Science and Football IV

This edited collection brings together the latest scientific research into the variety of sports known as *football*. With contributions by a large number of the leading international researchers in the field, the book aims to bridge the gap between the theory and practice of the various branches of football, and to raise the awareness of the value of a scientific approach to the various football codes.

The book contains nearly seventy papers, examining aspects ranging from match analysis and medical aspects of football to metabolism and nutrition, psychology and behaviour, and management and organisation in football. Containing a wealth of research data, and a huge range of examples of scientific applications in football, this book represents an invaluable reference for coaches, trainers, players, managers, medical staff, and all those involved in supporting performers in the many football codes.

Science and Football IV

Edited by W. Spinks, T. Reilly
and A. Murphy

London and New York

First published 2002
by Routledge
11 New Fetter Lane, London EC4P 4EE

Simultaneously published in the USA and Canada
by Routledge
29 West 35th Street, New York, NY 10001

Routledge is an imprint of the Taylor & Francis Group

© 2002 Routledge

Printed and bound in Great Britain by
The University Press, Cambridge

Publisher's Note

This book has been prepared from camera-ready copy provided by the authors.

British Library Cataloguing in Publication Data
A catalogue record for this book is available from the British Library

Library of Congress Cataloging in Publication Data
A catalog record for this book has been requested

Fourth World Congress of Science and Football
Sydney, Australia, 22 – 26 February 1999

Congress Organising Committee

Ross Smith (Australian Institute of Sport – Chair)
Warwick Spinks (University of Technology, Sydney – Congress Secretary)
Rod Carter (Australian Football League)
Peter Corcoran (Australian Rugby League)
Bill Ker (Australian Touch Association)
Warren Robilliard (Australian Rugby Union)
Jim Selby (Soccer Australia)
Nick Brooke (The Hotel Network)

Scientific Committee

W. Spinks (Chair)
R. Bower
A. Murphy
T. Reilly
R. Smith

Contents

PART FIVE Metabolism and Nutrition

PART SIX Paediatric Science and Football

Preface

The Fourth World Congress on Science and Football was held at the University of Technology, Sydney, 22–26 February, 1999. This event followed the inaugural Congress in Liverpool in 1987, the Second Congress at Eindhoven in 1991 and the Third Congress in Cardiff four years later. In all instances the Congresses were held under the aegis of the World Commission of Sports Biomechanics (re-named the World Commission of Science and Sports in January 2000) and in particular its Working Group on Science and Football (Chair: T. Reilly).

Patrons of the Congress included the International Council of Sports Science and Physical Education and the International Society of Biomechanics. The Congress was officially opened by Australia's Federal Minister for Sport, the Right Honourable Jacki Kelly. The organizers were fortunate to have generous support from Mars PLC, Gatorade, Hotel Mercure, Ansett Australia, the Australian Football League, the Australian Rugby Union, the Australian Rugby League, the Australian Touch Association, Soccer Australia, and the School of Leisure, Sport and Tourism of the University of Technology, Sydney. Delegates to the Congress also had the opportunity to witness at first hand the excellent preparations being made for the Sydney Olympics in 2000.

The philosophy underpinning the Congress is to bring together, every four years, scientists whose research is directly related to football and practitioners of football interested in obtaining current information about its scientific aspects. In this way it attempts to bridge the gap between research and practice so that scientific knowledge about football can be communicated and applied. The Congress themes are related to all the football codes, and the common threads among these are teased out in the formal presentations, workshops and seminars of the Congress programme.

The Congress programme included keynote addresses, oral communications, posters, demonstrations, symposia, and workshops. The detailed requirements of the scientific programme were adroitly handled by the local programme committee chaired by Warwick Spinks.

Delegates from all over the world attended the Congress. For its five days, football in its various forms was the topic of debate until the early hours of each morning. These debates will continue at the Fifth World Congress of Science and Football when it is convened at Lisbon, Portugal, in 2003.

The papers published in this volume constitute the proceedings of the Fourth Congress. They represent the material, either invited addresses or formal communications, that was submitted for publication and passed successfully through the peer-review process. Collectively they provide a flavour of the work currently under way in research in football and give an indication of the present state of knowledge in the area. As with previous volumes in the series, it is hoped that the content of this book will stimulate further research into science and football and encourage practitioners to implement some of the findings.

Thomas Reilly, February 2001

Chair, International Steering Group on Science and Football of the
 World Commission of Science and Sports
 (A service group of the International
Council for Sport Science and Physical Education
and the International Society of Biomechanics)

Introduction

This is the fourth volume in the Science and Football series. It is a testimony to the continuing growth and sophistication of sport and exercise science, which has matured sufficiently to generate a body of knowledge applicable to the football codes. The first three volumes confirmed the growing interest in bridging the gap between the theory and practice of the various branches of football and the increased awareness of the value of a scientific approach to these games. This volume, *Science and Football IV*, presents the Proceedings of the Fourth World Congress of Science and Football held at Sydney, Australia, 1999 and provides a record of selected research reports related to football.

The proceedings indicate current research work in football and provide markers of the topics that researchers are currently addressing. Fewer than half of the contributions to the conference programme are reproduced here. Researchers failed to meet either the deadlines set by the editors or the quality control standards set for publication. The manuscripts selected for publication provide a reasonable balance of the topics covered in the Congress programme.

The book is divided into ten Parts, each containing a group of related papers. Where possible, the titles of paper groupings used in previous volumes have been retained to facilitate future research. A feature of this volume are the numerous citations of papers published in the preceding volumes. Clearly the proceedings constitute a unique repository of data on football and footballers that would not otherwise be available to researchers and practitioners. The papers in each Part are related by theme or disciplinary approach. In a few cases, contributions cross disciplinary boundaries. Some others could sit comfortably in more than one Part with final location based on the main theme of a paper.

The editors are grateful to the contributors for their painstaking preparation of manuscripts to comply with the publisher's guidelines and our deadlines. We are equally indebted to Doug Moore at the University of Technology, Sydney, and Lesley Roberts at Liverpool John Moores University, Liverpool, for their skilful assistance with word-processing. The work of other members of the exercise and sport science community in refereeing papers was invaluable.

It is our aim that the Proceedings presented in this book should serve as a leading-edge reference for researchers in football and yield important current information for football practitioners. The material may motivate others to embark on research programmes, in time for the Fifth World Congress of Science and Football in Lisbon, Portugal, in April 2003.

Warwick Spinks
Thomas Reilly
Aron Murphy

PART ONE

Biomechanics of Football

1 THE TRACTION OF FOOTBALL BOOTS

E.B. BARRY* and P.D. MILBURN**
* RMIT University, Melbourne, Australia
** University of Otago, Dunedin, New Zealand

1 Introduction

There are many factors shared by the various football codes. They include the use of a ball, the playing surface and the need to change direction rapidly. Obviously, all players need to develop sufficient grip between the sole of the boot and the surface, to enable rapid change of direction, to accelerate, or to stop. Therefore, the purpose of this paper is to present research into the traction of football boots on natural surfaces including a non-linear regression model that calculates traction parameters. A prerequisite of the research was to design a suitable device for measuring traction. Traction tests were carried out under a number of conditions using combinations of: surfaces (3), moisture content (up to 10), boot types (4) and orientation (4). The test device and the protocols are described in the following sections.

2 Methods

2.1 Traction device and test protocols

The main components of the traction device (Figure 1) were (1) a weight-lifting frame and support table,(2) a vertical shaft holding the weights and boot that were fixed to a sliding plate, (3) the step motor drives that separately slid or rotated the vertical shaft, (4) the surface sample box clamped to a force plate (Kistler), and (5) the software and hardware control of the step motors, force plate and vertical displacement transducer (IBM PC).

Precise sliding displacements were applied to the vertical shaft and boot by a sliding plate connected to a screw thread shaft rotated by a step motor. Similarly, precise angular displacements could be applied to the vertical shaft through a reduction gear train rotated by another step motor. A switch on the step motor controller was used to set the device to operate either in translation or rotation. A PC (IBM) controlled both step motors and all data acquisition. The force plate range, the displacement rate, the direction and amount of the displacement, the centring of the shaft over the surface, when to lower and release the boot, and

the simultaneous triggering of the force plate and stepping motor for action, were all software selectable.

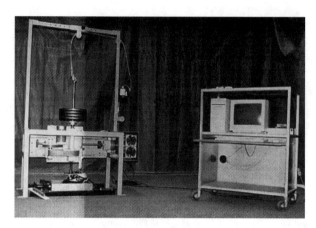

Figure 1. The RMIT boot/surface traction test device.

Pilot tests to establish reliable protocols were conducted using 'Blades Mudrunner' boots on sand with a 30-kg weight applied to the boot. The traction test conditions adopted for the remaining tests were as follows. The mass was increased to 35 kg producing a total load (weights + shaft) of about 400 N, a value used by Schlaepfer et al. (1983) and Beard and Sifers (1993) in related research. The forward and backward sliding rates were 50 mm·s^{-1} limited to a slide distance of 20 mm. For the selected rate and distance, the traction force was applied to the surface for 0.4 s since running, accelerating, braking or changing direction involve stance times of this order. The boot was fixed to the end of the shaft and lowered to just touch the surface of the sample. While in this position a transducer (Sakae) was positioned to measure the vertical displacement. The shaft was free to move as the device slid the boot horizontally over the surface. A clamping frame was bolted to a 900 mm x 600 mm force plate (Kistler) to hold the timber box and its surface sample rigidly and ultimately measure the traction force.

2.2 Exponential curve fit model
Figure 2 is one example of traction data fitted by a non-linear regression analysis technique using an exponential model. Exponential functions are used in terramechanics to study the traction behaviour and performance of vehicles on different terrains (Wong, 1989). They have been used in this research in a similar way to obtain the maximum traction force and other boot/surface interface parameters.

2.3 Surface samples and preparation

Three surfaces were tested. They were 1) ACI Sports 40 sand (sand), 2) StrathAyr Netlon turf made from ACI Sports 40 sand reinforced with Netlon mesh elements in the root zone and sown with perennial ryegrass (Netlon) and 3) StrathAyr natural river loam turf sown with perennial ryegrass (loam). Up to ten samples of each surface could be held in 400 mm x 300 mm x 100 mm marine timber boxes that were clamped to the force plate before testing with the traction device.

With sand, nine increasing moisture contents were held in sealed plastic containers before they were remoulded into the timber boxes. Remoulding of the sand was undertaken in a consistent manner by applying the same comparative effort to the sand as it was remoulded. Immediately after the completion of four traction tests per sample per boot, the sample was removed from the traction rig and covered with plastic to maintain the moisture content throughout the series of tests. A new sample at the next higher moisture content replaced the used sample and the tests repeated. Once all of the sand samples had been tested, another make of boot (see below) was fitted to the test rig and the above procedure repeated.

For turf, square pallets of each were pre-cut into 400 mm x 300 mm rectangular sods ready to place into the timber sample boxes. To maintain their healthy condition, the sods were placed side by side on a bed of sand in the gardener's compound and watered regularly. Five sods of the same turf were carefully fitted into the sample boxes. Before testing commenced, water was added in increasing amounts. Immediately after the completion of four tests per sample, one at the next higher moisture content replaced it. Once all of the samples at five different moisture contents had been tested, a second make of boot was fitted to the test rig and the procedure repeated. For the turf samples, traction testing was restricted to two makes of boot per sample to minimise the destruction of the grass and soil matrix of the surface. Furthermore, the stud lengths and arrangement of the two boots paired for test on the same sample were chosen on the basis of having different stud patterns and lengths. The two remaining makes of boots were tested in the same manner using a second set of samples at five similar moisture contents.

2.4 Verification of moisture contents

After the series of traction tests, two representative soil samples were placed in small aluminium tins to determine the percentage moisture content of the turf in accordance with Australian standard AS 1289.2.1.1 (1992). The two samples were taken from near the rootzone in the case of the turf surfaces. For sand, one sample was taken from near the surface and the other from near the base of the box. All moisture contents reported were the average of the two results.

2.5 Football boots and orientations

Four different types of boots were tested to obtain braking, propelling, and sideways traction forces. They were 'Adidas Flanker' rugby shoes with six front and two heel studs; 'Puma King Pro' with four front and two heel studs; two 'Blades' boots, namely 'Mudrunner', with six front blades and four-crossed heel blades, and 'Grassrunner', with ten front blades and four-crossed heel blades. Each boot was tested in the heel-off position. Only three boots, 'Adidas Flanker' rugby, 'Blades Mudrunner' and 'Puma King Pro' were tested on the sand at the nine moisture contents. However, all four boots were tested on the turf samples. As mentioned above, to decrease the damage to the surface and soil matrix of the turf samples, the paired boots tested on the same turf sample, had a different pattern and length of stud, with the shorter studded boots being tested before the longer studded boots. The boots were paired and tested in the following order, 'Blades Mudrunner' paired with 'Adidas Flanker' rugby followed by 'Blades Grassrunner' paired with 'Puma King Pro'.

Four different boot orientations were tested. They were braking traction involving a forward motion from heel to toe; propelling traction where the motion was backward; and two sideway tractions, where the boot's long axis (heel to the toe) was rotated at 45° and 135° to forward motion.

3 Results

The average moisture contents ranged between 1.7% and 19.0% in the sand. The ranges were more restricted and the values were higher in both the Netlon and the loam. They were respectively, 16.9% to 28.7% and 21.4% to 28.9%.

The coefficients of determination for the exponential model fitted to the data ranged between 0.6039 and 0.9983. The closer the coefficient is to one the better the fit. Most of the data had coefficients greater than 0.9. The traction in sand gradually increased up to maximum value after a slide distance of about 10 mm and remained constant from there (see Figure 2). However, in turf a maximum value was not exhibited even after a 20 mm slide distance because it continued to increase. The maximum traction forces for all surfaces, determined from the model, ranged between 123 and 1115 N.

With an increase in moisture content (1.7% to 19%) in sand, the traction force increased from approximately 150 N to 200 N depending on the direction of slide. For Netlon it ranged between approximately 300 N and 700 N apparently unrelated to the increase in moisture content (16.9% to 28.7%). Finally, for loam the range was approximately 300 N to 1000 N, also apparently unrelated to moisture content (21.5% to 28.9%).

In summary, each turf resulted in higher traction values than the sand and the river loam developed the greatest traction force. Finally, the type of surface, moisture content, and the slide direction affected traction.

4 Conclusions

The exponential model fitted the data for sand. For Netlon and loam data the model was limited, because the traction force did not reach a maximum (i.e. did not exhibit a failure plateau). With both of these turfs, the traction force continued to increase beyond the 20-mm slide distance and the loam developed slightly greater traction than the Netlon, which could perhaps increase the likelihood of injury to players. The traction force increased over the range of moisture contents for sand; however, there was no obvious relationship between traction force and the moisture content range tested for Netlon and loam turf. Each make of boot developed different traction forces on each surface and the orientation of the boot with respect to the boot's motion influenced the traction developed generally being smaller at 45° and 135°. Boots with blades developed a larger traction at 45° particularly in loam.

Variability in traction forces under varying conditions might be explained by a number of factors: preparation and test sequence; the influence of the Netlon plastic mesh elements introduced into the root zone; the variation in density of the ryegrass and their roots in the root zone; the probable variation in moisture content throughout each turf sample; the edge support received by the turf sods from end panels of the timber sample boxes.

5 Acknowledgments

This research was supported by grants received from RMIT University, Otago University and the Sports Science and Technology Board in New Zealand. Nike Sports Research Laboratory (NSRL) assisted with the preliminary design of the traction test device. ACI Industrial Minerals Division donated the Sports 40 sand. StrathAyr Pty Ltd reduced the cost of Netlon and the natural river loam turfs. Adidas, Blades and Puma donated their boots.

6 References

Australian standard AS 1289.2.1.1-1992 Methods of testing soil for engineering purposes, Method 2.1.1: Soil moisture content tests - Determination of the moisture content of a soil-Oven drying method (standard method).

Beard, J.B. and Sifers, S.I. (1993) Stabilization and enhancement of sand-modified root zones for high traffic sports turfs with mesh elements. A randomly oriented, interlocking mesh element inclusion system. The Texas Agricultural Experimental Station.

Schlaepfer, F., Untold, E. and Nigg, B.M. (1983) The frictional characteristics of tennis shoes, in *Biomechanical Assessment of Sports Protective Equipment: Proceedings of the International Symposium on Biomechanical Aspects of Sport* (eds B.M. Nigg and B.A Kerr), pp. 153-160.

Wong, J.Y. (1989) *Terramechanics and Off-Road Vehicles*, Elsevier, Amsterdam.

2 METHODOLOGY FOR GRAPHICAL ANALYSIS OF SOCCER KICKS USING SPHERICAL CO-ORDINATES OF THE LOWER LIMB

S.A. CUNHA[*], R.M.L. BARROS[**], E.C. LIMA FILHO[**] and R. BRENZIKOFER[***]
* Departamento de Educação Física,UNESP, Rio Claro, Brasil
** Faculdade de Educação Física, UNICAMP, Campinas, Brasil
*** Instituto de Física "Gleb Wataghin", UNICAMP, Campinas, Brasil

1 Introduction

Kicking is one of the most studied soccer skills (Lees and Nolan, 1998). The analysis of lower limb behaviour while kicking is an important procedure to determine individual soccer players' characteristics. The graphical analysis of kinematic data makes it easier to identify motor patterns and each athlete's skill level. The angle between thigh and shank, which expresses a relative measure and does not depend on the absolute position of each segment, is among the most used variables. This angle is usually calculated by measuring the plane defined by the vectors of both segments. However, for some movements, this kind of analysis does not allow distinguishing different situations. Soccer kicks are good examples since they can present identical knee angles (generated by the thigh-shank plane) for different angles from the ground plane when the kicking foot strikes the ball, which characterises different skills when performing the action. Further information about each segment position is therefore needed in order to determine motor patterns. Although research describing different kinds of kicks traditionally used the Cartesian co-ordinates system to determine kinematics variables (Asami and Nolte, 1983; Isokawa and Lees, 1988; Luhtanen, 1988; Rodano and Tavana, 1993), Redfern and Schumann (1994) developed a co-ordinate system centred on the hip that analyses foot trajectory in human locomotion and permits the production of an accurate human movement model through a spherical co-ordinates system.

In an attempt to find a better way to analyse this situation, this study refers to the spherical co-ordinates obtained from the transformation of the raw Cartesian co-ordinates as an alternative way to describe the same phenomenon. This transformation did not alter the characteristics of the movement studied (soccer kick). Choosing spherical co-ordinates guarantees the motor pattern analysis will be independent of the local reference system.

This study describes a methodology for analysing the motor pattern produced when kicking a stationary soccer ball using spherical co-ordinates to indicate the angular position of the lower limbs.

2 Methods

The methodology was applied to 12 stationary ball kicks registered by 2 S-VHS video cameras (30 Hz). Once the A–D conversion determined the tridimensional co-ordinates of the hip, knee and ankle joints, these Cartesian co-ordinates were transformed into spherical co-ordinates after a robust local weighted regression by continuous function (Cunha, 1998). With the latitude and longitude variables of each segment represented by continuous curves, a stereograph projection was performed onto a sphere of unitary radius and centred on the hip and knee joints.

A 37-year-old, highly experienced subject carried out the kicks. The ball was placed 11 m away from the goal and the subject tried to kick the ball into the corners formed by the crossbar and the goal posts. The two S-VHS video cameras focused on the subject's movement, more precisely on the kicking lower leg. After the images were digitised, the position of the lower limb joints (hip, knee and ankle) was measured from the supporting foot, in contact with the ground, to the end of the kick. From these data, a 3D reconstruction was carried out to obtain the cited real joint co-ordinates (Barros, 1997).

The chosen reference system had its $\mathbf{x'}$ axis defined by the direction opposite to that of the ball, its $\mathbf{y'}$ axis was vertical and its $\mathbf{z'}$ axis, whose orientation was defined by the outer product of $\mathbf{x'}$ and $\mathbf{y'}$, was orthogonal to both $\mathbf{x'}$ and $\mathbf{y'}$.

With three-dimensional co-ordinates in function of time $(\mathbf{X(t)}, \mathbf{Y(t)}$ and $\mathbf{Z(t)})$ for each one of the three joints (hip = $\mathbf{CF}$, knee = $\mathbf{JO}$, ankle = $\mathbf{TO}$), mathematical procedures allowed the determination of the normalised vectors of thigh and shank. These procedures are described as follows:

normalised vector of thigh $\rightarrow$ $\mathbf{COX} = (\mathbf{CF} - \mathbf{JO}) / (\|\mathbf{CF} - \mathbf{JO}\|_2)$

normalised vector of shank $\rightarrow$ $\mathbf{PER} = (\mathbf{TO} - \mathbf{JO}) / (\|\mathbf{TO} - \mathbf{JO}\|_2)$

with $\| \ \|_2$ defined as in the Euclidean norm of the vector.

This allowed the definition of a sphere of unitary radius, whose centre is at the knee and on whose surface the thigh and shank normalised vectors movements occur during the kick. The new system of references $(\mathbf{x}, \mathbf{y}$ and $\mathbf{z})$ whose origin is at the knee maintains its axis parallel to the laboratory references system $(\mathbf{x'}, \mathbf{y'}$ and $\mathbf{z'})$ (Figure 1).

The present study defined the variables of each segment angular position through the following co-ordinate transformations.

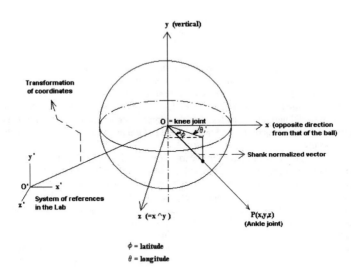

Figure 1. Transforming Cartesian co-ordinates (**x, y, z**) into spherical
co-ordinates (θ, φ).

$\phi_{COX} = \arcsin(Y_{COX})*180/\pi$ (latitude of the thigh)

$\phi_{PER} = \arcsin(Y_{PER})*180/\pi$ (latitude of the shank)

$\theta_{COX} = |\arctan(X_{COX}/Z_{COX})*180/\pi|$ (longitude of the thigh)

$\theta_{PER} = |\arctan(X_{PER}/Z_{PER})*180/\pi|$ (longitude of the shank)

The latitude describes the angle between the segment and a vertical line and ranges from
0°–90°. A value of 0° means the segment is on a horizontal plane while a value of 90°
indicates it is on a vertical plane. Positive values describe an up-oriented segment while
negative values design a down-oriented one. For each latitude, a circumference constitutes
the base of a cone whose vertex stands at the sphere centre.

The longitude indicates where the segment is located on this circumference and ranges from zero to a value of 180°. A value of 0° means the end of the normalised vector sits at the intersection of the spherical surface and on the plane constituted by the positive semi-axis **x** and the **y** axis (as, for example, in thigh extension). A value of 180° indicates the vector lies at the intersection of the spherical surface and of the plane constituted by the negative semi - axis **x** and the **y** axis (as, for example, in thigh flexion) A value of 90° describes the vector when at the intersection of the spherical surface and of the plane constituted by the **z** and **y** axes (as, for example, in thigh abduction).

It should be noted that this study did not translate the reference system, fixing the origin at the knee joint so that the angles (ϕ and θ) can describe the thigh and shank positions on a unitary radius sphere. Then, instead of three variables, as for the segments position in a Cartesian system (**x**, **y**, and **z**), both can be represented using only two variables (ϕ and θ), since the vectors are normalised, which permits the exclusion of the radial variable.

These two angles, which represent the segment spherical co-ordinates, allowed a stereographic projection linking each point (**P'**) on the sphere surface σ to a corresponding point **P** on the plane α by means of a semi-line originating at the North Pole (**N**) and 'perforating' the sphere at a point named **P'** (Figure 2). This projection presents two characteristics: a 'one to one' correspondence and the angle preservation (Cexeter, 1969).

The stereographic projection allows us to see both the latitude and longitude variables on a plane. It therefore yields graphics that can be used to analyse a kick pattern visually. Graphical analysis facilitates the interpretation of the real action with one figure for each segment.

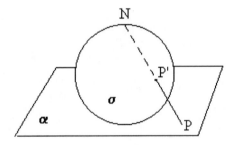

Figure 2. Schematic representation of the stereographical projection.

3 Results and Discussion

Figures 3 and 4 show the shank latitude and longitude as a function of time for a single kick, so that the next one (Figure 5), which brings longitude in function of latitude, can be better understood. Figure 6 presents the overlaid graphics of the 12 kicks studied and displays the motor pattern for this movement.

For the period of this kind of kick it can be seen that, as for skill pattern, the kicking leg presents a rough reduction in latitude values during the swinging phase, until the foot hits the ball, followed by an increase in these values after the subject kicked the ball (see Figure 3). With regard to the kick longitude, Figure 4 shows an enhanced curve increase after the contact with the ball. Figure 5, which brings longitude in function of latitude, describes what a kick pattern may look like for an experienced, mature subject.

The overlaid graph of longitude in function of latitude for thigh and shank (Figure 6) shows that the subject demonstrated a mature pattern for this movement with little variability among kicks. The variations that appear by the end of the movement are a consequence of the way the subject completed the movement. However, this is not relevant for the accomplishment of the action.

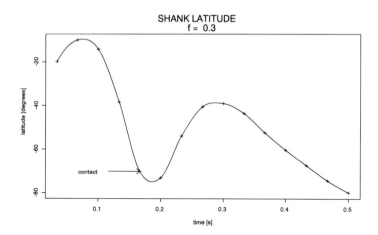

Figure 3. Graph of a kicking shank latitude in function of time.

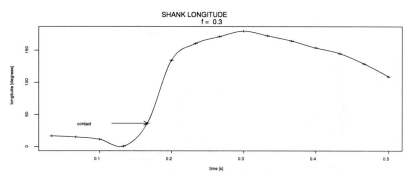

Figure 4. Graph of a kicking shank longitude in function of time.

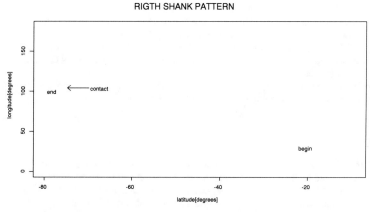

Figure 5. Graph of longitude in function of latitude for one kick (right shank).

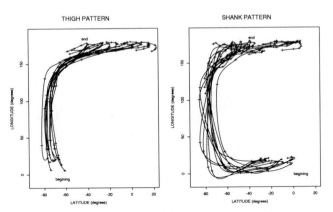

Figure 6. Patterns of the thigh and shank for a 12-kick series.

4 Conclusion

The analysis described in this report allows the use of only two variables (latitude and longitude) to identify the behaviour of each segment during the performance of a movement (soccer kick). The movement described by this methodology enables the analysis of different types of kicks, individual player characteristics, different skill levels and physical conditions and even age differences.

It is therefore possible to classify any kick within patterns which accurately represent the movement and which can be used to improve the players' technique.

5 References

Asami, T. and Nolte, V. (1983) Analysis of powerful ball kicking, in *Biomechanics VIII-B* (eds H. Matsui and K. Kobayashi), Human Kinetics, Champaign IL, pp.695–700.

Barros, R.M.L. (1997) *Concepção e implementação de um sistema para análise cinemática de movimentos humanos*, Doctoral dissertation, Faculdade de Educação Física, UNICAMP, Campinas, Brazil.

Coxeter, H.S.M. (1969) *Introduction to Geometry*, John Wiley and Sons, New York.

Cunha, S.A. (1998) *Metodologia para a suavização de dados biomecânicos por função não paramétrica ponderada local robusta*, Doctoral dissertation, Faculdade de Educação Física, UNICAMP, Campinas, Brazil.

Isokawa, M. and Lees, A. (1988) A biomechanical analysis of the instep kick motion in soccer, in *Science and Football* (eds T. Reilly, A. Lees, K. Davids and W.J. Murphy), E. and F. N. Spon, London, pp. 449–455.

Lees, A. and Nolan, L. (1998) The biomechanics of soccer: a review, *Journal of Sports Sciences*, 16, 211–234.

Luhtanen, P. (1988) Kinematics and kinetics of maximal instep kicking in junior soccer players, in *Science and Football* (eds T.Reilly, A. Lees, K. Davids and W.J. Murphy), E. and F.N. Spon, London, pp. 441–448.

Redfern, M. S. and Schumann, T. (1994) A model of foot placement during gait, *Journal of Biomechanics*, 27, 1339–1346.

Rodano, R. and Tavana, R. (1993) Three-dimensional analysis of the instep kick in professional soccer players, in *Science and Football II* (eds T. Reilly, J. Clarys and A.Stibbe), E. and F.N. Spon, London, pp. 357–361.

3 THREE-DIMENSIONAL KINEMATIC ANALYSIS OF THE INSTEP KICK UNDER SPEED AND ACCURACY CONDITIONS

A. LEES and L. NOLAN
Research Institute for Sport and Exercise Sciences, Liverpool John Moores University, Liverpool, UK

1 Introduction

The kicking skill and its variations have been well defined in the literature but mostly through 2D sagittal plane kinematic studies (Lees and Nolan, 1998). During the kicking motion, the kicking leg rotates about both the medio-lateral and longitudinal axes of the body, and so the true kinematics of the kicking skill can only be fully defined from a 3D analysis. A small number of 3D studies have been conducted but most have not reported on the movements which occur specifically in the frontal or transverse planes, such as hip and trunk rotations. Where these have been reported, they have provided an insight into the additional variables which may be associated with successful performance.

Rodano and Tavana (1993) reported data on lower limb kinematics for 10 professional soccer players performing an instep soccer kick. They used a 100 Hz opto-electronic system for recording 3D marker positions placed on the joints. Unfortunately, they did not report 3D movement data. Brown et al. (1993) investigated the free kick over a defensive wall and collected 3D data at 100 Hz from four highly skilled male soccer players. Their study was concerned only with the orientation of leg segments at impact and reported little kinematic data, which can be compared with other studies. Parassas et al. (1990) were interested in the means by which a soccer ball can be projected with a high and low trajectory. They investigated the 3D motion of low and high trajectory kicks from one subject filmed at 200 Hz. They noted that a high trajectory kick is commonly explained by a using a backward body lean, while the opposite is thought to occur for a low drive but they did not find significant differences in lean angle between the two kick types. The higher trajectory kick appeared to be achieved by contact on the ball at a lower point.

Browder et al. (1991) used cine film at 200 Hz to describe selected 3D characteristics of the instep kick of seven female players. They measured motions of hip and knee flexion and extension, and also reported data for pelvic rotation and hip adduction. They demonstrated that when comparing a fast to a slow kick, the pelvis showed greater range of motion (ROM) for the faster kick than for the slower kick (18° compared to 13°) although the phase of the movement that this covered was not defined. This suggests an important role for rotation of the hips in producing high velocity kicks. The ROM of the hip and the knee decreased as speed increased, with a ROM of 58° and 38° for the hip, and 95° and 75° for the knee for slow and fast kicks respectively. They suggested that pelvic rotation might be a method by which female players enhance the speed of their kick, rather than relying on joint extension at the hip and knee. Tant et al. (1991) focused on the temporal characteristics of the kick and claimed that in the performance of the maximal velocity instep kick, males used greater ROM at the hip and knee while females exhibited greater pelvic rotation.

All of the above studies suffer from a lack of clarity in defining terms such as reference systems from which angles are measured, and starting and finishing points for ROM. In several cases there are clear errors in the data reported and several important aspects of the data are not commented upon. In addition, these studies (with the exception of Browder et al. 1991) have not managed to identify any critical characteristics of the 3D nature of the kicking action, namely hip, trunk and shoulder rotations. Further, these studies have not related their findings to appropriate 'principles of movement' (Bunn, 1972) in order to explain changes in the characteristics of the kick as a function of the outcome of the kick, that is, ball speed and accuracy of ball placement.

The aim of this study was to investigate changes that occurred in the kinematic characteristics of the stationary instep kick under a speed-accuracy paradigm. It is known that in order to increase accuracy of outcome the speed of movement will reduce. Therefore this paradigm imposed a change of speed on the kick as the demands of accuracy increased. This was thought to be a more appropriate manipulation of kicking speed than by asking subjects to make voluntary changes. The objectives of the study were (1) to quantify selected 3D kinematic characteristics of the kick including hip, trunk and shoulder motion, (2) to determine the variability in selected 3D kinematic characteristics of the kick as the demand of accuracy increased, (3) to identify changes in selected 3D kinematic characteristics which occur as a result of a reduced speed of kick imposed by the demands of higher accuracy and (4) to develop a 3D model of kick performance and to relate aspects of this model to appropriate 'principles of movement'.

2 Methods

Two professional soccer players acted as subjects. Each took 10 instep kicks at a stationary ball placed in front of the goal mouth so as to simulate a penalty kick. They were required to hit a 1.0 m^2 target placed in the top right corner of the goal mouth. The requirement of the first five kicks was to hit the target with an emphasis on speed of kick. Subjects were asked to make a fast kick as if they were taking a penalty with the intention of placing the ball in the top right hand corner of the net. The requirement of the second five was for accuracy of placement to ensure the ball hit the target. Subjects had ample time to practise before filming took place. Each kick was filmed at 100 Hz using two cine cameras placed with their optical axes at approximately 90^0 to each other and the 3D co-ordinates of 18 body landmarks were used to reconstruct the 3D motion using standard DLT procedures. The x–y (sagittal) plane was in the direction of the kick and perpendicular to the line joining the two goal posts. From each kick, a variety of linear and angular displacement and velocity measures were computed and used to identify key aspects of performance. Accuracy was recorded as the number of hits on target.

Angles were obtained from the 3D displacement data of three joint markers. The angle reported is the angle made by these markers in the plane in which they lie. The joint reference system used was such that at full extension, the hip, knee and ankle joints were at 180°, trunk angles were measured to the vertical, and hip and shoulder lines in the transverse plane were measured so that they are zero when horizontal and facing forward. Ranges of motion were measured from the peak value produced from initiation of the kick to impact.

Comparisons were made between each condition using a Student's paired t-test and applied to each subject separately. The variability was quantified by the variance in the data for each subject's data and expressed as the standard deviation from which the coefficient of variation was computed. A probability level of $P<0.05$ was used to indicate statistical significance.

3 Results

Hip, knee, ankle and toe joints showed temporal patterning of velocity peaks in a classical proximal to distal sequencing of joint action except for the toes, which peaked slightly before the ankle. This was due to the plantar flexion of the foot in preparation for impact. Group mean data are given in Table 1 for peak values of selected kinematic characteristics. It can be seen for both subjects that speeds decreased as the demand for accuracy increases. The data in Table 2 represent the values of selected kinematic characteristics at the moment of impact. The data in Table 3 represent the ranges of motion of selected variables from their initiation, minimum or maximum value (as appropriate) until impact. These are presented in a way that reflects a model of kicking and best illustrate how variables change with the demands of change of speed.

Table 1. Mean (±SD) peak values for selected kinematic descriptors of the kick.

Subject	JM		PN	
Kick condition	Accuracy mean ± SD	Speed mean ± SD	Accuracy mean ± SD	Speed mean ± SD
Speed (m·s⁻¹)				
Approach	2.42 ± 0.11	3.32 ± 0.16^{+++}	2.54 ± 0.09	3.46 ± 0.23^{++}
Hip	2.38 ± 0.22	2.64 ± 0.05^{NS}	2.06 ± 0.27	3.12 ± 0.17^{+++}
Knee	4.44 ± 0.26	5.40 ± 0.30^{+}	4.52 ± 0.21	5.80 ± 0.14^{+++}
Ankle	13.16 ± 0.51	15.52 ± 0.17^{++}	11.94 ± 0.48	14.30 ± 0.18^{+++}
Toe (5th metatarsal)	15.56 ± 0.82	20.24 ± 0.25^{+++}	13.84 ± 0.43	18.36 ± 0.21^{+++}
Ball	20.4 ± 0.77	26.6 ± 1.51^{+++}	18.1 ± 0.77	24.3 ± 1.52^{+++}
Ball/toe speed ratio	1.31 ± 0.09	1.31 ± 0.04^{NS}	1.32 ± 0.08	1.32 ± 0.06^{NS}
Joint angular velocity (rad·s⁻¹)				
Ankle	1.1 ± 0.8	4.5 ± 1.3^{+++}	1.8 ± 0.9	1.9 ± 1.5^{NS}
Knee	20.5 ± 1.3	23.8 ± 1.4^{++}	13.8 ± 0.8	18.5 ± 1.3^{++}
Hip (thigh-trunk)	6.8 ± 0.7	8.6 ± 1.1^{NS}	7.6 ± 0.7	11.6 ± 0.01^{+++}
Accuracy (max=5)	5	3	5	4

NS = non-significant   + P<0.05   ++ P<0.01   +++ P<0.001

Table 2. Mean (±SD) angular kinematic descriptors of the kick at impact.

Subject	JM				PN			
Kick condition	Accuracy		Speed		Accuracy		Speed	
	mean	± SD	mean	± SD	mean	± SD	mean	± SD
Angle (degrees)								
Ankle	117.0	± 7.9	127.0	± 2.0$^+$	107.4	± 4.3	127.8	± 2.6^{++}
Knee	128.8	± 6.4	124.2	± 5.2^{NS}	118.4	± 5.1	126.2	± 3.4^{NS}
Hip (thigh–trunk)	166.2	± 1.6	158.0	± 3.9^{NS}	155.0	± 1.0	149.2	± 1.3^{++}
Thigh–pelvis angle	106.6	± 2.7	103.6	± 2.1^{NS}	108.2	± 1.6	108.2	± 2.7^{NS}
Trunk lean*	25.9	± 2.0	11.9	± 4.4^{+++}	21.7	± 3.0	0.0	± 3.3^{+++}
Trunk inclination**	12.1	± 1.8	9.6	± 1.8$^+$	15.4	± 1.3	16.4	± 1.6^{NS}
H–S separation***	1.2	± 0.7	3.5	± 5.3^{++}	3.6	± 2.8	14.6	± 6.1^{++}
H–S inclination	5.8	± 3.2	2.5	± 2.6^{NS}	3.3	± .0	1.8	± 1.8^{NS}
Angular velocity (rad·s^{-1})								
Ankle	1.0	± 3.8	-0.8	± 0.9^{NS}	0.4	± 0.9	-2.3	± 1.2$^+$
Knee	21.8	± 4.0	23.5	±1.3^{NS}	13.8	± 0.6	18.5	± 1.3^{++}
Hip (thigh-trunk)	-0.8	± 0.7	-5.2	± 0.3^{+++}	-5.1	± 0.8	-5.9	± 0.5^{NS}

* backward in sagittal plane   ** to support foot side in frontal plane   ***H–S = hip-shoulder
NS = non-significant;   + P<0.05;   ++ P<0.01;   +++ P<0.001

Table 3. Mean (± SD) ranges of motion. Units are degrees unless otherwise stated.

Subject	JM		PN	
Kick condition	Accuracy	Speed	Accuracy	Speed
	mean ± SD	mean ± SD	mean ± SD	mean ± SD
Last stride length (m)	0.55 ± 0.05	0.81 ± 0.06^{+++}	0.53 ± 0.04	0.72 ± 0.07^{+++}
Pelvic rotation	22.2 ± 3.3	33.8 ± 3.8^{+++}	22.4 ± 2.4	30.4 ± 6.7$^+$
H–S separation *	31.6 ± 2.1	38.4 ± 6.0$^+$	30.2 ± 4.2	41.6 ± 8.0$^+$
H–S inclination*	1.8 ± 1.5	9.0 ± 2.3^{+++}	3.4 ± 1.5	10.8 ± 4.0^{++}
Hip(thigh-trunk)**	34.2 ± 3.1	36.9 ± 5.5^{NS}	34.3 ± 1.6	51.3 ± 2.4^{+++}
Knee **	49.3 ± 6.3	52.1 ± 4.3^{NS}	32.8 ± 1.2	42.9 ± 4.8^{+++}

NS = non-significant   + P<0.05   ++ P<0.01   +++ P<0.001
*from initiation of forward motion to impact   **from min/max value to impact

4 Discussion

The ball speeds reported here compare well with those values reported in the literature (Lees and Nolan, 1998) with a maximum ball speed of 28.4 m·s^{-1} by subject JM and 26.1 m·s^{-1} by subject PN. Similarly, joint linear speeds compared well with that reported elsewhere for similar types of kick. Joint angular data were slightly lower than the 2D data reported in the literature (maximum values reaching up to 40 rad·s^{-1}), but similar to Rodano and Tavana who also used a 3D analysis method. Mean ball-to-toe speed ratio was uniformly 1.3 for each kick and each player, suggesting that the two conditions did not affect impact mechanics. The timing of peak joint speeds confirmed that these subjects demonstrated a proximal to distal sequencing in their performance as expected, except for the toes as previously noted.

The speed-accuracy paradigm led to a significant reduction in ball speed of about 6 m·s^{-1} for each subject. There was also a reduction in peak approach speed, peak linear and angular joint speeds for both subjects with the exception of the peak ankle angular velocity for subject PN and peak hip linear and angular speed for subject JM. There was little difference in orientation of the limbs at impact between the two types of kick as indicated by the joint angles and angular velocities at impact, although subject PN showed a slightly more flexed hip, plantar flexed ankle and more rapidly extending knee. Only two subjects were used but they were both highly skilled. Where differences in performance between the two conditions were found, it is likely that these would also be demonstrated by other skilled players.

There were also individual differences between the two subjects. Subject JM produced a faster kick than subject PN under both conditions. The approach speed, hip and knee joint speeds were similar, but JM produced higher knee extension angular velocities, which led to higher ankle and toe speeds. Differences were also apparent between the orientation of each player's kicking leg at impact, with PN being more upright in the sagittal plane but leaning more to the support leg side than JM.

The variability associated with each series of kick was generally small with coefficients of variation ranging between 1 and 10% with the majority closer to the lower limit. These indicate a repeatability of movement and generally both players were able to reproduce their kicks consistently. This was emphasised by the high success rate associated with hitting the required target. It was noted, however, that both subjects appeared to be more consistent in their whole body and proximal segment movements but less consistent in their distal segments during the slower kick. This suggested that the basic movement was repeated more precisely in the slow kick but that late modifications to the movement were being made as evidenced by greater variability in knee and ankle variables. This aspect of skilled kicking is worthy of further investigation.

The data here support the model of the mature soccer kick defined in the literature (see Lees and Nolan, 1998) but extend it with the inclusion of 3D information. The data from Table 3 suggest that to increase kick speed players increase their last stride length. This has the effect of opening out the hips (rotating the hip of the kicking leg backwards) which in turn enables the hip-shoulder separation angle to increase. This allows the musculature of the trunk to contribute to the effort of kicking by rotating the pelvis. The greater range of pelvis rotation, assisted by greater ranges of motion at the hip and knee joints, leads to greater foot speed and ultimately to a greater ball speed. The data support an important 'principle of movement' (Bunn, 1972) which is that increased end-point speed of a limb (e.g. the foot) can be achieved by using a greater range of motion at the joints. In particular, the 3D nature of the movement supports the notion that rotations about the longitudinal axis of the body are important in a full understanding of this principle. As well as greater range of motion at the knee and hip, a major contributor is the increased range of rotation between the pelvis and shoulder lines, a previously unreported finding. This supports and extends the findings of Browder et al. (1991) who reported that increased kick speed for female soccer players was associated with just the increased range of motion of the pelvis. The verification of this movement principle for soccer kicking should be useful to coaches and players, and suggests that performance outcome may be influenced by appropriate flexibility and strength training (which will increase range of motion and strength at the ends of ranges of motion) as well as skill training to implement these movement characteristics. It is not known at what point increased joint range of motion can no longer be an aid to performance, but once this is reached it is likely that performance will only increase further if some other physical characteristics of the player, for example strength, develops.

5 Conclusion

It is concluded that (1) the kinematics of the soccer kick can only be fully defined through a 3D study (2) movement patterns were very consistent with greater consistency being associated with the slower more precise action except for the distal segments, (3) the increases in kicking speed are associated with increases in range of motion at the pelvis, hip and knee joints and (4) the link between increased range of motion and increased toe and ball speed is a postulated 'movement principle' supported for the soccer instep kick by the data reported in this study.

6 References

Browder, K.D., Tant, C.L. and Wilkerson, J.D. (1991) A three dimensional kinematic analysis of three kicking techniques in female players, in *Biomechanics in Sport IX* (eds C. L. Tant, P. E. Patterson and S. L. York), ISU press, Ames, IA, pp. 95–100.

Brown, E.W., Wilson, D.J., Mason, B.R. and Baker, J. (1993) Three dimensional kinematics of the direct free kick in soccer when opposed by a defensive wall, in *Biomechanics in Sport XI* (eds J. Hamill, T.R. Derrick and E.H. Elliott), University of Massachusetts, Amherst, pp. 334–338.

Bunn, J.W. (1972) *Scientific Principles of Coaching*, Prentice Hall, Englewood Cliffs, New Jersey.

Lees, A. and Nolan, L. (1998) Biomechanics of soccer – a review, *Journal of Sports Sciences*, 16, 211–234.

Parassas, S.G., Terauds, J. and Nathan, T. (1990) Three dimensional kinematic analysis of high and low trajectory kicks in soccer, in *Proceedings of the VIIIth Symposium of the International Society of Biomechanics in Sports* (eds N. Nosek, D. Sojka, W. Morrison and P. Susanka), Conex, Prague, pp. 145–149.

Rodano, R. and Tavana, R. (1993) Three-dimensional analysis of instep kick in professional soccer players, in *Science and Football II* (eds T. Reilly, J. Clarys and A. Stibbe), E. and F.N. Spon, London, pp. 357–361.

Tant, C.L., Browder, K.D. and Wilkerson, J. D. (1991). A three-dimensional kinematic comparison of kicking techniques between male and female soccer players, in *Biomechanics in Sport IX* (eds C. L. Tant, P.E. Patterson and S.L. York), ISU Press, Ames, IA, pp. 101–105.

4   RELATIVE TIMING OF EMG PROFILES FOR NOVICE AND ELITE SOCCER PLAYERS

M.D. McDONALD
Queensland University of Technology, Brisbane, Australia

1 Introduction

Relative timing, a construct often used by motor control theorists to account for skilled motion, is considered invariant if the times allotted to the various events or phases of the movement remain in a constant ratio with respect to the total movement time, even if the movement time varies in accordance to different task dynamics. There have been a number of research reports arguing against the notion of invariant relative timing (IRT). Most notably, Gentner (1987) has argued that the evidence of IRT has resulted from inappropriate analysis techniques and statistical tests. His re-analyses of data, as well as the work by Neal et al. (1990) on gait kinematics, argues against the concept of IRT. Heuer (1988) claimed that the absence of IRT in the kinematics of movement doesn't rule out its existence at another higher level of the motor system. Heuer argued that the inertial characteristics of the system and delays between neural impulse arrival and muscle tension development are major causes of the lack of IRT at the kinematic level. Heuer went on to argue that the IRT aspects could be preserved at the electromyographic (EMG) level rather than at the kinematic level. If the neural commands for muscular contraction have a set time base, then one would expect to observe an adherence to temporal proportionality for movements initiated about proximal rather than distal joints, and for EMG rather than kinematic measures of motion. Winter (1983) has previously suggested that temporal proportionality exists within EMG correlates of gait which were not evident from gait kinematics. For the notion of IRT to hold, then one would expect that the total time to complete a movement such as a soccer instep place kick for either maximal or submaximal conditions would exhibit varying times to complete the movement but the various phases within the movement should remain in proportion. The purpose of this study was to determine if IRT, as evidenced from EMG patterns, exists at the neuromuscular level as opposed to the kinematic level within the instep soccer place kick.

2 Methods

Fifteen novices, drawn from two local swimming clubs (mean age = 19.3 years) and 15 expert players drawn from both the senior and junior national soccer league (mean age = 25.6 years) served as voluntary subjects for this study. The EMG and high speed video data were simultaneously collected for 12 trials under two different kicking conditions (fast and slow). The order of conditions was randomly assigned and sufficient practice was allowed so that subjects felt comfortable with the experimental procedure. Data collection was conducted within the biomechanics laboratory at QUT. The EMG signals were recorded during each trial using four pre-amplified (1000X) surface electrodes. The electrodes were connected to a transmitter module attached to the subject; from this source the EMG signals were transmitted to an FM recorder and transferred onto an AMLAB (Associated Measurements Pty Ltd) data acquisition unit. The sampling frequency for the EMG signals was 2000 Hz. The moment of impact was detected by an infra-red light which generated a pulse recorded on the AMLAB unit. Muscle activity was recorded from the following muscles of the right leg: rectus femoris, biceps femoris, tibialis anterior and lateral head gastrocnemius. The location of the electrodes on these muscles followed the guidelines of Delagi and Perotto (1974). The EMG records were analyzed by reference to the different phases of the kicking motion as identified by the kinematic data. The raw EMG records were full-wave rectified and then linear enveloped (filter time constants: 25 ms) as outlined in Winter (1990). The durations of muscle activity and inactivity were then expressed as a percentage of the total movement time.

The evaluation of the kicking action was divided into four events as depicted in Figure 1. The division of the kicking action into the various phases (times between events) is similar to the work done by DeProft et al. (1988) on muscle activity for the soccer place kick. The four events were defined as follows: MT began at toe off of the right foot during the last step prior to ball contact, and finished at maximum thigh back position of the kicking leg; HC began at maximum thigh back position and ended at left foot heel contact next to the ball; MS began at left foot heel contact and ended at maximum shank back position of the kicking leg; BC began at maximum shank back position and ended one frame prior to ball contact. The Motion Analysis Expert Vision Flextrak system (Motion Analysis Corp.) was used to digitise automatically the edges of retroreflective markers placed on anatomical landmarks of the kicking leg for each subject. Sampling frequency for the Flextrak system was 200 Hz. The centroid of each marker was determined then joined as paths which were smoothed using a Butterworth second order digital filter prior to calculating the temporal data for the four identified events by customised software. For comparison purposes between the two kicking conditions and subjects, the temporal values were normalised to percentage of the total movement time.

To investigate the notion of IRT, the statistical test of Gentner (1987) was used on individual subject EMG data. The test requires that the relative times spent in each of the phases of the kicking motion be regressed against the total movement time. If IRT is preserved, the slope of the regression equation should not be significantly different from zero. If, however, the gradient is different from zero, IRT cannot be supported for that subject. Gentner has proposed that if 10% or more of the subjects exhibit regression slopes that are different from zero then the notion of IRT must be rejected. Student's t-test was used to determine if the slope was different from zero for the data drawn from each of the individual subjects.

MT   HC   MS   BC

Figure 1. Schematic representation of the four distinct events within the kicking movement pattern.

3 Results

Table 1 indicates that not one muscle or individual phase of the movement sequence could fit into Gentner's notion of relative timing invariance. The data imply that the EMG activity time spent in the various phases of the kicking motion are not kept in proportion as the movement is scaled and modified to meet changing task demands. This finding casts doubt upon claims that relative timing is an invariant feature for various motor skills. Figure 2 exhibits a typical scatter diagram of the percentage of time spent in MT phase of the kicking sequence (plotted against the total movement time) for both an expert and novice player (for the EMG activity of biceps femoris) in an attempt to illustrate the regression procedure. It is evident that there is considerable variation in the percentage of time spent in this particular phase of the movement sequence for the two kicking conditions. Inter-subject variability was high, even amongst the expert players, indicating that there are many combinations of muscle action that can produce kinematics which are consistent with the goal of the task.

Table 1. Proportionate phase times regressions against total movement time (the number indicates the number of subjects out of 30 that had a regression greater than zero).

Muscle	Phases of the kicking motion			
	MT	MS	HC	BC
Rectus femoris	5	9	9	4
Biceps femoris	6	5	7	8
Gastrocnemius	5	7	8	6
Tibialis anterior	9	9	4	4

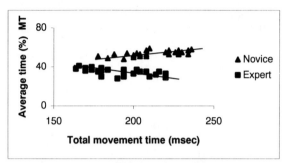

Figure 2. Average percentage time in MT against total movement time for one expert
and one novice subject for biceps femoris (for all trials of both kicking conditions).

4 Conclusions

Comparisons between the expert and novice subjects' EMG patterns lend support to the two-stage model of motor learning proposed by Newell et al. (1985). According to Newell et al. (1985) the novice has acquired the appropriate set of relative motions for the movement task. During advanced stages, the pattern of relative motions is scaled or invariant, with optimal scaling representing an expert level of performance. The similarities in the topological EMG characteristics between the expert and novice subjects' kicking patterns suggest that the novice subjects have acquired the appropriate set of relative motions.

5 References

Delagi, G. and Perotto, B. (1974) *Anatomic Guide for the Electromyographer,* C.C. Thomas, Champaign, IL.

DeProft, E., Clarys, J.P., Bollens, E., Cabri, J. and Dufour, W. (1988) Muscle activity in the soccer kick, in *Science and Football* (eds T. Reilly, A. Lees, K. Davids, and W.J. Murphy), E. and F.N. Spon, London, pp. 434–440.

Gentner, D.R. (1987) Timing of skilled motor performance: Test of the proportional duration model, *Psychological Review*, 94, 255–276.

Heuer, H. (1988) Testing the invariance of relative timing: Comment on Gentner (1987), *Psychological Review*, 95, 552–557.

Neal, R.J. et al. (1990) Invariant features of human gait, in *Proceedings of the 1990 Commonwealth and International Conference of PESHDR and Leisure*, Auckland, 124–137.

Newell, K.M. et al. (1985) Augmented information and the acquisition of skill in physical activity, *Exercise and Sport Science Reviews*, 13, 235–261.

Winter, D.A. (1983) Biomechanical motor pattern in normal walking, *Journal of Motor Behavior*, 15, 302–330.

Winter, D.A. (1990) *Biomechanics and Motor Control of Human Movement*, Wiley and Sons, New York.

5 THREE-DIMENSIONAL KINETICS OF IN-SIDE AND INSTEP SOCCER KICKS

H. NUNOME*, Y. IKEGAMI*, T. ASAI** and Y. SATO*
 * Research Centre of Health, Physical Fitness and Sports, Nagoya University,
    Nagoya, Japan
** Faculty of Education, Yamagata University, Yamagata, Japan

1 Introduction

Among the various skills of soccer, kicking is the most important. The in-side kick is the most frequently used when a shorter and precision pass is required. On the other hand, the instep kick is normally used for the generation of a faster ball speed. However, as most of the previous studies were executed with two-dimensional procedures, three-dimensional natures of these kicks are still unknown. In particular for the in-side kick, to hit the ball by the medial aspect of the foot, several complicated rotational motions may be combined to make a swing of the leg. Recently, Levenon and Dapena (1998) reported kinematic natures of the in-side and instep kicks, however, these kinetic aspects were still not quantified yet. To understand how the leg swing is produced for these kicks, the kinetic aspects may supply significant insight of the kicking techniques. Therefore, this study was designed to investigate the kinetic features of the two representing soccer kicks: in-side and instep kicks using a three-dimensional video analysis technique.

2 Methods

Five male elite high-school players (height = 174.6 ± 4.9 cm; weight = 67.6 ± 4.8 kg) volunteered to participate in this study. After an unlimited period of warm-up, the subjects performed instep and in-side kicks to a goal located 11 m ahead using a regulation soccer ball. Two electrically synchronised high-speed video cameras (Nac MEMRECAM C2) were used to sample the kicking motion at 200 fps (shutter speed 1/2000 s) from the rear and kicking leg (right) side. To calibrate the performance area, a calibration frame (1.5 x 1.5 x 1.8 m) with 16 control points was videotaped before the trials.

Prior to videotaping, reflective markers were placed at appropriate anatomical landmarks on each subject to aid in digitising. A personal computer was used to digitize body landmarks to define a four-segment body model including: right foot, right shank, right thigh and lower torso. Each trial was digitised from 10 frames before the take-off of the right toe to the impact of the foot with ball. The centre of the ball was also digitised in its initial stationary position and in all the available frames after it left the foot (typically five frames). The direct linear transformation method (Abdel-Azis and Karara, 1971) was used to obtain the 3D coordinate of each landmark.

Hip   Knee   Ankle

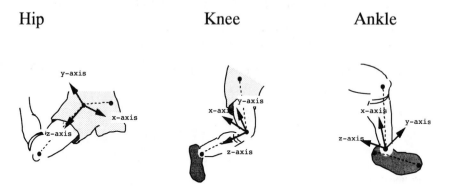

Figure 1. Right-handed orthogonal reference frame fixed at each joint.

2.1 Joint kinetics
To compute joint torque generating at each joint, the kicking leg was modelled as a three-link kinetic chain (Feltner and Dapena, 1986) composed of the thigh, shank and foot. Mass of the shoe was assumed to be negligible. The joint torques were separated into right-handed orthogonal axis fixed at hip, knee and ankle joints (Figure 1). The joint powers were also computed by taking the product of joint torques and joint angular velocities.

2.2 Data smoothing
The impact of the foot with the ball produces a sudden deceleration of the kicking leg, which causes a serious distortion of the kinetic data near the impact. To avoid such systematic errors, the time-dependent coordinates of the landmarks were digitally smoothed in a forward order (toward the impact) by a second-order Butterworth low-pass filter (Winter, 1990) at 12.5 Hz. Although the smoothing procedure followed in this study may have minimised the risk of systematic errors steaming from the impact as its reverse order of filtering was cancelled, the data was still prone to include phase distortion. However, the effect would not significantly change the quantitative parameters.

2.3 Statistics
All the kinetic variables were compared between the two types of kicks using Student t-tests. The criteria of statistical significance was $P < 0.05$ for all analyses.

3 Results and Discussion

The initial ball velocity (24.3 ± 0.8 m·s^{-1}) and the toe velocity immediately before the impact (17.7 ± 0.5 m·s^{-1}) of the in-side kick were significantly slower (P < 0.01) than those of the instep kick (ball; 28.0 ± 2.4 m·s^{-1} and toe; 24.2 ± 2.2 m·s^{-1}). In this study, the ball velocity of both kicks was similar to the values previously reported (Levanon and Dapena, 1998; Robertson and Mosher, 1985; Zerinke and Roberts, 1978).

Figure 2 shows the changes in average joint torques for the hip adduction/abduction, flexion/extension and internal/external rotation, knee flexion/extension, lower leg internal/external rotation, and ankle dorsi/plantar flexion and inversion/eversion. Although the general patterns of the hip flexion and knee extension torques were similar between the two types of kicks (Figure 2), their magnitudes were significantly different. The values of the in-side kick (hip flexion; 87.4 ± 21.2 Nm and knee extension; 74.8 ± 9.3 Nm) were significantly smaller (P < 0.01) than those of the in-step kick (hip flexion; 130.2 ± 42.3 Nm and knee extension; 106.4 ± 27.3 Nm). Moreover, a marked difference was observed for the hip external rotation torque between the two types of kicks. In the in-side kick, all the subjects exhibited the hip external rotation torque, and its magnitude (78.8 ± 7.8 Nm) was comparable to those of the hip flexion and knee extension.

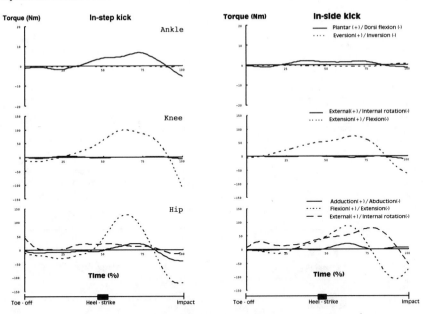

Figure 2. Average joint torques for the hip, knee and ankle joints.

Figure 3 shows the changes in average joint powers for the hip and knee joint motions. In the instep kick, the general patterns of positive joint power due to the hip flexion and knee extension torques were similar to those of Roberts and Mosher (1985). However, the magnitude of positive joint power due to the knee extension torque seemed rather larger than that of the previous study. In the previous study, as the deceleration produced by the ball impact was not treated properly, its magnitude would be underestimated. In the in-side kick, although the positive joint power due to the knee extension torque was not dominant, the positive joint power due to the hip flexion torque was dominant near the impact.

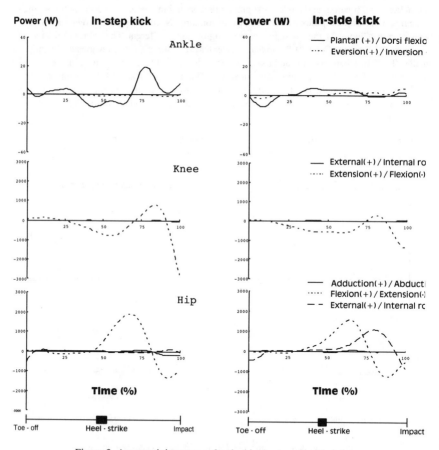

Figure 3. Average joint powers for the hip, knee and ankle joints.

These results indicated that a relatively complicated series of rotational motions are required in the in-side kick, suggesting that the hip external rotation motion is a significant contributor for making the leg swing. This finding may also help to estimate the possibility of lesion at the hip joint related to over-practice of these kicking motions.

4 References

Abdel-Azis, Y.I. and Karara, H.M. (1971) Direct linear transformation from computer coordinates into object space coordinates in close-range photogrammetry, in *Proceedings ASP Symposium on Close-Range Photogrammetry*, pp. 1–18.

Feltner, M.E. and Dapena, J. (1986) Dynamics of the shoulder and elbow joints of the throwing arm during a baseball pitch, *International Journal of Sports Biomechanics*, 2, 235–259.

Levanon, J. and Dapena, J. (1998) Comparison of the kinematics of the full-instep kick and pass kicks in soccer, *Medicine and Science in Sports and Exercise*, 30, 917–927.

Putnam, C.A. (1991) A segment interaction analysis of proximal-to-distal sequent segment motion patterns, *Medicine and Science in Sports and Exercise*, 23, 130–141.

Robertson, D.G.E., Zernicke, R.F., Youm, Y. and Huang T.C. (1974) Kinetic parameters of kicking, in *Biomechanics IV* (eds R.C. Nelson and C.A. Morehouse), University Park, Baltimore, pp. 157–162.

Robertson, D.G.E. and Mosher P.E. (1985) Work and power of the leg muscles in soccer kicking, in *Biomechanics VII-B* (eds D.A. Winter, R.W. Norman, R.P. Wells, K.C. Hays, and A.E. Patla), Human Kinetics, Champaign, IL, pp. 533–538.

Winter, D.A. (1990) *Biomechanics of Human Movement*, John Wiley and Sons, New York.

Zernicke, R.F. and Roberts, E.M. (1978) Lower extremity forces and torques during systematic variation of non-weight bearing motion, *Medicine and Science in Sports and Exercise*, 10, 21–26.

6 MUSCLE ACTIVITY DURING THE DROP PUNT KICK

J. ORCHARD*, S. WALT**, A. McINTOSH** and D. GARLICK*
*   School of Physiology and Pharmacology, University of New South Wales, Australia
**   Department of Safety Science, University of New South Wales, Australia

1 Introduction

Australian Rules Football is one of the country's most popular spectator and participant sports, but the biomechanics of its techniques have not been extensively studied, perhaps because the sport is not played internationally. The most important means of ball progression is kicking and the drop punt is the standard kicking technique in most situations, due to a combination of accuracy, distance and speed of execution. Punting is also a primary technique in Gaelic football and a secondary technique used by specific positional players in Rugby Union, soccer and American football. Therefore, some relevant research has included a study of the kinematics of punt kicking (Putnam, 1991) and electromyography of the soccer kick (Bollens, De Proft and Clarys, 1987).

The most common injuries in the Australian Football League (AFL) are hamstring strains, knee injuries and groin injuries. Most injuries are equally distributed between sides of the body, although quadriceps strains are more common on the dominant kicking side (Orchard et al., 1998). Hamstring, quadriceps and groin injuries occur more commonly than in Australia's other football codes (Seward et al., 1993). Factors that may explain the relatively higher incidence of hamstring and groin injuries include greater distance of sprinting during average effort, the less predictable flight of the ball, greater length of the playing arena, repetitive loads in kicking and longer duration of games. Kicking almost certainly increases the rate of quadriceps strains. It is hoped that a knowledge of when the muscle groups are working in the kicking cycle will help to explain this unique injury pattern of Australian Rules Football and ultimately lead to injury prevention.

The aims of this study were to examine the kinematics of the drop punt kick and the role of major lower limb muscle groups in powering the kick.

2 Methods

2.1 Subjects
Four professional AFL players were recruited as subjects (after two other players had been used as pilots). All four final subjects were right foot dominant.

2.2 The kicking task
Each subject was filmed while performing six right-foot and six left-foot drop punt kicks. The players were running in to kick on a hard surface and therefore wore their club issue Puma running shoes. They were instructed to kick a football to an imaginary target 40 m straight ahead of them.

2.3 Measurement and analysis of lower limb kinematics
A two-dimensional analysis of the lower limb kinematics was undertaken. Each subject was filmed in the sagittal plane using a high speed NAC video camera operating at 200 Hz. An array of 17 retroreflective markers was placed on each subject (Table 1). The limb closer to the camera (which had more markers) was the kicking foot in half of each of the left-foot and right-foot trials. The array of markers defined the limb segments, joint centres and local coordinate axis systems. After acquiring the video data, the coordinates of each marker were digitised using the BIOVISION system (UNSW, Sydney). Every fourth frame was digitised, resulting in an effective sampling rate of 50 Hz.

Ground reaction forces were measured on the stance leg during the kick. The acquisition of video and force platform data was synchronised. Data sets consisting of the time histories of marker coordinates and ground reaction forces were used to calculate the dynamics of the lower limbs during each kick. An inverse dynamics approach was used. The analysis was undertaken using programs written in ASYST.

2.4 Measurement and analysis of electromyography (EMG)
Surface EMG of major muscle groups was acquired using a Flexcomp system (Thought Technology, Montreal). The electrodes were amplified at the source and connected to the receiver which was worn around the subject's waist and connected to the data acquisition system by a fibre optic cable. The EMG data were acquired at a sampling rate of 1000 Hz. After acquisition, data were filtered with a dual-pass fourth order digital low-pass Butterworth filter with a cut-off frequency of 15 Hz. The input impedance of the electrode units is 1,000,000 MΩ and the Common Mode Rejection Ratio over 20–500 Hz is greater than 130 dB. Seven channels of EMG data were acquired, (i.e. right hamstring, left hamstring, right quadriceps, left quadriceps, right gluteal, left gluteal and rectus abdominus muscle groups). A foot switch was attached to the heel of the non-kicking leg. This was used to synchronise the data with values obtained from the force platform. Initial heel contact of the non-kicking leg (which occurred near the end of the kicking leg backswing) was chosen as the point of synchronisation.

Table 1. Location of markers.

Marker	Name	Location
1	Forehead	Front of headband worn
2	Rearhead	Rear of headband worn
3	Acromion	Acromion of near side
4	PSIS	Posterior Superior Iliac Spine of near side
5	ASIS	Anterior Superior Iliac Spine of near side
6	Greater trochanter	Greater trochanter of near leg
7	Mid thigh	Middle of vastus lateralis, near leg
8	Femoral condyle	Lateral femoral condyle, near leg
9	Head of fibula	Head of fibula, near leg
10	Lateral malleolus	Lateral mallelous, near leg
11	Calcaneous	Most posterior aspect of heel of shoe, near leg
12	5th metatarsal	Lateral aspect of shoe, over base of 5th metatarsal, near leg
13	Ball large	Middle of one half of football
14	Ball small	Middle of other half of football
15	Thigh	Middle of vastus medialis, far leg
16	Knee	Medial joint line of knee, far leg
17	Ankle	Medial malleolus, far leg

Maximum voluntary contractions (MVC) for each muscle group were obtained prior to the kicking trials. Standard muscle testing protocols were used that placed each muscle approximately at its resting length. Resistance was applied by an investigator and verbal encouragement was given. For all channels and each subject the isometric MVCs were less than the amplitude of the conditioned EMG signal obtained during the kicking trials. For the purposes of analysis the conditioned EMG signals were expressed as a percentage of the maxima recorded during the kick.

2.5 Examination of relationship between kinematics and muscle activation
Kinematic results were examined and the kicking motion for each subject was divided into six phases using hip and knee joint angles, and the position of the ball. The average EMG activity for each muscle group in each phase was pooled for the right-foot and left-foot kicks of each player. The timing of each phase for the EMG signal was determined by correlating heelstrike on the foot switch channel (t =0) with the kinematic and kinetic output.

The relationships between muscle activation, as indicated by EMG, and lower kinematics were examined qualitatively. The type of muscle contraction, isometric, eccentric or concentric, was interpreted based on an examination of the direction of the angular motion of the appropriate joint. For example, if the knee was extending and hamstrings were active, this was interpreted as eccentric hamstrings contraction. The effect of muscle force moments applied by two joint muscles can confound this method of interpretation.

3 Results

The phases of kicking were divided as follows: (a) Run-up and approach; (b) Backswing; (c) Wind-up; (d) Forward swing; (e) Follow-through; (f) Recovery. All phases are named with respect to the kicking foot, with (a) being a pre-swing phase, (b), (c) and (d) being the swing phases before ball contact and (e) and (f) the swing phases after ball contact. The phases are shown diagrammatically in Figure 1. The following section describes these phases in detail for a right foot kick.

(a) Run-up and approach (finishes – right foot toe off)

Figure 1(a). End of approach.

A six to eight step approach was used. Kinematic and EMG activity in this phase appeared to be consistent with normal running, which has been previously described (Mann, Moran and Dougherty, 1986). This phase was considered to finish with toe-off of the right leg. The last 0.2 s of the approach was designated the push-off period, in which the right hamstring sustained a strong concentric contraction to commence the backswing. The ball was held during the run-up by both hands and released by the left hand before the push-off and then by the right hand at the end of the approach. The left arm was abducted during the remainder of the motion to balance the body. The progression of the ball after release was downward (due to gravity) and forward (due to maintenance of momentum from the hand carrying it) (Hay, 1993).

(b) Backswing (starts – right foot toe off; finishes – right hip maximum extension)

Figure 1(b). Right foot toe off.

Backswing started with both feet off the ground and finished when the right hip was fully extended, lasting approximately 0.1 s. Left foot contact generally occurred near the end of the backswing. Initial stance contact of the left foot was with the heel in three subjects and the forefoot in the other. The activity of the right hamstring started to reduce in the backswing and the left gluteal was active to coincide with left leg support.

(c) Wind-up (starts – right hip maximum extension; finishes – right knee maximum flexion)

Figure 1(c). Right hip maximum extension.

The wind-up phase was characterised by right hip flexion and right knee flexion and also lasted approximately 0.1 s. The hip flexion was initiated and was continued for the remainder of the kicking motion, whilst the knee flexed to wind itself up for a rapid extension during the next phase. Although the right knee was still flexing, the right quadriceps took over from the hamstring as the most active muscle group. Presuming that all quadriceps muscles were active at this time, the right rectus femoris assisted the hip flexors with this motion, while the other quadriceps muscles were performing an eccentric contraction to decelerate knee flexion and initiate the forward swing.

(d) Forward swing (starts – right knee maximum flexion; finishes – initial ball contact)

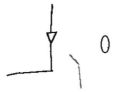

Figure 1(d). Right knee maximum flexion

The forward swing was a phase of continued hip flexion and rapid knee extension, which lasted approximately 0.05 s and was considered to finish at the time of ball contact. There was continued substantial activity of the right quadriceps during this phase and increasing activity of the rectus abdominus. At the time of ball contact, the right knee was still flexed approximately 50°, the lower leg was moving forward with an angular velocity of approximately $1400°.s^{-1}$ (24.4 rad s^{-1}) and the right ankle was held fixed in plantar-flexion.

(e) Follow through (starts – initial ball contact; finishes – right knee maximum extension)

Figure 1(e). Ball contact

The initial follow-through was considered to last until the knee reached full extension, which was slightly less than 0.1 s after initial ball contact. Overall muscle activity was much lower in this phase, with two of the four subjects exhibiting a significant eccentric contraction of the right hamstring. From this study, it was not possible to determine the exact duration of ball contact, although it was less than 0.015 s.

(f) Recovery (starts – right knee maximum extension; finishes – right hip maximum flexion)

Figure 1(f). Right knee maximum extension

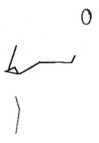

Figure 1(g). Right hip maximum flexion

The right hip continued to flex during the recovery part of the follow-through, for approximately 0.2 s after the right knee had finished extending. The total apparent range of right hip movement was up to 150°, over a period of 0.45 s. This range was only seen in two dimensions and some pelvic rotation may have contributed to this. The left hip appeared to move through less than 20° over the same time period. During most kicks, the left foot left the ground during the recovery phase. Muscle activity for all groups was low during this phase.

The average EMG values for left and right foot kicks in each phase are presented in Figures 2 and 3. Error bars presented represent standard deviations between subjects. Standard deviations within subjects were generally lower except for the rectus abdominus channel, which was possibly affected by artefact with increased signal appearing late in the kicking motion possibly corresponding to compression of the electrode by the receiver. Between subjects, results were still fairly consistent with the exception of kicking-leg hamstring activity in the follow-through, which had high activity in two subjects and low activity in the other two.

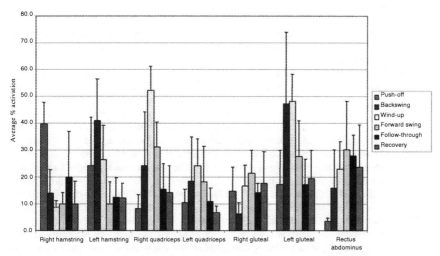

Figure 2. EMG of right foot kick.

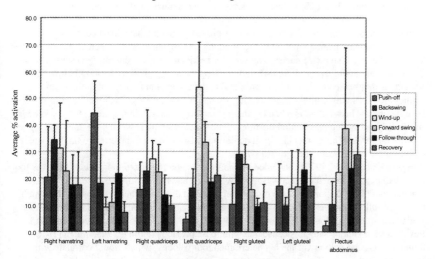

Figure 3. EMG of left foot kick.

4 Discussion

A study of the dynamics of kicking was undertaken using four first grade AFL players. While the number of subjects was low, the number of kicking repetitions was high. Altogether 48 trials were studied.

The aim of punt kicking is to propel the ball in a forwards and upwards direction as accurately (and often as far) as possible. Momentum is transferred from the kicking foot/ankle complex to the ball when the angular velocity of the leg is at its maximum. Kicking has been called a 'throw-like' pattern of movement, in that the movement of proximal segment is initiated early with the distal segment lagging behind. Momentum is then transferred from the decelerating proximal segment to the distal segment, which undergoes a rapid acceleration (Robertson and Mosher, 1985; Putnam, 1991; Kreighbaum and Barthels, 1996).

McCrudden and Reilly (1993) have compared EMG findings in punt kicking of a soccer ball with drop kicking finding similar muscle activity. Bollens, De Proft and Clarys (1987) divided the soccer kick into six phases and measured EMG activity for six muscles in the kicking leg during these phases. They called their phases: (1) first step; (2) second step; (3) loading phase; (4) swinging phase; (5) ball impact; (6) follow-through. They found that vastus medialis and lateralis contractions were maximal during the loading phase, at which time the knee was still flexing. They called this part of the 'soccer paradox' that a substantial amount of muscle work appears to be done eccentrically during soccer kicking. This corresponds to the 'wind-up' phase in the above description of punt kicking, where maximal quadriceps activity was also found.

Previous naming conventions for the various phases of kicking have been confusing as certain phases were named with respect to what the kicking leg is doing whereas others referred to the stance leg. The phase names refer only to the kicking leg. The terms backswing, wind-up and forward swing are preferred when describing the swing phase of kicking before ball contact, as they correctly describe the proximal to distal segment transfer of momentum. During the backswing both the thigh and leg are moving backwards (i.e. hip extending and knee flexing). The wind-up is when the thigh is moving forwards (hip flexing) but the knee is still flexing, so the leg lags behind the thigh. The forward swing is when both thigh and leg are moving forwards. The momentum built up during the wind-up is transferred to the leg, which rapidly accelerates during the forward swing.

Alexander and Holt (1974) concluded that more efficient transfer of momentum from the foot to the ball occurred when contact was made with the ankle region rather than the metatarsals. Plangenhoef (1971) reported a case of a punter who lost distance when wearing a shoe which prevented full plantar flexion at the ankle, suggesting that this was also important with respect to efficient transfer of force. The subjects kept their ankle joints relatively fixed in plantar flexion during kicking which is consistent with these previous findings. Macmillan (1976) found that angular velocity of the leg determined foot velocity, which would correlate with distance kicked given efficient transfer of force. Large angular velocities of the leg are achieved by rapid extension of the knee, assisted by a more gradual flexion of the hip. The prime movers of these movements are the quadriceps muscles with the chief antagonists being the hamstrings. Theoretically, the momentum imparted to the ball could be increased by increasing the tangential velocity at the foot, e.g. increasing the shank length or increasing the shank's angular velocity. The stabilisation of the ankle in plantar flexion effectively increases the shank length.

Overall the amount of activity in all muscle groups was high during kicking. In the current study, muscle activity was expressed as a percentage of the maximum value recorded at any stage, rather than the maximum seen on an isometric MVC. For all muscle groups, the maximum values recorded during kicking were greater than in manual contractions. This might reflect the recruitment patterns necessary for high velocity muscle contractions, although this aspect requires investigation before further research is undertaken.

The combined EMG profile for a left-foot kick was very similar to a right-foot kick (with the sides reversed). All subjects practised kicking with both feet but were right-foot dominant and felt more confident kicking on this side. The consistency of results between sides suggests similar muscle activation for dominant and non-dominant sides and also that the results of this experiment are reliable and reproducible. The only exception between sides was the stance gluteal muscle group, which showed considerably more activity during a right-foot kick (left gluteal) than a left-foot kick (right gluteal).

The quadriceps of the kicking leg showed much more activity than the stance leg. They were most active in the wind-up and forward swing phases, contracting eccentrically then concentrically. Quadriceps activity decreased substantially towards the end of the forward swing (ball contact). Robertson and Mosher (1985) also found that this occurred during soccer kicking. All subjects exhibited a concentric contraction of the kicking hamstring at the start of the backswing. There was a variation of hamstring activity during the follow-through, where two subjects showed minimal activity and the other two significant eccentric hamstring activity.

It is possible that the kicking set-up of the trials caused less eccentric hamstring activity during the follow-through phase than would be expected in some other kicks. The subjects generally showed knee angles which were quite flexed at ball contact (45° or greater) compared to previous descriptions (Elliott, Bloomfield and Davies, 1980; Baker and Ball, 1993). It has been suggested that the time of ball contact in the swing is related to the intended height of kicks (McDavid, 1985). Because the subjects were kicking in an indoor laboratory, they may have deliberately tried to kick with a low trajectory. This would mean ball contact would have occurred earlier in the forward swing (with the knee more flexed) and the hamstrings would not be required to contract as much eccentrically to decelerate the leg in the follow through. Further study could confirm whether the EMG pattern is different for the different types and trajectories of kicks made during a game.

The adductor and iliopsoas muscles were not studied due to a limitation of EMG channels. From the pilot trials, the adductor muscles appeared to have a similar profile to the hamstrings, possibly including some cross-talk. The adductors of the kicking leg would presumably show increased activity in around-the-corner kicks. The gluteal muscles also showed a similar activity profile to the hamstrings, although the gluteals of the stance leg contracted around the time of heel contact, presumably to abduct the stance hip and prevent dropping of the pelvis to the opposite side. The iliopsoas muscles of the kicking leg would presumably be active in the backswing and wind-up to flex the hip.

In terms of injury mechanisms some observations and comments can be made. The hip joint of the kicking leg moves through a large range over a short time. As the hip joint of the stance leg is relatively stationary large joint reaction forces would be applied through the pelvis, in particular the pubic symphysis. This may explain why osteitis pubis is commonly seen in Australian football (Fricker, Taunton and Ammann, 1991). The individual differences observed in hamstring activation during the follow through phase may represent a difference in hamstring function which may correlate with injury; however, it might simply demonstrate the different muscular effort involved in each kick.

The injury in Australian football that most commonly has the mechanism of kicking is the quadriceps strain (Orchard, Wood, Seward and Broad, 1998). It is likely that during a kick, the quadriceps muscle tears at the time of ball contact, when the ball transmits a retarding torque on the extending thigh. If this is the case, the muscle (rectus femoris) strain occurs during a concentric movement when the muscle is inactive (during the follow through phase) – not during an eccentric contraction. There is potential for this torque to be lowered, and hence the rate of injury to be reduced, by changing the mechanics of the ball (such as lowering the pressure of inflation, which may increase contact time and decrease peak retarding torque).

In conclusion, drop punt kicking in Australian Rules football has similarities with soccer kicking and other 'throwlike' motion patterns. The quadriceps (particularly of the kicking leg) and hamstring muscle groups are both highly active and exhibit both concentric and eccentric contractions during the various phases of kicking.

5 Acknowledgements

This research was funded by an Australian Sports Medicine Federation-Syntex research grant.

6 References

Alexander, A. and Holt, L.E. (1974) Punting, a cinema-computer analysis, *Scholastic Coach*, 43, 14–16.

Baker, J. and Ball, K. (1993) Biomechanical considerations of the drop punt (abstract), *Australian Conference of Science and Medicine in Sport*, Melbourne, Sports Medicine Australia.

Bollens, E.C., De Proft, E.and Clarys, J.P. (1987) The accuracy and muscle monitoring in soccer kicking, in *Biomechanics X-A* (ed B. Johnsson), Human Kinetics, Champaign, IL, pp. 283–288.

Elliott, B.C., Bloomfield, J. and Davies, C.M. (1980) Development of the Punt Kick: A Cinematographic Analysis, *Journal of Human Movement Studies*, 6, 142–150.

43

Fricker, P.A., Taunton, J.E. and Ammann, W. (1991) Osteitis pubis in athletes: Infection, inflammation or injury? *Sports Medicine*, 12, 266–279.

Hay, J.G. (1993) *The Biomechanics of Sports Techniques*, Prentice-Hall, Englewood Cliffs, New Jersey, pp. 272–274.

Kreighbaum, E. and Barthels, K.M. (1996) *Biomechanics: A Qualitative Approach for Studying Human Movement*, Allyn and Bacon, Needham Heights, MA, pp. 338–345, 377–378.

Macmillan, M.B. (1976) Kinesiological determinants of the path of the foot during the football kick, *Research Quarterly*, 47, 33–40.

Mann, R.A., Moran, G.T. and Dougherty, S.E. (1986) Comparative electromyography of the lower extremity in jogging, running, and sprinting, *American Journal of Sports Medicine*, 14, 501–510.

McCrudden, M. and Reilly, T. (1993) A comparison of the punt and the drop-kick, in *Science and Football II* (eds T. Reilly, J. Clarys and A. Stibbe), E. and F.N. Spon, London, pp. 362–366.

McDavid, R.F. (1985) Football kicking, in *Human Performance: Efficiency and Improvements in Sport, Exercise and Fitness* (ed T.K. Cureton), American Alliance for Health, Physical Education, Recreation and Dance, Reston, Virginia, pp. 415–421.

Orchard, J., Wood, T., Seward, H. and Broad, A. (1998) Comparison of injuries in elite senior and junior Australian Football, *Journal of Science and Medicine in Sport*, 1, 82–88.

Plangenhoef, S. (1971) *Patterns of Human Motion: A Cinematographic Analysis*, Prentice-Hall, Englewood Cliffs, New Jersey, pp. 98–105.

Putnam, C.A. (1991) A segment interaction analysis of proximal-to-distal sequential segment motion patterns, *Medicine and Science in Sports and Exercise*, 3, 130–144.

Robertson, D.G.E. and Mosher, R.E. (1985) Work and power of the leg muscles in soccer kicking, in *Biomechanics IX-B* (eds D.A Winter, R.W. Norman, R.P Wells, K.C. Hayes, A.E. Patla), Human Kinetics, Champaign, IL, pp. 533–538.

Seward, H., Orchard, J., Hazard, H. and Collinson, D. (1993) Football injuries in Australia at the elite level. *Medical Journal of Australia*, 159, 298-301.

7 THE EFFECTS OF SOLE CONFIGURATION ON GROUND REACTION FORCE MEASURED ON NATURAL TURF DURING SOCCER SPECIFIC ACTIONS

N. SMITH, R. DYSON and T. HALE
University College Chichester, Chichester, UK

1 Introduction

Until recently the starting point of the design of any new soccer shoe was based on the low cut, soft leather upper boot with polyurethane or aluminium studs designed in the 1960s. In recent years moulded blade sole units have evolved, some claiming to increase performance, some to reduce injury risk. However, these claims are not reinforced by data in the scientific literature.

Footwear performance needs to be addressed by consideration of the shoe–surface interface, rather than isolated properties of either one of these variables. Data regarding the performance of soccer players on natural turf with soccer footwear remain scarce. Only Saggini and Vecchiet (1994) have reported measurements of ground reaction forces from soccer players during straight running on natural turf. Furthermore, the ecological measurement of forces at the shoe–surface interface during soccer specific movements remains absent.

The importance of the shoe–surface relationship in soccer has been stated (Ekstrand and Nigg, 1989; Inklaar, 1994) due to the specific footwork associated with ball skills. With the effect of soccer specific movements such as turning and shooting on the ground reaction force unreported, it was the aim of the present study to measure such movements on a natural turf surface whilst also investigating the effect of using the modern, redesigned soccer sole unit.

2 Methods

2.1 Subjects
Eight male collegiate soccer players (mean age 24.4 ± 3.1 years, mass 78.3 ± 9.1 kg) volunteered for the study. All subjects wore shoe size UK 8 or 9, were right foot dominant and reported no musculoskeletal injuries at the time of testing. Soccer players were preferred due to their familiarity with the soccer specific moves required. The experimental protocol was explained, with each subject required to provide informed consent. Subjects were reminded of their right to cease participation in the study at any time.

2.2 Procedure

Each subject performed three soccer specific moves, each repeated five times. The moves were the shot, Cruyff turn, and drag-back turn. Turns were performed according to Football Association coaching guidelines (Hughes, 1994) with the ball placed beside the force platform, and each movement performed with the left foot on the force platform. The mean approach velocity for all subjects was 3.3 ± 0.3 m·s^{-1}. After completion of the turn, each subject was required to accelerate away with the ball under control, to replicate soccer match conditions. For the shot, subjects were instructed to contact the ball with the instep and strike hard and low to a target area (as if taking a penalty kick) and land on their striking foot. Subjects repeated the series of trials in a pair of standard six-studded soccer boots (Mizuno Pro Model) and a pair of moulded sole boots (Adidas Equipment Velez Traxion). Force data were collected for 3 s at a sampling rate of 1000 Hz.

2.3 Instrumentation

A Kistler force platform (type 9281B) was covered in a natural turf surface. The force platform was connected via an analogue to digital converter to an IBM compatible computer running Kistler Bioware 3.0 data acquisition and analysis software. Subject approach velocity was measured using infrared timing gates (Cla-Win Timer, Chichester Institute, UK) placed 3 m apart at hip-height. The approach to the force platform was also natural turf to maintain soccer pitch conditions. The force platform natural turf covering was prepared some days prior to testing and then kept irrigated to the appropriate level until testing commenced. Turf moisture was assessed with a soil wetness meter (Rapitest, UK), with tests proceeding only if a minimum value of 3 (scale 1–4) was attained.

The performance of all the trials was recorded on videotape (Panasonic VHS Supercam AG–DP800E, EG) to allow the subsequent slow motion review of all data. One camera was positioned to film foot contact with the force platform, whilst the second synchronised camera recorded the approach to assess performance technique.

2.4 Data analysis

From the five trials, the four most technically correct in performance were analysed further. Contact time with the platform was identified using a 10 N threshold in the Bioware software. Vertical, anterior–posterior, medial and lateral forces were analysed, and normalised for each subject's body weight. Values of maximal and average coefficient of friction in the horizontal plane were also measured. Frictional values were computed from the modulus of the resultant force in the horizontal plane divided by the vertical force. For some trials, this computation gave large values of friction when the vertical force was low. A threshold of 10 N was used in the present study to gather maximum data, as the effect on friction of studded footwear on natural grass was previously unknown.

3 Results

Mean maximal values of force were computed and compared between the two sole configurations. Overall, greater forces occurred in all three planes when studded footwear was worn.

Table 1. Comparison of mean maximum forces (± SE) occurring during the shot for 8 subjects.

	Studded sole	Moulded sole
Vertical	3.95 ± 0.14 BW	3.74 ± 0.10 BW
Anterior–posterior	1.27 ± 0.11 BW	1.22 ± 0.10 BW
Mediolateral	0.80 ± 0.06 BW	0.67 ± 0.06 BW

For each movement the mean graphical force data for one subject (who was representative of the subject group) are presented. Graphical data show forces recorded in the vertical (Fz), anterior–posterior (Fy) and mediolateral (Fx) planes, with the initial peak indicating heel strike.

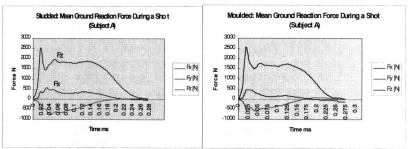

Figure 1. Ground reaction force during a shot in moulded and studded footwear.

Trials using the standard six-stud configuration yielded similar traces to the moulded sole. However, consistent differences were evident in the mediolateral (Fx) and anterior–posterior (Fy) traces, possibly corresponding to the discrete medial and lateral stud placement at the heel and forefoot of the studded boot. Such a difference was presumed to be a function of the footwear sole configuration, possibly giving an increased force loading capacity as the medial studs also interfaced with the turf. During the execution of a shot the body inclines toward the left supporting leg to enable the striking foot to obtain the correct position for ball strike. A medial (Fx) force occurred throughout the stance phase as a reaction to the lateral body incline.

Table 2. Comparison of mean maximum forces (± SE) in the Cruyff turn for 8 subjects.

	Studded sole	Moulded sole
Vertical	2.31 ± 0.15 BW	2.05 ± 0.10 BW
Anterior–posterior	1.02 ± 0.09 BW	0.94 ± 0.06 BW
Mediolateral	0.31 ± 0.04 BW	0.28 ± 0.04 BW

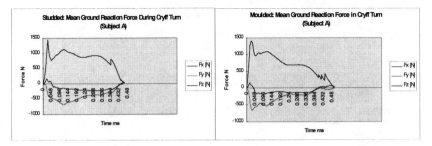

Figure 2. Ground reaction force during a Cruyff turn.

Overall, greater forces occurred in all three planes when studded footwear was worn during the Cruyff turn. The vertical (Fz) force showed a three peak trace, with the greatest initial force corresponding to heel strike. A third peak was evident in the vertical (Fz) trace when approaching ball contact as the swinging leg decelerated as it passed the support leg, creating more vertical force. Slight differences in contact time towards the end of the move caused peak artefacts during late stance. Mediolateral forces were smallest.

To enable the Cruyff turn to be executed correctly the body movement takes the body centre of mass medially, away from the supporting leg towards the right leg, resulting in forces directed predominantly towards the lateral border of the supporting sole.

The drag back movement was performed with two distinct techniques. Some subjects used one foot contact, whilst others preferred an initial foot contact to halt forward motion and initiate ball contact, and a second foot contact to accelerate away from the platform. As the two techniques enabled a technically correct turn to be executed, both styles were analysed.

Table 3. Comparison of mean maximum forces (± SE) in the drag-back turn for 8 subjects.

	Studded sole	Moulded sole
Vertical	2.28 ± 0.07 BW	2.06 ± 0.05 BW
Anterior–posterior	1.02 ± 0.08 BW	0.99 ± 0.05 BW
Mediolateral	0.31 ± 0.12 BW	0.13 ± 0.02 BW

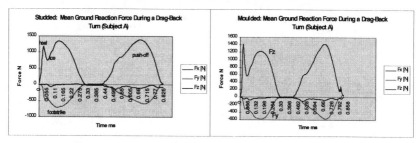

Figure 3. Ground reaction force of the supporting leg during a two contact drag-back turn.

The force data for the drag-back showed two-foot contacts of the supporting leg, with maximum values again greater for studded footwear. The large amount of external rotation of the support leg requires a different foot placement for the propulsive phase of the movement. Therefore, for the second contact, Fy becomes indicative of mediolateral force, and Fx of anterior–posterior force. For this subject the vertical force (Fz) showed a large impact peak in moulded footwear. This occurred because one trial generated an exceptionally large initial peak in the moulded sole, which acted to increase the magnitude of the mean vertical force trace. However, such a trial would not influence results overall, as reported mean maximal vertical forces occurred during the second foot contact.

Table 4. Average and maximal friction coefficient for the studded and moulded sole.

	Studded sole		Moulded sole	
	Average	Maximal	Average	Maximal
Cruyff	0.57	3.04	0.59	5.25
Shot	0.37	3.31	0.39	4.68

During the shot and Cruyff turn the average friction values were similar for both the moulded and studded sole configurations as shown in Table 4. The high maximal values occurred at the extremes of foot contact, where soccer players require high levels of traction to prevent slippage and loss of control. Poor foot fixation generally occurs close to foot strike or close to toe-off, therefore data analysis encompassing these end points was undertaken in order to better reflect the movement for ecological validity.

4 Discussion and Conclusions

For the shot, Cruyff turn and drag-back turn mean maximal vertical forces for all 8 subjects were greater by 0.26, 0.21 and 0.22 BW respectively when the standard six-studded soccer boot was worn. These mean maximal forces were reduced when wearing the moulded boot by 5%, 11% and 10% in the shot, Cruyff turn and drag-back turn, respectively. The horizontal mean maximal forces were also considerably less in the mediolateral plane (16%; 10%; 58%) with a smaller reduction evident in the anterior-posterior plane (4%; 8%; 3%). These results suggested that the moulded sole with its larger bladed contact area allowed these three soccer specific movements to be performed while exposing the body to reduced levels of ground impact. This property of the moulded sole, with its blade configuration should reduce the risk and severity of impact related injury and appears advantageous over the standard studded sole. The results indicate that the average coefficient of friction was similar for both soles while the moulded sole boot has the ability to allow greater horizontal force generation at the beginning and end of the ground contact when slippage is most likely. This was reflected in the much larger maximal coefficient of friction values obtained for the moulded sole boot than the standard six-studded boot.

During the performance of three soccer movements the wearing of the moulded sole boot with a blade configuration (Adidas Equipment Velez Traxion) resulted in reduced maximal forces during ground contact, yet allowed the generation of relatively high horizontal forces (and derived friction) during initial and final ground contact, acting to reduce the chance of slippage at these times.

5 References

Ekstrand, J. and Nigg, B.M. (1989) Surface-related injuries in soccer, *Sports Medicine*, 8, 56–62.
Hughes, C. (1994) *Soccer Tactics and Skills,* Queen Anne Press, Herts, London.
Inklaar, H. (1994) Soccer injuries II: aetiology and prevention, *Sports Medicine*, 18, 81–93.
Saggini, R. and Vecchiet, L. (1994) The foot–ground reaction in male and female Soccer players, in *Biomechanics in Sport XII* (eds A. Barabas and G. Fabian), University Park Press, Baltimore, pp. 213–215.

PART TWO

Fitness Test Profiles of Footballers

8 ASSESSMENT OF THE PHYSIOLOGICAL CAPACITY OF ELITE SOCCER PLAYERS

J. BANGSBO and L. MICHALSIK
Institute of Exercise and Sport Sciences, August Krogh Institute,
University of Copenhagen, Denmark

1 Introduction

The mean distance covered by male elite outfield players during a soccer match has been estimated to be about 11 km for a 90-min game. This value does not represent the total energy demand on players during a match since, in addition to walking and running, the players perform many other energy demanding activities (e.g., acceleration, change of direction, deceleration, jumping, static muscle contraction and getting up from the ground). The energy cost during a match may be better expressed by measurements performed during or immediately after the match such as heart rate, rectal temperature, muscle glycogen use and fluid loss. Based on these determinations the energy contribution from aerobic sources can be estimated to be about 70% of maximal oxygen uptake (VO_2 max) for elite players (Bangsbo, 1994). Exercising at such an average exercise intensity for 90 min puts an emphasis on the ability to perform prolonged intermittent exercise (endurance). In addition, a player should be able to exercise repeatedly at a high intensity, to sprint and to develop a high power output (force) in single match situations such as kicking, jumping and tackling (Figure 1). The basis for performance within these categories are the characteristics of the cardiovascular system and of the muscles, combined with the interplay of the nervous system. These characteristics are to some extent determined by genetic factors but can also be developed by training.

The players have to adapt to the requirements of the game, i.e. players at top level need to have a physical capacity that relates to the physiological demands required in soccer. This review will discuss how the physiological capacity of soccer players can be determined with a focus on aerobic and anaerobic performance in soccer. Data for elite male soccer players will be given in relation to position in the team.

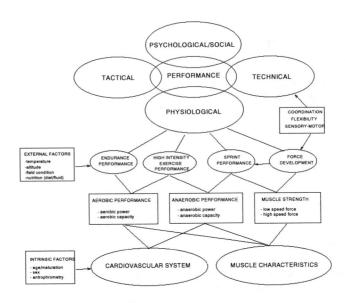

Figure 1. Physiological factors within a holistic model of soccer performance.

Performance during a soccer match is determined by a player's tactical, psychological/social, technical and physiological capacity. These areas overlap and influence each other. The physiological factors can be divided into several match performance abilities (upper part). These are dependent on variables which in part can be evaluated separately (middle part). Cardiovascular capacity, neural factors and muscle characteristics comprise basic components of physiological performance that are determined by both intrinsic biological make-up and training status (lower part). Performance during match-play is also influenced by various external factors, including environment anutrition.

2 Aerobic Performance

Aerobic performance is determined both by aerobic power and aerobic capacity. The former component reflects the ability to produce aerobic energy at a high rate and is characterised by VO_2 max. Aerobic capacity expresses the ability to sustain exercise for a prolonged period and is synonymous with endurance.

Several studies have determined the VO_2 max for male elite adult players, and mean values between 56 and 69 ml·kg⁻¹·min⁻¹ have been found (see Reilly, 1994). These values are similar to those obtained in other team sports, but are considerably lower than values for elite athletes within endurance sports, where VO_2max levels higher than 80 ml·kg⁻¹·min⁻¹ are observed (Reilly and Secher, 1990). Most studies of soccer players report a large variation in VO_2 max, which is partly associated with the different positions of the players within the team. Based on results obtained for Danish elite players, full-backs and midfield players appear to have the highest VO_2 max values, and goalkeepers and central defenders the lowest, but it has to be noted that a broad range was found within each positional group (Fig. 2). It has been observed that players in a Norwegian international top class team (Rosenborg) had higher VO_2 max values than players from a lower ranked team playing in the same league (67.6 vs. 59.9 ml·kg⁻¹·min⁻¹; Wisløff et al., 1998). On the other hand, in a study of Danish top-class players no difference in VO_2 max was observed between regular and non-regular first team players, which indicates that this variable is not crucial for good performance in soccer (Bangsbo, 1994 a,b). In general VO_2 max does not appear to be a sensitive measure of performance. In a group of cyclists with almost the same VO_2 max, it was observed that the time to exhaustion in a cycle test varied from 16 to 90 min.

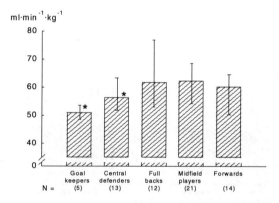

Figure 2. Maximum oxygen uptake (VO_2 max) of Danish elite male players. Means and range are given. (* Significantly different from full-backs, midfield players and forwards.)

3 Endurance Performance

Endurance performance is associated with the blood lactate response during submaximal continuous exercise. This makes blood lactate measurements useful for the evaluation of l ong-term exercise performance in groups of players that are performing the same test (see Jacobs, 1986).

Danish elite soccer players were tested during treadmill-running and it was observed that the oxygen uptake corresponding to a given blood lactate concentration was higher for full-backs and midfield players, than for central defenders and goalkeepers (Figure 3). In accordance with the VO_2max determinations, the forwards had values in between values for full-backs and midfield players. When the VO_2 for a given blood lactate concentration was expressed in relation to VO_2 max, slightly higher values were also obtained for the full-backs and midfield players (Bangsbo, 1994a). As for VO_2 max, there was no difference between regular and non-regular first team players in the relationship between submaximal treadmill speeds and blood lactate concentrations (Bangsbo, 1994a), suggesting that these determinations cannot be used to distinguish between players at a top level.

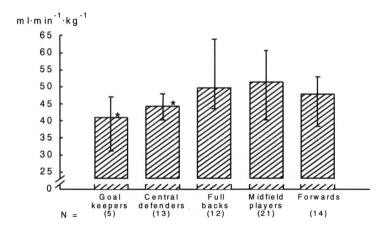

Figure 3. Oxygen uptake during treadmill running resulting in a blood lactate concentration of 3 mmol·l^{-1} ($VO_{2\ lac3}$) for Danish elite male soccer players. Means and range are given.
(* Significantly different from fullbacks, midfield players and forwards.)

4 Anaerobic Performance

When considering anaerobic performance a distinction has to be made between power and capacity. Anaerobic power represents the highest rate of anaerobic energy release, whereas anaerobic capacity reflects the maximum anaerobic energy production an individual can obtain in any exercise bout performed to exhaustion. It should be pointed out, however, that no proper measurement exists to determine anaerobic capacity (Bangsbo, 1997).

Several methods have been used to evaluate the maximal performance of soccer players during short-term exercise and thus, indirectly anaerobic power (see Reilly, 1994). Among these are the stair run test developed by Margaria et al. (1966). Using this test, di Prampero et al. (1970) found values for soccer players that were 5–15% lower than for middle-distance runners, sprinters and pentathletes. On the other hand, Withers et al. (1977) reported that soccer players had values about 20% higher than basketball players, walkers and runners. Similarly, Apor (1988) observed that Hungarian elite soccer players had a 15–30% higher level of anaerobic power than an age matched control group.

5 Intermittent Exercise Performance

The measurements discussed above express to a certain degree the physical capacity of a soccer player, and they can be used for comparisons with other sports. However, none of these determinations per se accurately express the ability to perform prolonged intermittent exercise with alternating intensities or repeated high intensity exercise, as in soccer (Bangsbo and Lindquist, 1992). A number of field tests of endurance performance with a higher validity for soccer have been used.

An intermittent field test has been developed to test soccer players (Bangsbo and Lindquist, 1992). The players carry out many activities (e.g. accelerations, decelerations, stops, turns, backwards and sideways runs) which are encountered in a soccer match. The players are working intermittently switching between 15 s of high speed running and 10 s of active rest. The test lasts 16.5 min and the test result is the performance during 10 min of high intensity running. Performance in this test was related to the distance covered during a combined intermittent endurance field and treadmill test to exhaustion lasting 2–2.5 h.

In a study of elite Danish players the mean distance covered during the test by full-backs and midfield players was significantly longer than for goalkeepers and forwards (Bangsbo, 1994b). It is worthwhile emphasizing that in contrast to the results obtained during treadmill running, the central defenders performed better than the forwards during this type of exercise. Thus, it seems that the intermittent field test adds valuable information to that which was obtained by testing in laboratory conditions.

Recently two other field tests of relevance for soccer have been developed namely, the Yo-Yo intermittent endurance test and the Yo-Yo intermittent recovery test. In the Yo-Yo intermittent tests the players perform repeated 20-m shuttle runs interspersed with a short recovery period during which the players jog. The time allowed for a shuttle, which is progressively decreased, is dictated by audio bleeps from a tape. The players should each complete as many shuttles as possible.

The Yo-Yo intermittent endurance test evaluates a player's ability to repeatedly perform intense exercise after prolonged intermittent exercise. Thus, a significant relationship ($r = 0.85$; $P < 0.01$) has been observed between the performance in the test and long-term intermittent exercise performed on a treadmill of a duration between 60 and 180 min (Michalsik, Søndergaard and Bangsbo, unpublished data). In the Yo-Yo intermittent endurance test the players have a 5-s rest period between each shuttle. The results of top class Portuguese soccer players performing the Yo-Yo intermittent endurance test are shown in Figure 4. It is clear that the full-backs and midfield players performed significantly better than the defenders and forwards. Similar differences were obtained when Danish top class players were tested (Michalsik and Bangsbo, 1994).

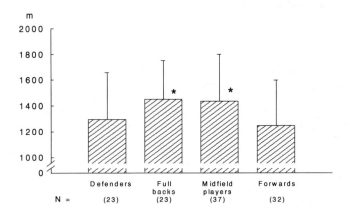

Figure 4. Performance of elite Portuguese male soccer players in the Yo-Yo intermittent endurance test (m). (* Significantly different from defenders and forwards.)

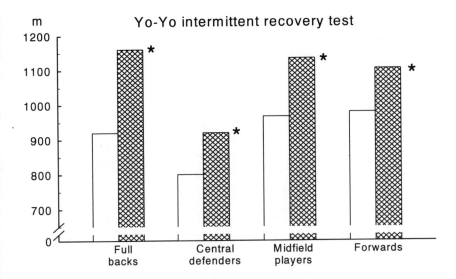

Figure 5. Performance of professional Danish male soccer players (N=27) in the Yo-Yo
intermittent recovery test (m) in different positions of a team at the start (□) and the
end (▨) of a six- week preparation period prior to the season. (* Significantly different
from defenders and forwards).

The aim of the Yo-Yo intermittent recovery test is to examine a player's ability to recover from
intense exercise. In this test the running speeds are higher than during the endurance test and
there is a 10 s period of jogging between each shuttle. The test appears to be sensitive to
changes in the performance level of top class players. When comparing performance of 44
professional players from two leading teams in the Danish top-league during the first and last
phase of the preparation period for the season, it was observed that all groups of players had
marked increases in performance during the period (Figure 5). The study also revealed that heart
rate during the test can be used to determine changes in performance, since the players had a
significantly lower heart rate at each speed after the preparation period (Figure 6). Restricting
the test so that the players only perform a certain number of shuttles during which heart rate is
monitored has the advantage that the players do not need to perform the test to exhaustion.

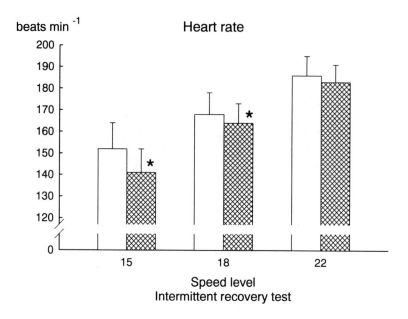

Figure 6. Heart rate of professional Danish male soccer players (N=27) at various speed levels of the the Yo-Yo intermittent recovery test at the start (□) and the end (▨) of a six-week preparation period prior to the season. Means and range are given. (* Significantly different from start of preparation period.)

The ability to sprint repeatedly during a game is often of great importance in soccer. This ability can be evaluated by having the players perform a number of sprints each separated by a break which is so short that the players can not fully recover before the next sprint. An example would be seven 34.2-m sprints with changes in direction and a 25-s recovery period during which the players jog a distance of 30 m (Bangsbo, 1994b). When players in two top-class professional Danish teams performed the test in the last phase of the preparation period and in the early phase of the season, it was observed that mean sprint time (Figure 7) and fatigue index were significantly lowered during the season. These findings suggest that the test can reveal changes in performance of the players.

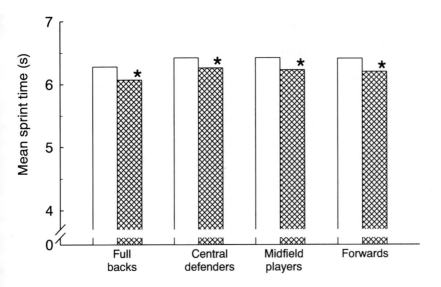

Figure 7. Mean time (s) of professional Danish male soccer players (N = 22) in a repeated sprint test at the start (□) and the end (⊠) of a six-week preparation period prior to the season. (* Significantly different from defenders and forwards.)

6 Summary

Various types of measurements have been conducted to evaluate specific aspects of the physical performance of soccer players. Several measurements obtained in the laboratory such as VO_2 max and power output during maximal cycling, may give a general picture of the physical capacity of a player. Such tests do not give a precise measure of physical performance in soccer. Instead, field tests for specific aspects of the game provide a better measure of performance in soccer. Nevertheless, it has to be recognized that no single method may allow for a representative assessment of a player's physical performance during a soccer match.

7 References

Apor, P. (1988) Successful formulae for fitness training, in *Science and Football* (eds T. Reilly, A. Lees, K. Davids, and W.J. Murphy), E. and F.N. Spon, London, pp. 95–107.

Bangsbo, J. (1994a) The physiology of football with special reference to intense intermittent exercise, *Acta Physiologica Scandinavica*, 151, 1–156.

Bangsbo, J. (1994b) *Fitness Training in Football – A Scientific Approach*, HO+ Storm, Copenhagen.

Bangsbo, J. (1997) Quantification of anaerobic energy production during intense exercise, *Medicine and Science in Sport and Exercise*, 47–52.

Bangsbo, J. and Lindquist, F. (1992) Comparison of various exercise tests with endurance performance during soccer in professional players, *International Journal of Sports Medicine*, 13, 125–132.

di Prampero, P.E., Finera Limas, F. and Sassi, G. (1970) Maximal muscular power, aerobic and anaerobic, in the athletes performing at the XIXth Olympic Games in Mexico, *Ergonomics*, 13, 665–674.

Jacobs, I. (1986) Blood lactate. Implications for training and sports performance, *Sports Medicine*, 3, 10–25.

Margaria, R., Aghemo, P. and Rovelli, E. (1966) Measurement of muscular power (anaerobic) in man, *Journal of Applied Physiology*, 21, 1661–1664.

Michalsik, L. and Bangsbo, J. (1994) Evaluation of physical performance of soccer players by the Yo-Yo intermittent endurance test, *Abstract from the 3rd World Congress on Science and Football*, Cardiff, Wales.

Reilly, T. (1994) The physiological profile of the soccer player, in *Soccer* (ed. B. Ekblom), Blackwell, Oxford, pp. 31–42.

Reilly, T. and Secher, N. (1990) Physiology of sports: an overview, in *Physiology of Sports* (eds T. Reilly, N. Secher, P. Snell and C. Williams), E. and F.N. Spon, London, pp. 465–487.

Wisløff , U., Helgerud, J. and Hoff, J. (1998) Strength and endurance of elite soccer players, *Medicine and Science in Sports and Exercise*, 30, 462–467.

Withers, R.T., Roberts, R.G. D. and Davies, G.J. (1977) The maximum aerobic power, anaerobic power and body composition in South Australian male representatives in athletics, basketball, field hockey and soccer, *Journal of Sports Medicine and Physical Fitness*, 17, 391–400.

9 ANTHROPOMETRIC AND PHYSIOLOGICAL DIFFERENCES BETWEEN GENDER AND AGE GROUPS OF NEW ZEALAND NATIONAL SOCCER PLAYERS

M.N. DOWSON, J.B. CRONIN and J.D PRESLAND
Auckland Institute of Technology, Auckland, New Zealand

1 Introduction

As reported by *NZ soccer* there are 100,000 registered players within New Zealand (NZ), which constitutes approximately 3% of the total population. Soccer was reported as the fastest growing sport in New Zealand in 1998, increasing registrations by over 8% between the ages 7–16 years and female numbers had one of the fastest growing rates in the world. Considering the increase in popularity of soccer in NZ and no recent studies of the physical condition of elite NZ soccer players, one purpose of this study was to compare physiological data between NZ and other National squads. In terms of the physiological requirements of soccer, Davis and Brewer (1993) suggested that soccer play has very similar requirements for males and females. Both assume high levels of aerobic power, muscular strength and endurance, speed endurance, flexibility and speed. Another purpose of this study was to compare the anthropometric and physiological data between gender and age of NZ elite players. Such comparisons provide information for personnel who are responsible for player development, setting training goals and arranging competition based on physiological similarity.

2 Methods

2.1 Subjects
Physiological and anthropometric assessments were conducted on five different NZ representative soccer squads (male U15, U17, U19, senior and female senior). A description of the characteristics of the players for each of the National squads is shown in Table 1.

2.2 Test procedures and equipment
The assessments of the players were performed during the NZ pre-season (January to March). All tests were performed in the morning and were preceded by a rest day. The order of the tests remained the same on each occasion.

2.3 Body composition
Each player was weighed (kg) using electronic scales (Seca Alpha Model 770). The height (cm) of each player was determined using a stadiometer (Seca Model 220). The sum of two skinfolds (Sk2) was used as a measure of fat distribution (Wilson, Russell and Wilson, 1993) for each player. A skinfold calliper (John Bull, England) was used to assess skinfold measurements (mm) from the abdomen and suprailiac crest for the males and triceps and suprailiac crest for females. The skinfold sites were taken as outlined by ISAK (International Society for the Advancement of Kinanthropometry).

2.4 Aerobic endurance
Maximal oxygen uptake (VO_2max) was determined by using the multi-stage fitness test (National Coaching Foundation, 1988). This test has been validated as a predictor of VO_2max by Leger and Lambert (1982) and by Ramsbottom, Brewer and Williams (1988).

2.5 Leg power
Following a structured warm up, the player's leg power was measured by performing a vertical jump (cm) on a contact mat system (Swift Sports Equipment, Australia). Comparisons between contact times from the contact mat system and force platforms have shown a difference of 3.6% and correlations between the two test methods (r=0.99, P<0.01) was very high (Young, Pryor and Wilson, 1995). The results reported from the mat are therefore considered accurate and reliable and the methods used for jump assessment were as outlined by Young (1994). The players started in a stationary position with both feet on the mat, but were allowed to crouch, and use their arms to assist in driving upwards. All attempts with incorrect technique and bent legs during flight time in the air were excluded with each player performing four attempts.

2.6 Sprint performance
Sprint performance was measured using dual beam electronic timing equipment (Speed Light Sports Timing System, Swift Sports Equipment). Players began from a standing start with the front foot 0.5 m from the first timing gate. The players' times for three 20-m sprints were recorded of which the two best times were averaged.

2.7 Statistical methods
The results from these tests were compared using analysis of variance (ANOVA) between the groups. A criterion alpha level of $P \leq 0.05$ was used and Bonferroni post-hoc comparisons were used to distinguish between the groups.

3 Results

The anthropometric measurements for the NZ soccer squads are shown in Table 1. No height measurements were recorded for the U19 squad. Mean height and body mass increased with age, although there were no significant differences between height and body mass for the U17 squad (175.1 ± 5.8 cm) and U19 (70.7 ± 6.8 kg) as compared to the senior men (178.8 ± 6.8 cm and 78.9 ± 6.0 kg). The women's body mass (65.9 ± 9.8 kg) was significantly higher than U15 (58.3 ± 8.9), but there was no difference in height between these groups. The Sk2 skinfold measurements showed that the U17 squad had the highest skinfold (17.6 ± 11.1 mm) and was significantly higher than senior men (13.3 ± 4.2 mm). Skinfolds for the senior women cannot be compared with the other male squads because of different measurement sites.

Table 1. Anthropometric measurements for the NZ male age groups, senior men's and senior women's soccer teams.

	Body mass (kg)	Height (cm)	SO2 (mm)
U15 (n=56)	$58.3 \pm 8.9^{*1,2,3,4}$	$168.6 \pm 8.6^{*1,2}$	14.8 ± 6.5
U17 (n=23)	$69.9 \pm 6.6^{*1,5}$	$175.1 \pm 5.8^{*1,3}$	$17.6 \pm 11.1^{*}$
U19 (n=25)	$70.7 \pm 6.8^{*2,6}$	–	14.7 ± 6.9
Senior Men (n=21)	$78.9 \pm 6.0^{*3,5,6}$	$178.8 \pm 6.8^{*2,4}$	$13.3 \pm 4.2^{*}$
Senior Women (n=20)	$65.9 \pm 9.8^{*4}$	$166.3 \pm 5.6^{*3,4}$	21.3 ± 9.3

($* P \leq 0.05$)
1 significant difference between U15 and U17
2 significant difference between U15 and U19
3 significant difference between U15 and senior men
4 significant difference between U15 and senior women
5 significant difference between U17 and senior men
6 significant difference between U19 and senior men

The results of the multi-stage fitness test (VO_2 max) are presented in Figure 1. Senior men had a higher VO_2 max (60.5 ± 2.6 ml·kg^{-1}·min^{-1}) than U15 (51.0 ± 4.2), U17 (56.1 ± 5.2), and senior women (49.1 ± 5.5). No significant differences were found between the other squads.

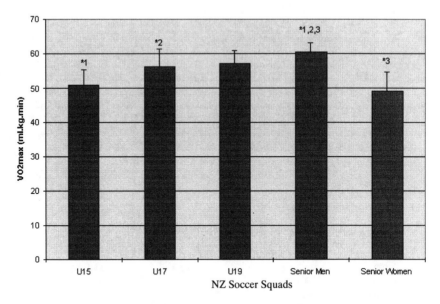

Figure 1. Mean VO₂ max values of the NZ soccer squads as predicted by the
multi-stage fitness test. (* P ≤ 0.05)
1 Significant difference between U15 and senior men
2 Significant difference between U17 and senior men
3 Significant difference between senior men and senior women

The results for leg power, as measured by a counter-movement vertical jump are shown in
Figure 2. Leg power for the male U19 squad was not measured. Jump heights increased with
age with senior men the highest (48.05 ± 4.63 cm), whereas the senior women had the lowest of
all groups (33.75 ± 3.84 cm). All groups were found to be significantly different from one
another.

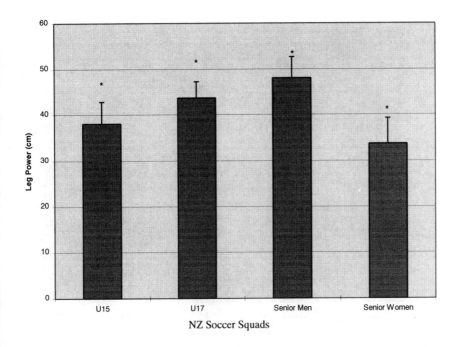

Figure 2. Counter-movement jump for NZ soccer squads as measured by
contact mat equipment. (* P ≤ 0.05)

The 20-m sprint times are presented in Figure 3. The senior men's 20-m sprint performance
(2.91 s ± 0.06 s) was faster than the U15 (3.23 ± 0.16 s) and senior women (3.29 ± 0.13 s).
The senior women were significantly slower than all the male squads with the exception of the
U15 squad.

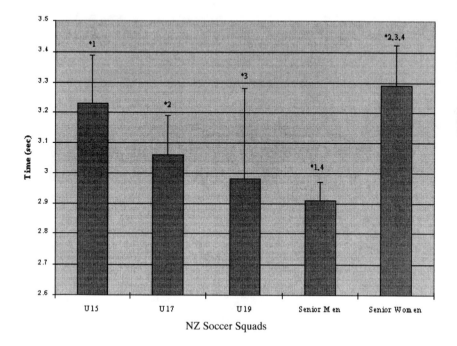

Figure 3. Mean sprint times over 20 m for the NZ soccer squads using
electronic timing equipment. (* P ≤ 0.05)
1 Significant difference between U15 and senior men
2 Significant difference between U17 and senior women
3 Significant difference between U19 and senior women
4 Significant difference between senior men and senior women

4 Discussion

4.1 Body composition

The results of this study have shown height and body mass, as expected, increase with age in
the male groups. There was no difference, however, in body mass between the U17 and U19
groups, an age span when growth is expected, but this could be explained by U17 having the
highest skinfold thickness results (17.6 mm ± 11.1). Although body fat should not necessarily
increase immediately after puberty from hormone changes, it may be anticipated that an

increase could be associated from lifestyle changes as this is the time many young adults begin socialising more, especially in sporting environments.

Both height and body mass for the age group squads are similar to other National youth squads (cited in Chin et al., 1994; Nyland et al., 1997). Senior men appear taller and heavier compared with senior elite soccer players from Asia (Chin et al., 1992; Adhikari and Das, 1993) but appear similar to European elite players (cited in Chin et al., 1992). Although there may be differences in body composition from different parts of the world, Ekblom (1986) stated body size and other anthropometric measures within normal ranges are of minor importance for physical performance.

4.2 Aerobic endurance

Soccer requires players to work at high intermittent intensities for distances of 10–14 km (Ekblom, 1986). Therefore aerobic endurance must play an important role in team performances. Hughes (1990) stated that a high level of fitness from all players in a team helps to allow for a high work rate and maintenance of good technique throughout a match. The multi-stage fitness test is a strong predictor of VO_2 max (Ramsbottom et al., 1988), a widely acknowledged measure of aerobic endurance. An increase in VO_2 max was evident with age although it appeared to display a plateau with the U19 squad. Such findings can be explained by Astrand and Rodahl (1986) who stated VO_2 max attained a peak between 18–20 years of age. Senior women have a significantly lower VO_2 max (49.1 $ml \cdot kg^{-1} \cdot min^{-1}$ ± 5.5) than senior men (60.5 ± 2.6). This suggests that senior male soccer players can work at a higher rate than senior females although total distances covered in a match may be similar for both groups (Davis and Brewer, 1993). There were no differences between women and male age group squads, although the women did produce the lowest average.

In comparison to other published literature, it appears VO_2 max results for a NZ age group are lower than that of other countries' players. The range of VO_2 max values found in this study between U15 and U19 was 51.0–57.0 $ml \cdot kg^{-1} \cdot min^{-1}$ compared with Hungarian Junior National squad (70.0), Hungarian junior club (66.0), United States Junior squad (61.8) and the Canadian Junior squad (58.3) (cited Chin et al., 1992). Elite women's physiological data cited in Davis and Brewer (1993) indicated VO_2 max values between 47.1–57.6 $ml \cdot kg^{-1} \cdot min^{-1}$. The present findings show that NZ women are within this range, although their VO_2 max was in the lower quartile. This may mean both NZ junior players and women should concentrate more on developing their aerobic endurance to be able to compete with other nations. Ekblom (1986) stated that a higher VO_2 max would improve soccer performance especially in the second half. The NZ senior men's VO_2 max found in this study appear similar in comparison to published results from other countries as listed in Table 2. Also listed in Table 2 are FIFA rankings as released in January 1999 which, although not directly corresponding to the time of testing, can provide an indication of the relationship between aerobic endurance and performance. There does not appear to be a strong relationship between international ranking and VO_2 max values. New Zealand's current ranking is 100, which may suggest rather than spending additional time of aerobic endurance, this time could be spent on other performance related activities.

Table 2. New Zealand senior men's mean VO_2max values compared with published elite senior soccer players from other Countries and their FIFA ranking, January 1999.

Country	$ml \cdot kg^{-1} \cdot min^{-1}$	World ranking (FIFA 1999)
Germany (1988, cited in Chin et al., 1992)	62	5
Czechoslovakia (Heller et al., 1993)	59.3	7
England (Davis, Brewer and Atkin, 1992)	60.4	11
Sweden (Ekblom, 1986)	62	18
Austria (1988, cited in Chin et al., 1992)	58.3	22
Kuwait (1987, cited in Chin et al., 1992)	51.9	49
Australia (1977, cited in Chin et al., 1992)	62	50
Canada (1986, cited in Chin et al., 1992)	58.7	90
New Zealand (1999)	60.5	100
India (Adhikari and Das, 1993)	59.3	116
Hong Kong (Chin et al., 1992)	59.1	131

4.3 Leg power and sprints

Counter-movement jumps are commonly performed in soccer whilst contesting for the ball in the air. Sprint times over short distances are important in soccer to reach a certain position or make contact with the ball before the opposition. Both leg power and sprints improve with age, thus reinforcing that senior players have an advantage over younger opposition. The women's squad was significantly lower than all male squads in leg power and slower than all male squads with the exception of the U15. Therefore, if arranging suitable competition for senior women against male squads, they appear best matched with the U15 male squad, in terms of power and speed.

5 Summary

In summary, senior men generally demonstrated greater aerobic endurance and were faster and more powerful than the U15, U17, U19 and women's groups. The data for the NZ senior men's soccer squad appear to be similar to that for other published national squads. Therefore, it may be concluded that players' skill and talent are possibly the limiting factor to NZ men's success in world soccer. However, both male age groups and senior women produced VO_2max results that were also lower than other published national squads, which suggests this is one area to improve on for international competitiveness.

6 References

Adhikari, A. and Das, S.K. (1993) Physiological and physical evaluation of Indian national soccer squad, *Hungarian Review of Sports Medicine*, 34, 197–205.

Astrand, P.O. and Rodahl, K. (1986) *Textbook of Work Physiology*, McGraw-Hill, New York.

Chin, M.K., So, R.C.H., Lo, Y.S.A. and Li, C.T. (1992) Physiological profiles of Hong Kong elite soccer players, *British Journal of Sports Medicine*, 26, 262–266.

Chin, M.K., So, R.C.H., Yuan, Y.W.Y., Li, R.C.T. and Wong, A.S.K. (1994) Cardiorespiratory fitness and isokinetic strength of elite Asian junior soccer players, *Journal of Sports Medicine and Physical Fitness*, 34, 251–257.

Davis, J.A. and Brewer, J. (1993) Applied physiology of female soccer players, *Sports Medicine*, 16, 180–189.

Davis, J.A., Brewer, J.,and Atkin, D. (1992) Pre-season physiological characteristics of English first and second division soccer players, *Journal of Sports Sciences*, 10, 541–547.

Ekblom, B. (1986) Applied physiology of soccer, *Sports Medicine*, 3, 50–60.

Heller, J., Prochazka, L., Bunc, V., Dlouha, R. and Novotny, J. (1993) Maintenance of aerobic capacity in elite football players during competitive period, *Acta-Universitatis-Carolinae*, 29, 79–87.

Leger, L.A. and Lambert, J. (1982) A maximal multi-stage 20 m shuttle run test to predict VO$_2$max, *European Journal of Applied Physiology*, 49, 1–12.

National Coaching Foundation. (1988) *Multistage Fitness Test*, White Line Press, Leeds.

Nyland, J.A., Caborn, D.N.M., Brosky, J.A., Kneller, C.L. and Freidhoff, G. (1997) Anthropometric, muscular fitness, and injury history: Comparisons by gender of youth soccer teams, *Journal of Strength and Conditioning Research*, 11, 92–97.

Ramsbottom, R., Brewer, J. and Williams,C. (1988) A progressive shuttle run test to estimate maximal oxygen uptak, *British Journal of Sports Medicine*, 22, 141–144.

Wilson, N., Russell, D. and Wilson, B. (1993) Body composition of New Zealanders, *Life in New Zealand Activity and Health Research Unit*, University of Otago.

Young, W. (1994) A simple method for evaluating the strength qualities of the leg extension muscles and jumping abilities, *Strength and Conditioning Coach*, 2, 5–8.

Young, W.B., Pryor, J.F. and Wilson, G.J. (1995) Effect of instructions on characteristics of counter movement and drop jump performance, *Journal of Strength and Conditioning Research*, 9, 232–236.

10 AN EXAMINATION OF LONGITUDINAL CHANGE IN AEROBIC CAPACITY THROUGH THE PLAYING YEAR IN ENGLISH PROFESSIONAL SOCCER PLAYERS, AS DETERMINED BY LACTATE PROFILES

G.M.J. DUNBAR
Nike Lab, London, UK.

1 Introduction

It is widely documented that the aerobic system is the main source of energy in soccer. However, in this high intensity intermittent sport, many other factors also contribute to performance. Elements such as speed, power, strength, agility, flexibility and anaerobic capacity all combine with aerobic capacity and indeed all of these were also measured in the group of soccer players to be discussed. The purpose of such monitoring was to construct individual fitness profiles to indicate strengths and weaknesses for subsequent training prescription. This report is focused on just the aerobic capacity analysed in players throughout a pre-season period and full playing season.

Traditionally, maximum oxygen uptake (VO_2 max) has been used to classify the fitness of soccer with elite outfield players registering values of around 60 ml·kg^{-1}·min^{-1} (Reilly, 1996). It has become increasingly common to assess players using progressive shuttle runs as a means of estimating this property. The advantages have instant appeal in that many players can be tested at the same time, testing is cheap and easy to administer and can be performed in a familiar environment. However, such maximal testing, whether in the laboratory or field, is not perfect. Being a motivation dependent means of assessment can be problematic in players who have performed the test on several occasions, bringing into question validity on a longitudinal basis. Furthermore, there can be difficulty in programming such tests in the training week, especially when playing two matches.

Dunbar and Doherty (1991) showed a blood lactate profile to be more sensitive to changes in aerobic fitness than VO_2 max when assessing games players. Such sub-maximal testing also has the advantage of being less disruptive to the training week and gives more information for training prescription in terms of setting individual training intensities. The testing principle here is similar to that we have used when testing National squad athletes from sports such as rowing and running (Dunbar et al., 1997a; Dunbar et al., 1997b).

Islegen et al. (1997) used blood lactate measurements in a three-stage test to monitor changes in fitness between the start and finish of pre-season. Brady et al. (1997) used an incremental treadmill test to monitor fitness in Scottish players over a two-season period, using lactate threshold as a criterion measure.

The particular purpose of this study was to analyse change in aerobic capacity in English professional soccer players through a pre-season period and playing season, using reference lactates in a blood lactate profile.

2 Methods

The soccer squad consisted of 33 players from a highly successful English First Division club, who played 46 League, 5 Cup and 3 play-off matches in the season. Eleven players completed all four tests at the same time during the testing period, comprising the sample. The playing positions of the eleven were: a Goalkeeper (1), Defenders (4) Midfield (3) Attackers (3). All testing took place during the 1997/98 season. Test 1 took place in early July, 8 days after players reported for pre-season training. Test 2 was 5 weeks later in August, just after the start of the season, whilst test 3 followed in January, 25 weeks after test 2; and test 4 was 16 weeks further on in May, between the semi-final and final of the play-offs.

Skinfold measurements were taken from bicep, tricep, subscapular and suprailiac sites and the sum of measurements calculated. All running took place on a Powerjog GX100 (Sport Engineering Ltd, Birmingham) running machine with a level gradient. After a controlled warm-up of 5-min running between 8.5 and 9.0 $km·h^{-1}$ each player performed a progressive incremental running protocol. The initial running speed was 11.5 $km·h^{-1}$ and increased by 1.1 $km·h^{-1}$ every 3 min. At the end of each running speed, heart rate was recorded by means of radiotelemetry (Polar, Finland) and players stood astride of the treadmill belt whilst blood was taken from the earlobe for blood lactate determination using an Analox GM7 (Hammersmith).

For each player a blood lactate profile was constructed using linear interpolation. This was used for a statement of the current level of aerobic fitness for each player, as well as for the purpose of endurance training prescription. The speeds at reference lactate concentrations of 2 and 3 mmol $·l^{-1}$ were determined for each player.

Statistical differences between the four testing times for skinfold measurements and running speeds at the two reference blood lactates were analysed using repeated measures ANOVA. Significance was accepted at the 0.05 level of probability.

3 Results

No significant difference was seen in skinfold measurements and running speed at both 2 and 3 $mmol·l^{-1}$ (Table 1) across the four sessions of testing in the playing season.

Table 1. Mean (± SD) sum of skinfolds and running speed at reference lactate levels.

	July	August	January	May
Sum of Skinfolds (mm)	31.3 (± 7.3)	28.6 (± 5.6)	27.7 (± 6.3)	28.5 (± 8.6)
V-2 $mmol·l^{-1}$ $(km.h^{-1})$	14.3 (± 1.4)	14.5 (± 1.2)	14.8 (± 1.5)	13.9 (± 1.7)
V-3 $mmol·l^{-1}$ $(km.h^{-1})$	15.4 (± 1.2)	15.4 (± 1.1)	15.7 (± 1.6)	15.0 (± 1.5)

4 Discussion

It was expected that aerobic capacity would increase between July and August in response to pre-season training, which is traditionally endurance based, and would decline throughout the season due to the large number (54) of games played. Indeed, seven players produced their worst lactate profile in May, at the end of the playing season. This trend was not statistically significant. Several players reported for pre-season training in good aerobic condition in response to self-selected base running programmes in the close-season period. This represents a current change in culture in English soccer, where many players voluntarily remain active during the close season, as opposed to 10–15 years ago, where players reported for pre-season training in a de-conditioned state (White et al., 1988).

Other factors may also account for the apparent lack of change in aerobic fitness through the playing year. There is the potential influence of injury and individual change in form of players. Although none of the eleven players included in this sample had any long-term injuries (> 4 weeks), they may have had minor problems which could have influenced fitness at any given time. Furthermore, players in large squads may have different training regimens according to whether they are playing first team or reserve team football. This is often a function of their current form.

Despite the fact that many researchers have previously examined the changes in fitness levels between the start and finish of the pre-season period, few have examined continued responses through the playing season as well. Brady et al. (1997) examined changes in aerobic fitness in Scottish soccer players over two seasons and found differences between in season and close season measures. However, they used lactate threshold (Tlac) as the criterion measure, a phenomenon we have had difficulty in consistently identifying in soccer players, who do not always display typical lactate curves. Furthermore, they did not analyse exactly the same players on each occasion, rather a different number out of an overall pool of 24 players.

The running speeds in relation to blood lactate concentrations were of similar levels to those displayed in the work of Brady et al. (1997). This, however, represents a good level of aerobic conditioning for soccer players in comparison to our unpublished measures from three other soccer squads in the same season.

Had the players been less conditioned at the start of the pre-season period, one might have expected a significant increase in aerobic fitness through the pre-season and playing season periods. This particular squad seemed able to maintain aerobic capacity throughout the year. It may well be that the fitness described here is different to specific match fitness, which is likely to be more varied throughout the year. Such match fitness is better reflected via an intermittent high intensity protocol (Bangsbo and Lindquist, 1992). However, the aerobic endurance was just one aspect of soccer fitness that was monitored throughout the year.

5 Conclusion

It was concluded that despite the trend for running speed at reference blood lactate values to be lower at the end of the playing season, there was no significant difference in aerobic capacity through the year in this sample of top class English professional soccer players, as determined by lactate profiles.

6 References

Bangsbo, J. and Lindquist, F. (1992) Comparison of various exercise tests with endurance performance during soccer in professional players, *International Journal of Sports Medicine*, 13, 125–132.

Brady, K., Maile, A. and Ewing, B. (1997) An investigation into the fitness of professional soccer players over two seasons, in *Science and Football III* (eds T. Reilly, J. Bangsbo and M. Hughes), E. and F.N. Spon, London, pp. 118–124.

Dunbar, G.M.J. and Doherty, M. (1991) An investigation of training effects: Multistage vs OBLA, *Journal of Sports Sciences*, 9, 413–414.

Dunbar, G.M.J., Warrington, G.D. and White, J.A. (1997a) Training prescription for junior rowers using blood lactates and heart rates, *Journal of Sports Sciences,* 15, 47.

Dunbar, G.M.J., Renfree, A. and White, J.A. (1997b) A comparison of speed and heart rate to control running at threshold intensity, *Journal of Sports Sciences,* 15, 47–48.

Islegen, C., Acar, M.F., Cecen, A., Erding, T., Varol, R., Tiryaki, G. and Karamizrak, O. (1997) Effects of different pre-season preparations on lactate kinetics in professional soccer players, in *Science and Football III* (eds T. Reilly, J. Bangsbo and M. Hughes), E. and F.N. Spon, London, pp. 103–105.

Reilly, T. (1996) *Science and Soccer,* E. and F.N. Spon, London.

White, J.A., Emery, T.M., Kane, J.L., Groves, R. and Risman, A.B. (1988) Pre-season fitness profiles of professional soccer players, in *Science and Football* (eds T. Reilly, A. Lees, K. Davids and W.J. Murphy), E. and F.N. Spon, London, pp. 164–171.

11 THE DEVELOPMENT AND EVALUATION OF A TASK SPECIFIC FITNESS TEST FOR ASSOCIATION FOOTBALL REFEREES

R.A. HARLEY, R. BANKS and J. DOUST
Chelsea School Research Centre, University of Brighton, Eastbourne, East Sussex, UK

1 Introduction

The referee plays a significant role in the game of Association Football. For referees to optimise their decision making, they must take up the correct position on the field of play and therefore, must have good knowledge of the laws of the game and the physical fitness to be in the right place at the right time. Match analysis data have revealed that referees at Football League level, cover, on average 9.4 km over the two 45-min halves of play (Catterall et al., 1993). The percentage of total distance covered by sprinting has been reported as being less than 10%, while on average the total distance covered by jogging, running backwards and walking are 50%, 20% and 20% respectively (Catterall et al., 1993; Johnson and McNaughton, 1994). It would appear therefore that refereeing is essentially an endurance task, which involves some short periods of high intensity activity interspersed with periods of lower intensity activity. A number of different fitness tests are employed to assess the fitness levels of Association Football referees in England. A 12-min continuous run is employed at County League level, the pass criterion being 2600 m. National League level Division 3 to Premier League referees have to undertake the multistage fitness test (Ramsbottom et al., 1988), the pass criterion being level 10 stage 5, which equates to a VO_2 max of 48 ml·kg⁻¹·min⁻¹. Although these tests are essentially tests of aerobic endurance, they fail to incorporate the movement patterns employed during game situations. The tests are performed in unfamiliar surroundings and, in the case of the 12-min run, involve pace judgement. Since the tests are judged on a pass/fail basis they offer limited information to referees on monitoring their fitness. Thus the aim of this study was to develop and evaluate a task specific fitness test to assess fitness levels of Association Football referees.

2 Methods

2.1 Designing the new football referees (FR) test
The test was designed so that it could be performed on any football pitch. The movement patterns and percentages of total distances covered on the new test were based upon published match analysis data (Catterall et al., 1993; Johnson and McNaughton, 1994). Audio instructions from a cassette player negated any problems with pace judgement and the incremental nature of the test allowed changes in fitness to be assessed. The test design can be seen in Figure 1.

Starting at the corners of the square, the test involves jogging for 50 m, a 6-s passive recovery, a jog backwards for 20 m, turn and sprint for 10 m and then walk for 20 m. This sequence is then repeated. Pace is determined by the audio instruction. The test consists of nine levels with each level lasting approximately 3 min. The time to complete each 100-m sequence is reduced at the end of each level requiring the referee to move a little faster in each phase of the test. The test is incremental to maximal volitional effort.

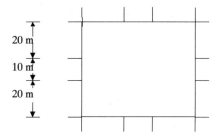

Figure 1. The layout of the new FR test.

2.2 Assessment of criterion validity
Sixteen Sussex County League referees volunteered to take part in the study (36.6 ± 7.7 years, 84.9 ± 10.3 kg). Each performed three maximal tests (incremental treadmill test with oxygen uptake analysis, MST to predict VO_2 max and the new FR test), on three occasions, over a two-week period following a Latin square design. All the tests were performed in the early evening.

The incremental oxygen uptake test was performed on a motorised treadmill (Woodway). After a warm-up the treadmill velocity was raised to that which elicited a heart rate about 25 beats.min[-1] lower than age predicted maximum. The slope was kept constant at 1% to best represent the energetics of outdoor running for subsequent exercise prescription (Jones and Doust, 1996) while velocity was increased 0.5 km.h[-1] every minute until the subjects reached volitional exhaustion. Expired air was collected in Douglas bags during the final 40 s of each minute and subsequently analysed using paramagnetic transducer (Servomex series 1100) O_2 analyser, and an infrared CO_2 analyzer (Servomex model 1490). Expired volumes were assessed using a Harvard dry gas meter.

The MST was performed on a football pitch using the protocol outlined by Ramsbottom et al. (1988). The new FR test was performed following the protocol described earlier. The total distance covered was recorded as the score.

2.3 Assessment of reliability and objectivity

Twelve Sussex County League referees study (34.2 ± 8.3 years, 82.6 ± 11.3 kg) who had previously not undertaken the new FR test, volunteered to take part in the study. Each performed two FR tests until volitional exhaustion, at the same time of day, one week apart. To examine the objectivity of the new test, three test administrators each marked separately the performance of subjects during the first test.

3 Results

Significant correlations were determined between distance covered on the FR test and predicted VO_2 max from the MST (r = 0.73, P < 0.0005) and peak VO_2 (r = 0.82, P < 0.0005) measured in the laboratory. The relationship between peak VO_2 and distance covered on the FR test can be seen in Figure 2.

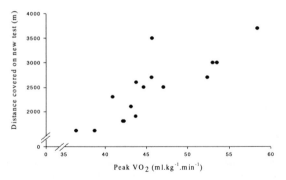

Figure 2. Relationship between laboratory measured peak VO_2 and distance covered on FR test.

No significant difference (t = 1.1) was found to exist between test retest data for the FR test, with the coefficient of variation using the least square method being 5.4%. Graphical representation of these data can be seen in Figure 2. The 95% confidence intervals for test-retest are ± 400 m which is ~3.5 ml·kg⁻¹·min⁻¹ expressed in terms of VO_2 max.

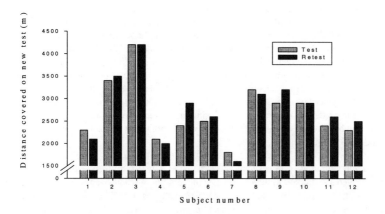

Figure 3. Test re-test data for the FR test.

No significant difference (F = 0.000) was found between the scores recorded by the three independent assessors.

4 Discussion and Conclusions

Catterall et al. (1993) concluded that the major physiological demands on referees are imposed on the oxygen transport system. They reported that high intensity exercise only accounted for 18.5% of the total distance covered and concluded that the major energy source for soccer refereeing is the aerobic system. It would appear that a certain level of aerobic fitness is required by association football referees in order to perform the tasks required of them. Accurate and specific fitness assessment is needed to ensure that a reasonable level of aerobic fitness can be determined. The time and expense involved with laboratory assessment of aerobic power is much greater than that of field assessment. Field assessment also has the advantage that relatively large numbers of subjects can be assessed at the same time. However, it is recognised that some degree of measurement accuracy will be lost. Hawley (1987) has pointed out that in order to obtain the most valid information possible about the maximal aerobic power of an athlete, it is important to select an appropriate test situation, which involves large muscle mass and optimal recruitment of specifically trained muscle fibres. The design of the new FR test was such that the movement patterns of referees were incorporated into an incremental endurance task which could be performed on any football pitch. It is therefore proposed that the test has ecological and face validity. Although not formally assessed, subject satisfaction with the new test was strong with most of the referees expressing greater satisfaction with the new test because of task specificity. A number of other criteria also need to be achieved if a test is to be deemed valid and reliable. Ramsbottom et al. (1988) and Grant et al. (1995) have both reported significant correlations between aerobic power predicted from the MST and measured directly on a treadmill using the oxygen uptake method. The results indicate a significant relationship in predicted and measured aerobic power by similar methods for this particular group of referees. The highly significant correlation between laboratory assessed aerobic power and distance covered on the new FR test demonstrates the high criterion validity of the test in assessing task specific aerobic power and that it could be used to predict VO_2 max of referees using linear regression. The results also demonstrate a high degree of repeatability of the new FR with no significant difference between repeated trials and a CV of 5.4%.

The new test has now been used for two seasons and a total of 100 referees tested. Test results have been used to help guide referees towards suitable individualized fitness programmes. The results have proved satisfactory to the County League in judging whether an individual referee's fitness is of a suitable level to be eligible for the official referees' list. The key advantageous features of the new test are: utilisation of football-specific movement patterns; control of pace; incremental in nature; high criterion validity; and good repeatability.

5 References

Catterall, C., Reilly, T., Atkinson, G. and Coldwells, A. (1993) Analysis of the work rates and heart rates of association football referees, *British Journal of Sports Medicine*, 27, 193–196.

Grant, S., Corbett, K., Amjad, A.M., Wilson, J. and Aitchison, T. (1995) A comparison of methods of predicting maximum oxygen uptake, *British Journal of Sports Medicine*, 29, 147–152.

Hawley, J.A. (1987) Physiological laboratory testing to identify athletic potential, *The New Zealand Journal of Sports Medicine,* 15, 7–10.

Johnson, L. and McNaughton, L. (1994) The physiological requirements of soccer refereeing, *The Australian Journal of Science and Medicine in Sport,* 6, 93–101.

Jones, A. and Doust, J. (1996) A 1% treadmill grade most accurately reflects the energetic cost of outdoor running, *Journal of Sports Sciences,* 14, 321–327.

Ramsbottom, R., Brewer, J. and Williams, C. (1988) A progressive shuttle run test to estimate maximal oxygen uptake, *British Journal of Sports Medicine,* 22, 141–144.

12 CORRELATIONS BETWEEN FIELD AND LABORATORY TESTS OF STRENGTH, POWER AND MUSCULAR ENDURANCE FOR ELITE AUSTRALIAN RULES FOOTBALLERS

C. HRYSOMALLIS *, R. KOSKI *, M. McCOY ** and T. WRIGLEY *
*   Centre for Rehabilitation, Exercise and Sport Science, Victoria University, Melbourne, Australia
** Richmond Football Club, Melbourne, Australia

1 Introduction

Field and laboratory tests are frequently used to evaluate the physiological capacities of athletes. Common field tests include sprint times, vertical jump height and long jump distance. Tests conducted in the laboratory with isokinetic dynamometers can determine the strength and anaerobic muscular endurance of particular muscle groups. Isokinetic measures have been shown to significantly correlate with many aspects of athletic performance (Wrigley, 1999), but data on Australian Rules footballers is lacking. Field tests are generally used where utility and cost are important, whereas laboratory tests are used where additional precision and control is required. Investigating the association between field and laboratory tests will assist in the determination of which field tests to conduct when laboratory access is limited. Additionally, relationships may be identified which have implications for training and enhancing performance. It may be found that strength of a particular muscle group (e.g. quadriceps versus hamstrings) is more related to a particular motor skill (e.g. sprinting). The aim of this investigation was to determine the association between some common field tests and laboratory tests for elite ARF players.

2 Methods

Twenty-two professional players from the Richmond Football Club provided voluntary informed consent and served as subjects. The subjects warmed up by jogging approximately 1 km followed by lower body stretches for about 10 min and moderate intensity striding (4–5 × 40 m). The field tests were conducted in one morning session, late in the pre-season. Sprint times were determined with photocell timing lights/gates placed at 0, 5, 10 and 20 m along an outdoor synthetic athletic track. Two sets of lights were placed at each distance at different heights (about knee and mid-torso) requiring both the upper and lower body to break the beams prior to triggering. Adopting a staggered standing start and wearing jogging footwear, the subjects positioned themselves 0.5 m from the first set of timing gates. These first gates (0 m mark) were activated with the first stride. As with all field tests, three trials were permitted with the best value recorded and later used in the analysis. Subjective evaluation of the wind conditions did not detect noticeable head- or tail-wind.

A vertical jump and reach test was executed using a five-step run up and one-legged take-off. A free standing device (Vertec®) was used. It comprised of indicators that swivelled when struck. Subjects were instructed to run, jump and reach up and strike the swivel indicators at the apex of their jump. Standing long jump was performed in a long jump pit. An arm swing was permitted and the distance was measured from the front of the feet to landing point closest to the take-off point.

The laboratory tests were conducted on a different day to the field tests. A Biodex® isokinetic dynamometer was used to measure strength and muscular endurance. Strength was defined as the peak torque value during maximal repetitions. Muscular endurance was defined as the total work during 20 maximal repetitions of a newly developed protocol. The subjects warmed up and stretched before being seated and stabilised on the Biodex® with a hip angle of 80°. The strength tests consisted of concentric knee flexion and extension at three angular velocities: 60, 240 and $360° \cdot s^{-1}$ (1.05, 4.19, and 6.28 $rad \cdot s^{-1}$). Subjects were allowed three practice repetitions and then three maximal test repetitions. Standardized instructions and strong verbal encouragement was given throughout all testing procedures. Gravity correction was implemented for all isokinetic tests.

Muscular endurance was assessed by eccentric/concentric stretch-shorten cycle (SSC) knee flexor and knee extensor tests. These two tests comprised 20 repetitions at $90° \cdot s^{-1}$ (1.57 $rad \cdot s^{-1}$). Twenty repetitions was chosen based on pilot work which indicated that it was enough to induce a fatigue response, but not a large enough volume to elicit considerable soreness. The testing velocity of 1.57 $rad \cdot s^{-1}$ was chosen as a suitable angular velocity in which the movements could be co-ordinated. The knee flexion test was conducted through a range of motion (ROM) of 100° flexion to maximal knee extension. A 1-s pause was integrated into the test after the completion of the concentric flexion phase. Subjects were asked to try to relax during this pause. This was incorporated in order to mimic the work-rest sequence of the SSC that occurs during running. The sequence during running is an eccentric muscle action followed immediately by a concentric muscle action and then recovery (as demonstrated by the knee flexors during late swing, early stance of sprinting). Subjects were given five practice repetitions and a 1 min recovery. They then completed the 20-repetition endurance test. At the completion of the knee flexor eccentric/concentric testing, subjects completed the knee extensor concentric/eccentric testing. The ROM for the knee extensor endurance testing was limited to 30° from full extension to 100° flexion (there can be a tendency to lock the knee at full extension, which may then result in the lever arm subsequently moving into flexion lifting the subject out of the chair). The 1-s pause this time occurred at the top of the movement (after concentric knee extension). After testing one limb, the same procedure was completed on the other limb.

Pearson correlation coefficients (r) were calculated to determine the statistically significant relationships ($P < 0.01$) between the field and laboratory test results.

3 Results

The mean (± SD) age, body mass and height of the subjects were 22.0 ± 3.5 years, 88.1 ± 9.8 kg and 1.90 ± .09 m, respectively. Table 1 depicts the significant relationships between the field and laboratory tests. It is indicated in the table whether the results for the laboratory tests apply to just one limb (left or right) or averaged for both and if they are expressed in absolute or relative (divided by body weight) terms.

Table 1. Significant correlations between field and laboratory test results.

Field test	Isokinetic test	r
Sprint time to 5 m	Hamstring strength, 4.19 rad.s^{-1}, left limb, absoluté	-0.556
	Hamstring strength, 6.28 rad.s^{-1}, left limb, absolute	-0.615
Sprint time to 10 m	Hamstring strength, 6.28 rad.$^{-1}$, left limb, absolute	-0.567
Vertical jump	Quadriceps strength, 4.19 rad.s^{-1}, right limb, absolute	0.591
	Quadriceps strength, 4.19 rad.s^{-1}, average, absolute	0.573
	Hamstring endurance, concentric, right limb, absolute	0.634
	Hamstring endurance, eccentric, right limb, absolute	0.666
	Hamstring endurance, eccentric, right limb, relative	0.593
	Quadriceps endurance, concentric, right limb, absolute	0.618
	Quadriceps endurance, concentric, average, absolute	0.708
	Quadriceps endurance, eccentric, left limb, absolute	0.690
	Quadriceps endurance, eccentric, average, absolute	0.696
Long jump	Hamstring endurance, eccentric, left limb, relative	0.653

Moderate correlations were detected, ranging in magnitude from 0.556 to 0.708. Only single limb absolute hamstring strength correlated with sprint time. Absolute quadriceps strength was related to vertical jump height, but surprisingly endurance measures had higher correlations. No strength measures significantly correlated with long jump performance.

4 Discussion

Most significant correlations with the field tests were with single limb absolute isokinetic values. It is unclear why both limbs did not correlate with field tests that involved both limbs. It could be speculated that any deficit in isokinetic values between limbs or previous injury were confounding factors. When the average of both limbs was used or when the isokinetic value was expressed relative to body weight, the correlations in most cases were no longer significant.

In contrast to previous findings, quadriceps strength did not correlate with sprint times. Some researchers have detected the relationship between sprint times and hamstring strength. Alexander (1989) determined the relationship between sprint times for 100 m and isokinetic strength (relative peak torque, average value for both limbs. For 14 male elite sprinters, a significant correlation (r = -0.71, P < 0.01) was found with quadriceps concentric strength at $230°·s^{-1}$. Neither the hip muscles nor hamstrings had a significant correlation with sprint time. Dowson et al. (1998) found significant correlations between sprint times for 0–15 m and 30–35 m and isokinetic strength (dominant limb) for both the quadriceps and hamstrings. The 24 subjects consisted of: eight rugby players, eight sprinters and eight competitive sportsmen. Concentric quadriceps torque at 4.19 rad·s^{-1} had the highest relationship (0–15 m time: r = -0.518, P < 0.01; 30–35 m time: r = -0.688, P < 0.01). These relationships were improved for 0–15 m, but not for 30–35 m when strength was expressed relative to body weight (0–15 m time: r = -0.581; 30-35 m time: r = -0.659). The correlations with hamstring strength were only slightly lower in magnitude compared with those involving the quadriceps. Nesser et al. (1996) found slightly higher significant (P < 0.05) correlations for the hamstrings (r = -0.561 at 3.14 rad.s^{-1}; r = -0.613 at $450°·s^{-1}$) than for the quadriceps (r = -0.546 at 7.86 rad·s^{-1}) with 40 m sprint time. The 20 male subjects tested were all involved in sprint-type sports such as gridiron, lacrosse and baseball.

A reason which may be put forward for the lack of significant association between sprint times and quadriceps strength is that when a homogeneous group is used, the range of scores may not vary enough for significant correlations to be detected. If this was the case, the same would be expected for the group of elite sprinters in the study by Alexander (1989).

Significant correlations between quadriceps strength and vertical jump height have previously been reported elsewhere (Wrigley, in press) and are not unexpected considering the role of knee extension in jumping. The vertical jump test has been traditionally used as an indicator of lower limb power. An unexpected finding of this study was the significant relationships between the measures of quadriceps and hamstring concentric and eccentric endurance and vertical jump height. In fact, the highest correlations detected in this study were for those parameters. Standing long jump performance only correlated significantly with single limb eccentric relative hamstring endurance.

5 Conclusion

For this group of ARF, sprint times to 5 and 10 m were significantly associated with hamstring isokinetic strength at 4.19 and 6.28 rad·s^{-1} (r ranged from -0.556 to -0.615). For short sprints from a standing start, hamstring strength appears to be important and strengthening this muscle group may enhance sprint performance. The vertical jump test is a reasonable indicator of quadriceps strength at 4.19 rad·s^{-1} (r = 0.591 and 0.573) and anaerobic muscular endurance of both the hamstrings and quadriceps acting concentrically and eccentrically (r values ranged from 0.593 to 0.708). Standing long jump performance had limited association with isokinetic measures.

6 References

Alexander, M.L. (1989) The relationship between muscle strength and sprint kinematics in elite sprinters, *Canadian Journal of Sports Science,* 14, 148–157.

Dowson, M.N., Nevill, M.E., Lakomy, H.K.A., Nevill, A.M. and Hazeldine, R.J. (1998) Modelling the relationship between isokinetic muscle strength and sprint running Performance, *Journal of Sports Sciences,* 16, 257–265.

Nesser, T.W., Latin, R.W., Berg, K. and Prentice, E. (1996) Physiological determinants of 40-meter sprint performance in young male athletes, *Journal of Strength and Conditioning Research,* 10, 263–267.

Wrigley, T.V. (in press) Correlation of isokinetic dynamometry with athletic performance, in *Isokinetic Enhanced Performance* (ed L.Brown), Human Kinetics, Illinois.

13 SEASONAL VARIATIONS IN THE FITNESS OF ELITE GAELIC FOOTBALLERS

T. REILLY and S. KEANE
Research Institute for Sport and Exercise Sciences, Liverpool John Moores University, Liverpool, UK

1 Introduction

Seasonal variations in fitness levels are anticipated to correspond with the phases of the competitive season. Such cycles in fitness characteristics have been described for soccer (Thomas and Reilly, 1979) and Rugby Union (Tong and Mayes, 1995) players. In Gaelic football the major competition, the All-Ireland inter-county championship, takes place in the summer, starting in May and culminating in the All-Ireland final in September. The inter-county League programme is held over the winter and spring months. The aim of the research was to monitor fitness and performance measures in elite Gaelic footballers over a complete season.

2 Methods

A senior inter-county male squad (N = 32) was measured on six occasions throughout the playing season from January to September. The players undertook a systematic training programme, on average three times a week, in addition to regular competition with their club as well as the county team.

Anthropometric measures included body mass and percent body fat, estimated from four skinfold thickness measures (Durnin and Womersley, 1974). The performance battery included sprints over 50, 100, 200, 400 m and a 12-min run. Maximal oxygen intake (VO_2 max) was estimated from a progressive 20-m shuttle run (Ramsbottom et al., 1988).

A one-way analysis of variance was used to examine changes in mean values for the squad of players over the course of the season from Test 1 (January) to Test 6 (September). A least significant difference test was used to assess where specific differences lay.

3 Results

Over the six tests there were improvements in 50 m (7.43 ± 0.2 to 6.56 ± 0.2 s), 100 m (15.3 ± 0.5, to 13.0 ± 0.5 s), 200 m (33.5 ± 0.9 s to 30.4 ± 0.9 s), 400 m (83 ± 2.4 to 73 ± 1.9 s) runs. Distance covered in 12 min increased from 2633 ± 151 m to 3028 ± 114 m but was shorter at Test 2 than at other times ($P < 0.05$) and longer at Tests 3 and 4 than in the first two tests ($P < 0.05$). The VO_2 max showed a non-significant change from 52.8 ± 2.1 ml·kg^{-1}min^{-1} to 53.3 ± 3.2 ml·kg^{-1}min^{-1}. These changes were accompanied by an average weight reduction of 3.3 kg.

Table 1. Mean (± SD) test results for the inter-county squad for the entire season from January (Test 1) to September (Test 6).

	Test 1	Test 2	Test 3	Test 4	Test 5	Test 6
Body mass (kg)	85.99 ±5.3	85.65 ±5.3	85.13 ±5.2	84.35 ±5.2	82.87 ±5.7	82.72 ±5.0
VO_2 max (ml·kg^{-1}min^{-1})	52.78 ±2.1	53.29 ±1.9	50.17 ±3.5	53.30 ±3.2	-	-
Cooper's 12-min run (m)	2633 ±151	2563 ±117	3024 ±90	3028 ±114	-	-
50-m sprint test (s)	7.43 ±0.2	7.19 ±0.2	6.95 ±0.2	6.96 ±0.3	6.78 ±0.2	6.56 ±0.2
100-m sprint test (s)	15.3 ±0.5	14.5 ±0.5	14.0 ±0.4	13.6 ±0.4	13.4 ±0.5	13.0 ±0.5
200-m sprint test (s)	33.5 ±0.9	33.7 ±0.7	31.3 ±0.7	32.7 ±1.5	30.6 ±1.1	30.4 ±0.9
400-m sprint test (s)	83 ±2.4	78 ±2.7	73 ±2.1	75 ±2.9	76 ±3.2	73 ±1.9

4 Discussion

The main improvements in the fitness profiles of the players were the reductions in sprint time which ranged from 50 m to 400 m. The improvement was associated with a reduction in body mass, itself largely attributable to loss of body weight as fat. The seasonal profile provides a good baseline for the Gaelic football players as this team won the all-Ireland championship during the year of the study.

The improvements in the anaerobic measures were not matched by their aerobic test counterparts. Performance in the 12-min run was enhanced by the systematic training programme. The estimated VO_2 max did increase but the difference was non-significant

(P > 0.05). Data were not obtained for this measure in the later stage of the season prior to the major championship matches when the aerobic fitness levels would be expected to reach their peak. It should also be emphasised that not all players were able to continue to improve during the course of the season, fitness levels regressing when players were injured and had to reduce the training load.

5 References

Durnin, J.V.G.A. and Womersley, J. (1974) Body fat assessed from total body density and its estimation from skinfold thickness: measurements on 481 men and women aged 16–72 years. *British Journal of Nutrition,* 32, 77–97.

Ramsbottom, R., Brewer, J. and Williams, C. (1988) A progressive shuttle run test to estimate maximal oxygen uptake, *British Journal of Sports Medicine,* 22, 141–144.

Thomas, V. and Reilly, T. (1979) Fitness assessment of English League soccer players throughout the competitive season, *British Journal of Sports Medicine,* 13, 103–109.

Tong, R.J. and Mayes, R. (1995) The effect of pre-season training on the physiological characteristics of international Rugby players, *Journal of Sports Sciences,* 13, 123–127.

14 FACTORS ASSOCIATED WITH PRE-SEASON FITNESS ATTRIBUTES OF RUGBY PLAYERS

K. QUARRIE and S. WILLIAMS
Department of Preventive and Social Medicine, Medical School, University of Otago, Dunedin, New Zealand.

1 Introduction

Knowledge of the fitness requirements needed for rugby is limited by the difficulty in obtaining physiological information from players during matches (Nicholas, 1997). Studies that have attempted to assess the physiological demands placed on players during games provide conflicting results (Docherty, Wenger and Neary, 1988; Jardine et al., 1988; McLean, 1992). The difficulty in ascertaining the physiological demands of Rugby Union is partly due to the large number of factors which exert an influence on the patterns of activity during a match (e.g. environmental conditions, level of play, officiating styles). Reilly (1990) stated that 'The game requires a mixture of fast reactions, speed, agility, muscular strength, anaerobic and aerobic power but not in a clearly definable way.'

Owing to the difficulties in measuring the fitness of players during play, fitness test batteries are commonly used to determine players' fitness levels. The items included in test batteries are generally regarded as 'proxy' measures of rugby fitness, which are thought to bear some relationship to the physical demands placed on players during games. Thus players are often assessed on tests designed to measure endurance (aerobic, anaerobic and muscular), speed, strength, power and agility (Maud, 1983; Maud and Shultz, 1984; Rigg and Reilly, 1988; Ueno et al., 1988; Bell et al., 1993; Quarrie et al., 1995). At present, however, the relationships among tests commonly used in fitness test batteries are unclear. If several of the tests within a battery of fitness tests measure similar fitness attributes, then more efficient fitness testing could be achieved by avoiding tests which are redundant.

The New Zealand Rugby Injury and Performance Project (RIPP) (Waller et al., 1994) was a prospective cohort study designed to yield information about factors associated with Rugby Union player performance and injury risks. It was found that the anthropometric and fitness attributes of Rugby Union players differ across various playing levels (grades), between backs and forwards and amongst the various positional groups within the backs and forwards (Quarrie et al., 1995). Players at higher grades generally performed better than those at lower grades. Backs tended to perform better on most measures of fitness when compared with forwards.

Other factors that may be associated with fitness levels of players have been the focus of less research attention. For example, little is known about the relative importance of anthropometric characteristics, ethnicity, amount of training, and health status, on the fitness levels of Rugby Union players. The information obtained through the RIPP study presented an opportunity to investigate these factors.

This study therefore had two purposes:

1. To examine the relationships among the various fitness tests in the RIPP fitness test battery. This was done to find out whether the fitness test variables used in RIPP could be well explained in terms of a smaller number of underlying 'factors'.

2. To investigate the importance of associations between selected demographic, anthropometric, health, training, and lifestyle variables and pre-season fitness in Rugby Union as assessed via an aerobic shuttle test and a 30 m sprint test.

2 Methods

Three hundred and fifty-six Rugby Union players (258 males and 92 females) were enrolled in the RIPP cohort during the pre-season club-training period before the 1993 Rugby Union season. Male players were drawn from the following grades: Senior A, Senior B, Under 21, Under 19/Under 18. As there were too few females for the purposes of multivariate analyses, they were excluded from the analyses for this report. One hundred and seventy four males (67%) completed all of the anthropometric and physical performance measures. Two hundred and fifty two players, however, completed the aerobic shuttle test and the 30-m sprint from a 5-m running start.

2.1 Questionnaire measures

Information on demographic characteristics, rugby and injury experience, training history, and health and lifestyle factors was obtained by means of a self-administered questionnaire. Where possible, measures that had been validated with other populations were used.

2.2 Anthropometric and physical performance measures

Anthropometric and physical performance assessments were conducted by a group of ten trained assessors. The anthropometric assessments taken included height, body mass, and skinfolds from the triceps, subscapular, supraspinal, abdominal, front thigh, and medial calf sites. The sum of skinfolds from these six sites was used in the subsequent analyses.

The physical performance assessments included:

- A 20-m multi-stage shuttle run test to gauge aerobic performance. The number of shuttles completed by each player was recorded.
- A modified Sargent vertical jump test. The best of three attempts was recorded.
- Push-ups at a constant cadence of 50 b·min^{-1} were performed as a test of upper body muscular endurance. Push-up position was standardized for all players.
- 30-m sprint times from a standing start and a 5-m running start were recorded.

- Two agility runs – one in which the player was required to perform a series of left hand turns around a set of cones, and one in which the player performed right hand turns.

Anaerobic endurance was assessed by a set of six repeated high intensity shuttles. Each repetition of the test consisted of sprinting a distance of 70 m. The players ran out to a 5-m mark and back to the start, out to a 10-m and back to the start, and then out to a 20-m mark and back to the start. A 'fatigue index' for each player was calculated. The players who were unable to complete all six repetitions were penalised by multiplying their score by six (the maximum possible number of shuttles) and dividing by the number they completed.

2.3 Factor analysis

A confirmatory factor model was fitted to the measures in the fitness test battery using LISREL 8 (Joreskog and Sorbom, 1993). One hundred and seventy four players completed all of the fitness tests, and were included in the model. It was hypothesized that the fitness variables measure two underlying dimensions or latent variables. The first, endurance, was measured by the aerobic shuttle test, push ups and the anaerobic 'fatigue index'. The second, power, was measured by the vertical jump, agility runs turning right and left, and sprint speed from both a standing and running start.

2.4 Prediction of aerobic endurance and sprinting performance

The aerobic shuttle test was chosen as the dependent variable to investigate factors associated with pre-season aerobic fitness of the players, and 30-m sprint time from a 5-m running start was chosen as the dependent variable for factors associated with pre-season speed and power attributes of players. Hierarchical regression analysis was used to investigate factors associated with performance on these tests. Grade and position, which have been shown to discriminate between Rugby Union player fitness levels through previous work were included in the models first. These were followed by variables postulated to be associated with the dependent variables. Potential variables for inclusion in the model were selected following the steps outlined by Noreau and Shephard (1994). Variables that had low correlations ($P < 0.05$) with the outcome variable were excluded from the models. The amount of variance explained by the grade and position is presented, as well as the additional variance explained by the exploratory section of the analysis. The stepwise method was used to select those included in the final model. This was done because apart from grade and position, there were no strong hypotheses as to which factors were likely to be strongly associated with the fitness of Rugby Union players. Final regression diagnostics and a residual analysis were run to ensure that the assumptions underlying multiple regression were not violated. Bootstrap analysis of the data was undertaken to examine the consistency of the models obtained. Analyses were run on one hundred sub-samples of the data set, and the frequency with which each variable entered the model was produced.

3 Results

The means, standard deviations and correlations for the fitness measures are shown in Table 1. The correlations between the measures are also shown.

Table 1. Means, standard deviations and correlations of fitness variables.

		Mean	SD	1	2	3	4	5	6	7
1	Aerobic Shuttle test	111.7	18.6							
2	Push Ups	26.1	7.10	0.38						
3	Fatigue index	62.1	2.51	-0.48	-0.21					
4	Vertical Jump	60.3	7.60	0.24	0.30	0.04				
5	Agility run – right	12.0	0.64	0.41	-0.32	0.11	-0.41			
6	Agility run – left	11.9	0.64	-0.38	-0.35	0.14	-0.39	0.79		
7	Sprint (standing start)	4.5	0.24	-0.23	-0.18	-0.05	-0.52	0.46	0.41	
8	Sprint (running start)	3.9	0.20	-0.38	-0.18	-0.02	-0.62	0.51	0.50	0.69

3.1 Factor analysis

The results of the factor analysis are shown in Figure 1. The loadings give an indication of the correlation between the observed item and the latent variable. The correlation between the aerobic shuttle test and endurance, for instance, was 0.84. Two minor modifications were made to the hypothesised model to improve its fit. The first was to allow the fatigue index to load on the latent variable for power as well as that for endurance and the second was to allow the error terms for the two agility runs to be correlated. With these modifications, the model fitted the data well [Chi square = 36.3 (17 df)] and adjusted goodness of fit index was 0.90. The endurance factor explained 71% of the variance of the aerobic shuttle test and the power factor explained 81% of the variance of the sprint speed from running start.

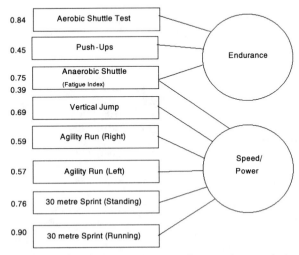

Figure 1. Factor loadings for the confirmatory factor analysis.

3.2 Prediction of aerobic performance

Table 2 shows the variables that were included in the final model for aerobic performance. The correlations between the dependent variable and each of the independent variables and the effect a change of one unit in the independent variable has on the dependent variable are provided. The final model accounted for 53% of the variance in the number of shuttles completed. Table 2 also contains the number of times the variable was entered into the model when the bootstrap analysis was undertaken. Examination of the bootstrap frequency shows that the variables included in the model were generally included in a higher number of models.

Table 2. Regression model for Aerobic Shuttle Test (n = 225).

Variables in model	r	Parameter estimate	SE	p	Bootstrap Frequency
Senior B Grade		0.3	2.8	0.91	100
Under 21 Grade		-2.8	2.6	0.27	
Under 19/18 Grade		-12.9	2.8	<0.01	
Position (Backs)		4.1	2.1	0.06	61
Sum of six skinfolds	-0.53	-0.4	0.0	<0.01	100
Months of endurance training	-0.24	1.7	0.5	<0.01	62
Ethnic origin (Maori or Pacific Islander)	-0.23	-8.1	2.5	<0.01	94
Hazardous drinking	0.09	0.8	0.4	0.06	58
Smoking					
(ex-smoker)	-0.04	-5.2	2.7	0.05	88
(current smoker)	-0.11	-7.0	2.5	<0.01	
Health status	-0.24	4.9	1.9	<0.01	82
Self rated ability	0.18	3.3	2.0	0.09	69
Attempting to lose weight	-0.20	-5.2	2.8	0.07	65
Variables not in model:					
Age	0.14				12
Body mass	-0.27				31
Amount of endurance training	0.23				44
Years played rugby	0.17				17
Off-season training programme	0.16				24
High fitness sports during off season	0.10				30

3.3 Prediction of sprinting speed

The correlations between the 30-m sprint and the variables included in the regression model are shown in Table 3. The final model accounted for 43% of the variance in sprint times. Once again, it can be seen that the variables included in the final model were included in a greater proportion of the models derived through the bootstrap analysis.

Table 3. Regression model for 30-m Sprint Test (n = 202).

Variables in model	r	Parameter estimate	SE	p	Bootstrap Frequency
Senior B Grade		0.088	0.039	0.03	95
Under 21 Grade		0.000	0.027	0.99	
Under 19/18 Grade		0.100	0.029	<0.01	
Position (Backs)		-0.128	0.025	0.01	100
Sum of six skinfolds	0.48	0.002	0.001	0.01	100
Current injury	-0.09	-0.042	0.023	0.07	61
Ethnic origin (Maori or Pacific Islander)	-0.11	-0.066	0.029	0.03	74
Representative rugby last season	-0.21	-0.052	0.025	0.04	77
Self-rated ability	-0.20	-0.031	0.016	0.05	58
Variables not in model:					
Body Mass	0.26				42
Played rugby in off-season	-0.07				24
Speed training in off-season	-0.19				42

4 Discussion

4.1 Factor analysis

The factor analysis confirmed that two factors termed 'endurance' and 'speed/power' provided an adequate representation of the eight variables considered. While the tests used in the RIPP fitness test battery are valid measures of their various fitness components, the factor analysis indicates that there is some redundancy among the tests. Although fitness assessment via the use of a battery of tests is common for Rugby Union, there is no reason to believe that these test batteries measure the 'combination' of fitness demands faced by players during games. Further investigations into the physiological demands of match-play, and development of fitness tests that mimic the requirements of the sport should help more specific training programmes for players to be developed.

4.2 Prediction of fitness attributes

The variables that were associated with both aerobic endurance and sprint performance were grade, position, sum of six skinfolds, ethnic origin, and self-rated ability in Rugby Union. The current analyses revealed that a decrease in the sum of six skinfolds was associated with better performance on both the aerobic and sprint tests. These findings are consistent with previous research into the performance characteristics of Rugby Union players. Bell et al. (1993) found that rugby players with a large fat free mass were capable of producing greater maximal power and Quarrie et al. (1996) showed that possessing a large overall body mass was important in generating momentum. Both of these are important in the body contact phases of play. A high level of aerobic fitness, which is presumably associated with the ability of players to last a match, and sprinting speed, which is important for both attacking and defense are also strongly related to body composition. It appears that maintaining a high total body mass, combined with relatively low adiposity confers an advantage to players.

Players who identified themselves as being of Polynesian ethnicity (New Zealand Maori and Pacific Islanders) completed on average eight fewer shuttles than those of European descent, but were faster on the sprint test. The regression equations take into account the influence of the other factors in the model (e.g. anthropometric characteristics, grade, and position). Given that these factors were controlled for, the differences in performance on the tests between the ethnic groups were large. From casual observation of Rugby Union in New Zealand, it appears that a large number of Pacific Island and Maori players possess the attributes required for successful performance in the sport. Little is known, however, about the exact nature of these attributes. Investigations into the factors that are associated with these differences in performance may enhance our understanding of the requirements for success in Rugby Union. These findings also raise issues relating to the validity of assessing the fitness requirements rugby performance by employing fitness tests that do not appear to relate well to performance during matches.

Both cigarette smoking status and months of off-season endurance training were significantly associated with performance on the aerobic shuttle test. Of the factors that remained in the model, these are two, which are directly modifiable by the players.

5 Conclusions

The fitness tests used in the RIPP test battery can be well explained in terms of two underlying factors, which related to endurance, and speed/power. The pre-season variables that were related to both the endurance performance and strength and power of players were grade, position, ethnicity, sum of six skinfolds, and self-rated ability. Also associated with aerobic endurance were: months of off-season endurance training, self-rated health status, hazardous alcohol use, trying to lose weight in the off-season and cigarette smoking status. Having played representative rugby during the previous season and entering the season carrying an injury were other variables associated with sprinting speed. A higher sum of six skinfolds was associated with decreased performance on both aerobic fitness and sprint tests. New Zealand

Maori and Pacific Island players generally performed better on measures of sprinting speed and worse on measures of aerobic fitness than their European counterparts of similar grade and position. It appears that to ensure their fitness for Rugby Union is optimal, players should choose an appropriate position, try to maintain their lean body mass at as high a level as possible, perform sufficient training in the off-season and abstain from smoking. The results also raise issues as to the validity of using tests that examine only one aspect of fitness for performance in Rugby Union.

6 References

Bell, W., Cobner, D., Cooper, S.M. and Phillips, S. (1993) Anaerobic performance and body composition of international rugby union players, in *Science and Football II* (eds T. Reilly, J. Clarys and A. Stibbe), E. and F.N. Spon, London, pp. 15–20.

Docherty, D., Wenger, H.A. and Neary, P. (1988) Time-motion analysis related to the physiological demands of rugby, *Journal of Human Movement Studies,* 14, 269–77.

Jardine, M.A., Wiggins, T.M., Myburgh, K.H. and Noakes, T.D. (1988) Physiological characteristics of rugby players including muscle glycogen content and muscle fibre composition, *South African Medical Journal,* 73, 529–32.

Jöreskog, K.G. and Sörbom, D (1993) *LISREL 8 User's Reference Guide,* Scientific Software, Inc. Mooreseville, Indiana.

Maud, P.J. (1983) Physiological and anthropometric parameters that describe a rugby union team, *British Journal of Sports Medicine,* 17, 16–23.

Maud, P.J. and Shultz, B.B. (1984) The US National Rugby Team: a physiological and anthropometric assessment, *Physician and Sports Medicine,* 12, 86–94.

McLean, D.A. (1992) Analysis of the physical demands of international rugby union, *Journal of Sports Sciences,* 10, 285–96.

Nicholas, C.W. (1997) Anthropometric and physiological characteristics of rugby union football players, *Sports Medicine,* 23, 375–96.

Noreau, L. and Shephard, R.J. (1994) Guarding against pitfalls in multivariate analysis. An illustration from fitness testing of the spinally-injured, *Journal of Sports Medicine and Physical Fitness,* 34, 192–198.

Quarrie, K.L., Handcock, P., Toomey, M.J. and Waller, A.E. (1996) The New Zealand Rugby Injury and Performance Project. IV. Anthropometric and physical performance comparisons between positional categories of senior A rugby players, *British Journal of Sports Medicine,* 30, 53–56.

Quarrie, K.L., Handcock, P., Waller, A.E., Chalmers, D.J., Toomey, M.J. and Wilson, B.D. (1995) The New Zealand Rugby Injury and Performance Project. III, Anthropometric and physical performance characteristics of players, *British Journal of Sports Medicine,* 1995, 29, 263–270.

Reilly, T., Secher, N., Snell, P. and Williams, C. (1990) *Physiology of Sports,* E. and F.N. Spon, London.

Rigg, P. and Reilly, T. (1987). A fitness profile and anthropometric analysis of first and second class rugby union players, in *Science and Football* (eds T. Reilly, A. Lees, K. Davids and W.J. Murphy), E. and F.N. Spon, London, pp. 194–200.

Ueno, Y., Watai, E. and Ishii, K. (1987) Aerobic and anaerobic power of rugby football players, in *Science and Football* (eds T. Reilly, A. Lees, K. Davids and W.J. Murphy), E. and F.N. Spon, London, pp. 201–205.

Waller, A.E., Feehan, M., Marshall, S. and Chalmers, D.J. (1994) The New Zealand Rugby Injury and Performance Project, I. Design and methodology of a prospective follow-up study, *British Journal of Sports Medicine,* 28, 223–228.

PART THREE

Match Analysis

15 AN ANALYSIS OF THE PLAYING PATTERNS OF THE JAPAN NATIONAL TEAM IN THE 1998 WORLD CUP FOR SOCCER

KUNIO YAMANAKA, TSUTOMU NISHIKAWA, TOMOYUKI YAMANAKA and M.D. HUGHES*
Institute of Health and Sport Sciences, University of Tsukuba, 1-1-1 Tennodai, Tsukuba-shi, 305-8574 Japan.
* Centre for Notational Analysis, Cardiff Institute, Cyncoed, Cardiff, XF2 6XD, Wales.

1 Introduction

In November 1997, the Japan national soccer team came third in the Asian final qualifying match in Malaysia, meaning they got a chance to participate in the 1998 FIFA World Cup in France. This was the first time for Japan to qualify for the World Cup since its Football Association (JFA) was founded in 1921. However, Japan lost all three games in the Group H matches, which meant that they could not go through to the final tournament (the final 16), but it seemed that Japan's players performed at their best. It is certain that the crucial factor for the improvement in Japanese football was due to the foundation of the J - League, now into its sixth year, and the introduction of many South American and European players and coaches which has had a good influence on the players. It is thought to be important for the future development of Japanese football to evaluate from a skill and tactical point of view, the way Japan played in, what was for Japan, the first World Cup in 1998.

The purpose of this study was to perform a computerized notational analysis of games in the 1998 World Cup so that the playing patterns and problems of the respective teams may be examined, with particular emphasis on Japan.

2 Methods

2.1 Subjects

Three games were selected for study from among the games played between Japan, Argentina, Croatia and Jamaica in order to analyse the playing patterns of the respective teams:
(1) Japan vs Argentina (0–1), (2) Japan vs Croatia (0–1), (3) Japan vs Jamaica (1–2).

2.2 Notational analysis system

The software used to input data was that developed by Hughes et al. (1988). The most recent version was used for this study. Performance data during games were entered by replaying the videotapes of games a number of times. Thirty-two different actions of players were entered into 'time', 'place', 'player' and 'action' categories by clicking the 'mouse'.

2.3 Data processing

Using data during a 90-minute game (excluding injury time), and by dividing the pitch into six areas horizontally and three areas vertically, the frequency of each action per area was recorded. For statistical processing, a chi-square test was used.

3 Results and Discussion

3.1 Team Performance

Using 12 items processed in this study, Table 1 shows the mean values per game for three matches; Japan vs Argentina, Japan vs Croatia and Japan vs Jamaica. From each item, the following significant findings were obtained.

First, Japan used passes more often than Croatia ($P < 0.01$), but passed the ball less frequently than Argentina ($P < 0.01$). Second, Japan dribbled the ball less frequently than Argentina ($P < 0.01$) and Croatia ($P < 0.05$). Third, Japan outnumbered Jamaica in the frequency of shots on goal ($P < 0.05$), but made shots on goal less frequently than Argentina ($P < 0.01$). Fourth, Japan outnumbered Jamaica in the frequency of crosses ($P < 0.01$). Fifth, Japan cleared the ball less frequently than Jamaica ($P < 0.01$), but cleared the ball more frequently than Argentina ($P < 0.01$). In other items, no significant differences were found statistically. As previously reported in Yamanaka et al. (1993), Japan characteristically used dribbling as a tactic, and made a large number of passes. This result is similar to the analysis of the playing patterns of Japan in the Asian Cup. Japan executed dribbling in the 1998 World Cup in the same way as the team had done in the Asian Cup: compared to the opponents, Japan used dribbling much less frequently (Japan vs Argentina 49:95, Japan vs Croatia 58:85, Japan vs Jamaica 54:62).

The findings illustrate the frequencies of passes, crosses and clearances of Croatia and Jamaica which characterise the defence-oriented performance. The frequencies of cross-passing and shots on goals are closely related to scoring. From these findings, it is inferred that the team performance of Japan in the three games was unstable, sometimes influenced by unskilled play. It seems that teams using dribbling more frequently, relied mainly on using individual skills in their attacking tactics. From this point of view, it seemed that Japan played more organized football by using passes rather than using individual attacking skills in the games of the 1998 World Cup. In other words, Japan's attacking tactics were based on text-book style soccer, but thoughtlessly.

3.2 The distribution of passes in different areas and the use of 'negative space'

Japan set a priority on controlling the ball by means of passing, while Croatia, with the use of fewer passes and more frequent dribbling, aimed for a more defensive game performance. The game performances of Jamaica may also be examined using passing as criterion. Using data from three games between Japan against Argentina, Croatia and Jamaica, the playing pattern of Japan was totally different between the match versus Argentina and the other two matches.

Table 2 shows the distribution of passes in different areas, obtained by a division of the pitch into six areas, using data from the three matches. Japan used more passes in the midfield

Table 1. The mean value for each variable for Japan and their opponent teams in the 1998 World Cup for soccer (frequencies/team game).

	Japan vs Argentina		Japan vs Croatia		Japan vs Jamaica	
Fouls	25	33	11	18	23	14
Dribbling	49	**95	58	*83	54	62
Clearing kicks	84	**50	37	45	35	**67
Passes	294	**410	379	**292	435	395
Loss of control	15	23	13	11	17	13
Throw-ins	17	16	14	14	15	7
Free kicks	28	24	17	16	25	13
Corner kicks	6	4	6	1	8	5
Shots on goal	10	*22	11	17	24	*11
Headers	54	58	42	44	47	43
Crosses	20	13	31	19	38	**9
End of poss.	197	196	162	162	185	186

* P < 0.05 ** P < 0.01

Table 2. The distribution of passes for Japan and their opponent teams in the 1998 World Cup. The pitch was divided into six vertical divisions. Direction of attack is A to F. (Frequencies/team game.)

	A	B	C	D	E	F
Japan	11	69	94	85	33	2
			*	**	**	**
Argentina	12	56	132	131	66	13
Japan	17	64	125	119	44	10
				**	*	**
Croatia	15	66	123	64	23	1
Japan	7	63	145	137	68	15
	**	**		**	**	*
Jamaica	21	113	150	74	33	4

* P < 0.05 ** P < 0.01.

area of the opponents and also in offensive areas than did Croatia and Jamaica (P < 0.05 – P < 0.01), but less in midfield and offensive areas than Argentina (P < 0.05 – P < 0.01). All teams made the most passes in the midfield area beyond the halfway line while Argentina made the same number of passes in midfield within the opponent's half. Furthermore, Japan made fewer passes in defence than did Jamaica (P<0.01). The frequencies of passes by Japan's players in their own half and the opponent's half of the pitch in three games were 174 (59.2%) and 120 (40.8%), 206 (54.4%) and 173 (45.6%), 215 (49.4) and 220 (50.6%), respectively. On the other hand, passes by Argentina, Croatia and Jamaica in their own half and the opponent's half of the pitch numbered 220 (48.8%) and 210 (51.2%), 204 (69.9%) and 88 (30.1%), 284 (71.9%) and 111 (28.1%), respectively. Furthermore, the number of passes used by Japan within the offensive areas in the three games were 3554 and 83, respectively.

Table two shows the frequencies of each pass in the opponent's court for Japan and the opposing teams in the 1998 World Cup. The overall number of forward or square passes and backward passes in the opponent's court for Japan in three games were 92 – 26, 129 – 47, 159 –52 per game, respectively. Against Japan, Argentina had 156 – 63, Croatia had 61 – 31 and Jamaica had 73 – 36 forward or square and backward passes. It seems that these results have a directly proportional relationship with the overall number of passes. Every team had more forward passes than backward passes, and the ratio was about 2.6 to 1.

4 Conclusions

The main findings were as follows:

(1) Japan used passes more often than Croatia, but less frequently than Argentina (P < 0.01). Japan dribbled the ball less frequently than Argentina and Croatia (P < 0.05 – P < 0.01). Japan outnumbered Jamaica in the frequency of shots at goal (P < 0.05), but had fewer than Argentina (P < 0.05).

(2) Japan used more passes in the midfield and offensive areas than did Croatia and Jamaica (P < 0.05 – P < 0.01), but fewer in the midfield and offensive area than Argentina (P < 0.05 – P < 0.01).

(3) The ratio of short and middle backward passes in the opponent's court for Japan's three games were 12.4 and 9.1 (21.5%), 13.1 and 8.0 (21.1%), 14.0 and 6.8 (20.8%), respectively. Against Japan, Argentina, Croatia and Jamaica exhibited 17.4 and 9.1 (26.5%), 12.0 and 15.2 (27.2%), 14.7 and 14.7 and (29.6%) respectively.

(4) We can conclude from the results of our study that it is important for Japan to develop an attacking quality by using 'negative space' effectively and to use more middle backward passes to develop attacking possibilities.

5 References

Hughes, M.D., Robertson, K. and Nicholson, A. (1988) Comparison of patterns of play of successful and unsuccessful teams in the 1986 World Cup for Soccer, in *Science and Football* (eds T. Reilly, A. Lees, K. Davids and W. Murphy), E. and F. N. Spon, London, pp. 363–367.

Yamanaka, K., Uemukai, K. and Matsumoto, M. (1993) An analysis of playing patterns of the Japan national team in the 1992 Asian Cup for soccer, (in Japanese) *Ibaraki Journal of Health and Sport Sciences,* 9, 17–27.

16 GOAL SCORING PATTERNS OVER THE COURSE OF A MATCH: AN ANALYSIS OF THE AUSTRALIAN NATIONAL SOCCER LEAGUE

G. A. ABT, G. DICKSON and W.K. MUMMERY
School of Health and Human Performance, Central Queensland University,
Rockhampton, Australia

1 Introduction

'If events recur in predictable ways (as days must follow nights, and new births
compensate old deaths), then life includes pattern amidst the flux.'
(Stephen Jay Gould, 1998)

The game of soccer can be differentiated in the world of contemporary sport for two reasons:
first, because of its global popularity and second, the low frequency (and therefore high value)
of scoring. Consequently, there is merit in the objective examination of goal scoring patterns
that may determine the factors which ultimately lead to successful scoring opportunities. Much
attention has been focussed on the analysis of goal scoring from a tactical perspective, such as
the attacking methods employed in scoring and the number of passes involved (Jinshan et al.,
1993; Garganta et al., 1997). These types of analyses are important for identifying the
characteristics of successful team play. However, there appears to be few studies of other
aspects of goal scoring, particularly the time at which goals are scored during match play
(defined as goal scoring patterns). An analysis of this kind would provide useful information to
both coaches and sports scientists, as the relationship between goal scoring and time would
appear to be linked to those aspects of play which inherently change as a match progresses, such
as physical conditioning and tactical play.

Previous research examining goal scoring patterns during soccer match play has generally
been limited to one tournament or season (Morris, 1981 [cited in Ridder et al., 1994]; Jinshan,
1986; Jinshan et al, 1993; Ridder et al., 1994; Reilly, 1996). Morris (1981), who was the first to
examine goal scoring patterns, reported an increase in the frequency of goals scored as a match
progressed. Jinshan (1986; 1993) examined the goal scoring patterns of the 1986 and 1990
World Cups respectively, reporting conflicting results. For the 1986 World Cup analysis,
Jinshan examined the number of goals scored each 15 min, revealing that most goals were
scored during the 60–75 min period, with all other periods resulting in a similar number of
goals. Conversely, the 1990 World Cup saw goal scoring patterns essentially increase over
time, with the most number of goals scored in the final 15 min of play. Ridder et al. (1994) also
examined goal-scoring patterns per 15 min interval of play for 340 matches during the 1991–92
Dutch professional season, reporting the frequency of scoring increased monotonically during
match play. Reilly (1996) examined goal scoring patterns during the 1991–92 Scottish League

season, reporting a higher than average scoring rate in the final 10 min of play. Although not reported, there also appears to be a gradual increase in scoring frequency when the data are expressed per 15 min of play. Collectively, these studies suggest a trend towards an increase in the frequency of goal scoring as a match progresses.

The Australian National Soccer League (NSL) is still relatively young. Introduced in 1977, the NSL is now played during the summer months, frequently resulting in hot and humid playing conditions for most of the season. Although some players competing in the NSL are full-time professionals, the majority are still semi-professional. Together, these factors create a playing environment quite different to that observed in Europe, possibly resulting in a change in the goal scoring patterns observed over the course of a match. Therefore, given the European perspective and the limited sample size of previous research, an extended analysis of NSL matches appeared to be warranted.

2 Methods

Four seasons of NSL matches, commencing in 1994–95 through to and including 1997–98 were analysed. For each match, the time at which all goals were scored was recorded. The total number of goals scored in each minute was then calculated for each season. Seasonal totals were then merged to determine the total number of goals scored in each minute over all four seasons. Data were examined by 45-, 15-, and 5-min time periods. A limitation of the analysis was the inability to determine, for example, if a goal scored in the 47th minute was actually scored in the 2nd minute of time added on to the first half, or the 2nd minute of the second half. This may have resulted in a slight reduction in the frequency of goals scored just prior to half time. Additionally, goals scored after the 90th minute were included in the 90th minute figure. Given the non-parametric nature of these data, chi-square analysis was utilised to determine significant differences between the 1st and 2nd half. Non-parametric trend analysis (Ferguson, 1965) was utilised to determine the relationship between 15-min and 5-min periods across all four seasons. Alpha was set at 0.05.

3 Results

Altogether 2065 goals from 703 matches were analysed. Chi-square analysis revealed a significant ($P < 0.001$) difference between 1st and 2nd halves. This accounted for a 34% increase in the frequency of goals scored in comparison to the first half (see Figure 1). Trend analysis revealed a significant ($P < 0.01$) upward trend across the six, 15-min periods. This represents an approximate 10% increase between each successive period (see Figure 2). Trend analysis also revealed a significant ($P < 0.01$) upward trend across the 18, 5-min periods. This represents an approximate 6% increase between each successive period (see Figure 3).

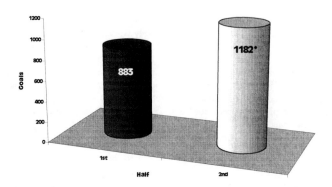

Figure 1. The frequency of goals scored in the 1st and 2nd halves of 703 NSL matches between 1994–95 and 1997–98. * Significantly different from 1st half at P < 0.001.

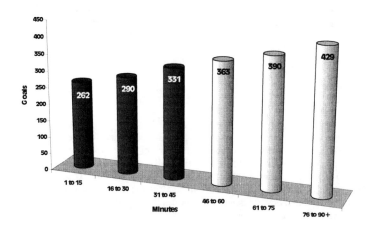

Figure 2. The frequency of goals scored in each of the six, 15-min periods of 703 NSL matches between 1994–95 and 1997–98. Significant upward trend across periods at P < 0.01.

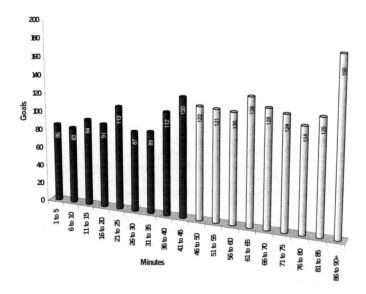

Figure 3. The frequency of goals scored in each of the 18, 5-min periods of 703 NSL matches between 1994–95 and 1997–98. Significant upward trend across periods at P < 0.01.

4 Discussion

This study has shown that the frequency of goals scored during a match is time dependent. A systematic and significant upward trend was observed in the number of goals scored as time progressed. Given the large sample size employed, the results of this study lend strong support to previous research suggesting an increase in the frequency of goals scored over the course of a match (Morris, 1981; Jinshan et al., 1993; Ridder et al., 1994; Reilly, 1996). The cause of this phenomenon is most probably multifactorial. Reilly (1996) suggested a number of possible explanations, including a greater deterioration in physical condition among defenders (thereby providing attackers with an advantage) and lapses in concentration. Certainly, a 'fatigue' factor would appear to be present, although the magnitude of this is unknown. From a purely physiological perspective there is a strong body of knowledge supporting a reduction in physical condition over the course of a match leading to a state of fatigue and reduced physical performance (Saltin, 1973; Smaros, 1980; Jacobs, Westlin et al., 1982; Bangsbo, 1994). However, it appears that physical condition may not influence goal-scoring ability. Recent studies by Zeederberg et al. (1996) and Abt et al. (1998) have shown that neither carbohydrate depletion nor supplementation appears to influence the performance of game related skills such as shooting. As such, maintenance of shooting ability as a match progresses would further aid attackers in gaining an advantage over defenders. Reilly (1996) also suggested that play may become more urgent towards the end of play as teams chase a result. Although urgent play is difficult to quantify, it would appear that players are willing to take greater risks towards the end of a match in order to effect an outcome. It is also possible that the losing team pushes players forward in order to create scoring opportunities, thereby scoring themselves or conceding further goals (Reilly, 1997).

Lapses in concentration may help to explain the peak in scoring frequency observed in the period immediately prior to half time (see Figure 3). Given that the number of goals scored in this period is probably under-represented in the results of this study, it would appear that the few minutes immediately prior to half time are very important for both scoring and conceding goals. During this period players may be concentrating more on getting to the dressing room than on playing the match. Similarly, the frequency of goals scored in the final 5-min period of the second half attests to the importance of this period for scoring goals.

Fluid balance may also be a factor influencing the goal scoring patterns in those games studied. From an Australian perspective, the hot and humid conditions under which many games are played increase the risk of players becoming hypohydrated. Hypohydration caused by hot and humid conditions has been shown to impair endurance capacity during prolonged high intensity exercise (Coyle and Hamilton, 1990; Sawka and Pandolf, 1990). Moreover, cognitive function is diminished in the hypohydrated state (Reilly and Lewis, 1985; Gopinathan et al., 1988), possibly leading to a reduction in decision making ability and/or skill performance. Conversely, Hoffman et al. (1995) reported no decrease in shooting ability during a simulated basketball game, despite fluid losses approaching 2% of body mass.

The practical implications arising from the results of this study centre around the ability of players to endure the full 90+ minutes of a match. Players who are able to minimise the effects of fatigue, especially towards the end of a match, should gain an advantage over their opponent. Strategies including dietary manipulation and ensuring adequate hydration prior to and during matches will certainly aid this process. Coaches could also consider tactics such as the rotation of players in order to minimise the effects of fatigue. Additionally, the manner in which substitutes are utilised could possibly be addressed.

The results of this study have shown that the frequency of goals scored over the course of a match is time dependent. Future research will be directed towards examining questions such as how teams respond successfully when opponents score early in a match and what goal scoring patterns those teams use. Additionally, useful information could be gained from an examination of the changes in goal scoring patterns across a season. Are goal scoring patterns different at the start of the season compared to the end?

5 Acknowledgments

The authors would like to acknowledge the contributions of Dr Brendan Humphries, Dr Robert Ho, Dr Grant Schofield, Thomas Esamie, John Punshon and Andrew Howe to this research.

6 References

Abt, G., Zhou, S. and Weatherby, R. (1998) The effect of a high carbohydrate diet on the skill performance of midfield soccer players after intermittent treadmill exercise, *Journal of Science and Medicine in Sport,* 1, 203–212.

Bangsbo, J. (1994) The physiology of soccer - with special reference to intense intermittent exercise, *Acta Physiologica Scandinavica,* 151(Suppl. 619), 1–155.

Coyle, E. F. and Hamilton, M. (1990) Fluid replacement during exercise: effects on physiological homeostasis and performance, in *Perspectives in Exercise Science and Sports Medicine* (eds C.V. Gisolfi and D.R. Lamb), Benchmark Press, Carmel, pp. 281–308.

Ferguson, G.A. (1965) *Nonparametric trend analysis.* McGill University Press, Montreal.

Garganta, J., Maia, J. and Basto, F. (1997) Analysis of goal-scoring patterns in European top level soccer teams, in *Science and Football III* (eds T. Reilly, J. Bangsbo and M. Hughes), E. and F.N. Spon, London, pp. 246–250.

Gopinathan, P.M., Pichan, G. and Sharma, V.M. (1988) Role of dehydration in heat stress-induced variations in mental performance, *Archives of Environmental Medicine,* 43, 15–17.

Hoffman, J.R., Stavsky, H. and Falk, B.(1995) The effect of water restriction on anaerobic power and vertical jumping height in basketball players, *International Journal of Sports Medicine,* 16, 214–218.

Jacobs, I., Westlin, N., Karlsson, J., Rasmusson, M. and Houghton, B. (1982) Muscle glycogen and diet in elite soccer players, *European Journal of Applied Physiology,* 48, 297–302.

Jinshan, X. (1986) The analysis of the techniques, tactics and the scoring situations of the 13th World Cup, *Sandong Sports Science and Technique* (April), 87–91.

Jinshan, X., Xiaoke, C., Yamanaka, K. and Matsumoto, M. (1993) Analysis of the goals in the 14th World Cup, in *Science and Football II* (eds T. Reilly, J. Clarys and A. Stibbe), E. and F.N. Spon, London, pp. 203–205.

Reilly, T. (1996) Motion analysis and physiological demands, in *Science and Soccer* (ed T. Reilly), E. and F.N. Spon, London, pp. 65–81.

Reilly, T. (1997) Energetics of high-intensity exercise (soccer) with particular reference to fatigue, *Journal of Sports Sciences* 15, 257-263.

Reilly, T. and Lewis, W. (1985) Effects of carbohydrate feeding on mental functions during sustained exercise, in *Ergonomics International 85* (eds I.D. Brown, R. Goldsmith, K. Coombes and M.A. Sinclair), Taylor and Francis, London, pp. 700–702.

Ridder, G., Cramer, J.S. and Hopstaken, P. (1994) Down to ten: estimating the effect of a red card in soccer, *Journal of the American Statistical Association,* 89, 1124–1127.

Saltin, B. (1973) Metabolic fundamentals in exercise, *Medicine and Science in Sports,* 5, 137-146.

Sawka, M.N. and Pandolf, K.B. (1990) Effects of water loss on physiological function and exercise performance, in *Perspectives in Exercise Science and Sports Medicine* (eds C.V. Gisolfi and D.R. Lamb), Benchmark Press, Carmel, pp. 1–38.

Smaros, G. (1980) Energy usage during football match, *1st International Congress on Sports Medicine Applied to Football,* Rome.

Zeederberg, C., Leach, L., Lambert, E.V., Noakes, T.D., Dennis, S.C. and Hawley, J.A. (1996) The effect of carbohydrate ingestion on the motor skill proficiency of soccer players, *International Journal of Sport Nutrition,* 6, 348–355.

17 REGAINING POSSESSION OF THE BALL IN THE DEFENSIVE AREA IN SOCCER

J.F. GREHAIGNE*, D. MARCHAL* and E. DUPRAT**
* University of Franche-Comte, Besançon, France
** UFR-STAPS Evry, France

1 Introduction

In its evolution, soccer was first characterized by attack in play; then, progressively, it evolved more defensive options. Nowadays, there appears to be a balance between the two phases of play. Tactical play is completely immersed in this attack – defence rapport and the role shift it implies following an attack on goal, recovery of the ball, its retention or loss. Téodorescu (1965) defined defence as 'a tactical situation during which a team fights for regaining possession of the ball while avoiding penalties called for by the rules and preventing opponents from scoring points'. Using quantitative analysis, Hughes and Franks (1997) show that one misses more passes and loses the ball more often if one is part of the team dominated. They also show that a team in trouble loses the ball significantly more often in the attacking third.

Gréhaigne's (1989) study of goals scored during soccer World Cups showed that a greater offensive efficiency is related to a fast and high (in the opponent's area) recovery of the ball, to an accelerated shift from defence to attack and to an almost instantaneous exploitation of the ball. Such results emphasize the importance of time and speed in this type of sequence of play.

After a brief overview of the offensive aspects of the game (Gréhaigne 1992 a,b; Bouthier 1993; Gréhaigne, Bouthier and David 1997), we shall proceed to an analysis of active recovery of the ball while in defence.

In this report, a recovery is called active if it has these characteristics; (a) the ball is recovered from the opponents (attackers); (b) recovery is made by the defenders, in the defensive half of the pitch; (c) the defence (now on attack) must make at least two consecutive passes or two sequences of play related to an offensive action. We chose these characteristics because we wanted to study true regain of the ball, not passages of play in which continual change of possession does not allow identification of the real ball holder (either team or player).

2 Research Design

As already discussed in the case of offensive actions (Gréhaigne and Bouthier, 1994), the setting evolves from a given state 1 to a state n. The configuration of play determines the players' positions at a given time t. Considering players' tactical choices in a collective attack, we shall analyse 'dynamical states' of all players who are part of the attack/defence system (Gréhaigne, 1989). To do so, we shall take note, for each player, of his position, the direction of his moves and his speed, thus determining possible changes in direction and the amount of ground than can be covered.

2.1 Sectors of play and sectors of intervention.

To represent these kinetic data in the appropriate plane, we propose the notions of a 'sector of play' for the attackers and a 'sector of intervention' for the defenders. These sectors spatially define the limits of possible action for the different players, given the three variables mentioned above. Consequently, we hypothesise that for a recovery of the ball to occur, there must be bad choices on the ball holder's part and/or good choices by defenders. Rules applied by defenders should be contrary to elements involved when goals are successful.

How can one define such types of action? From a preliminary analysis, we shall submit that the following configurations of play must be encountered in a play setting for the ball to be recovered. Figures 1 to 4 illustrate the four ways of recovering the ball. The shape of the sectors is a function of the players' speed.

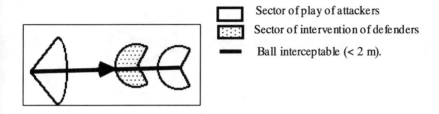

Figure 1, Case 1: For an interceptable ball (< 2 m), a sector of intervention must act as a block between two sectors of play where attackers are exchanging the ball. This is the classical case of a ball recovery with a static defender in block.

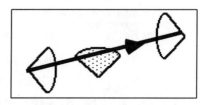

Figure 2, Case 2 (on a pass which lasts approximately one second): If an interceptable ball crosses a sector of intervention, the time between the ball being kicked and its clearing the sector must not be less than the time taken by the defender to cut the trajectory of the ball. The defender moves in block in order to intercept the ball.

Figure 3, Case 3: A non-interceptable ball (> 2 m) must leave a sector of play, free or not, and reach another sector of play occupied by one or many opponents. Then, a defence on the receiver applies.

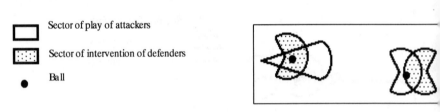

Figure 4, Case 4: The sector of play of an attacker who decided to dribble the ball is occupied by a defender or the latter enters the former's sector of play.

Besides these parameters, one may presume that, when an active recovery occurs, there are constants with regards to:

- the attack/defence rapport
- the recovery area
- the kind of gesture and action selected by the defender in order to take control of the ball.

3 Methods

This study was conducted durnig the 1994 World Cup in USA. From all matches played, four were chosen: Switzerland–Romania; Brazil–Russia; South Korea–Spain; and Italy–Mexico. These matches show different playing styles but are typical of different ways of recovering the ball as shown by Duprat (1996) in his study of tackles in that World Cup. In these matches, we considered all actions where the defence actively regained possession in its half, and carried on with at least two consecutive passes or a dangerous manoeuvre. Any ball called out of play in favour of the defence was rejected from the analysis. Interventions by the goalkeeper to recover the ball were not deemed active recoveries because they involve a player who, by rules of the game, has privileges that make him a special case in the defensive system. In the soccer literature, this goalkeeper's particular task is usually considered part of the goalkeeper distribution section.

3.1 Data collection

To determine the players' positions, direction of movement and speed, we used a video recorder with stabilised alimentation (capable of moving forward or rewinding one picture at a time). The beginning of each sequence was determined depending upon the situation as follows:

- when the ball was passed within the offensive half of the pitch, the beginning of the sequence of analysis corresponded to the time the ball was kicked;

- when the ball was dribbled or when it was passed from the defensive part of the pitch, the beginning of the sequence was set at the time the ball crossed the midline of the pitch. Each sequence lasted until recovery of the ball and was analysed on a 1-s interval basis and calculated speeds were average speeds.

On the basis of the observation grid (see Figure 2), the distance travelled by the different players from one configuration of play to the next one (given that 7 mm on the grid represented a 2 m distance on the pitch) was known; thus the distance travelled was equal to $2 \cdot \alpha$ (distance in mm covered on the grid). Using the video recorder, one could determine the time interval between two configurations of play (knowing that 25 pictures correspond to 1 s). The player's speed was then determined by dividing the distance travelled by the time. For the modelling of sectors of play and the sector of intervention, the protocol for collecting the data was identical to the one used by Gréhaigne and Bouthier (1994) for their analysis of attack configurations. Thus, the areas for sectors of play and sectors of intervention were functions of the players' speed.

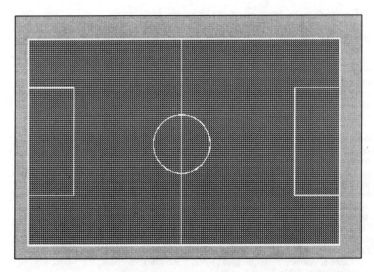

Figure 5. Observational grid (reduced size).

4 Results and Analysis

The results for the four selected matches show (a) variations in the number of recoveries (from 20 to 80) depending upon the opposition encountered but also (b) a relative homogeneity between half-times of each match (Table 1).

Table 1. Results.

	Recoveries first half	Recoveries second half	Total
Switzerland–Romania	39	33	72
Brazil–Russia	43	37	80
South Korea–Spain	22	22	44
Italy–Mexico	15	14	29
			225

All of the 225 active recoveries analysed satisfy the parameters discussed earlier. The distribution of the recoveries from case 1 to case 4 is as follows:

Case 1 = 17 ; Case 2 = 85 ; Case 3 = 77 ; Case 4 = 46

In the study considering 102 offensive actions (Gréhaigne, Bouthier and David, 1997), the results for corresponding cases were the following:

Case 1 = 22 ; Case 2 = 17 ; Case 3 = 17 ; Case 4 = 46

These results seem to be logical and complement elements provided by Gréhaigne et al. (1997). Indeed, it is better, from the offensive point of view, to proceed, using cases 1 and 4, in which the sector of play of the attacker is free or partially free. From a defensive point of view, recoveries are essentially obtained in case 2 settings, with interceptions where players' speed is important, or in case 3 settings, when attackers play long passes forward or try to displace the play, thus allowing defenders to position themselves and regain the ball before intended receivers. Let us examine more closely the characteristics of the 225 active recuperations registered in our study.

4.1 Positioning of defence
At first, one can examine whether the defence is in block or in pursuit of the attack at the time of the recovery.
  • The defence may be considered in 'block' when it is positioned between the ball holder, the attackers and its own goal.
  • The defence is considered in 'pursuit' when it is positioned behind the ball holder, the attackers and its goal.
The tools used to classify in this way the positioning of the defence are the main axes of inertia (representing the strong dimension) and the axes of dispersion of the attack and the defence. More specifically, at the interaction of these axes one finds the team's centre of gravity (bary centre of the players' positions). These data are automatically computed by a specific computer program developed by Gréhaigne (1992b). Results concerning the positioning of the defence are presented in Table 2.

Table 2. Positioning of defence according to cases.

	Case 1	Case 2	Case 3	Case 4	Total
Defense 'in block'	17	85	77	45	224
Defense 'in pursuit'	–	–	–	1	1

The results are clear. All but one active recoveries, the four cases combined, are performed with a defence in block. Thus, if the defence is in pursuit, it will rarely be able to recover the ball and thereafter resume an action. The exception registered here corresponds to a rare case where, using a correctly executed tackle (i.e., without touching the attacker), a defender regains the ball and immediately resumes with a counter-attack.

4.2 Number of players involved
The number of players involved in the configuration of play represents another indicator for looking at regularities in the game. Results presented in Table 3 show that numerical superiority is a fundamental element in defence.

Table 3. Numerical ratio of defence and attack.

	Case 1	Case 2	Case 3	Case 4	Total
Defence in numerical superiority	15	76	74	39	204
Defence in numerical balance	2	9	5	5	20
Defence in numerical inferiority	–	–	–	–	1
					225

Results indicate that for an active recovery to be obtained, the defence must generally be in numerical superiority (204/225). If numerically balanced, there appears to be an 8.8% chance of success (20/225). Such results imply that defenders are likely marking attackers and that other defending players are :
(a) either in support or covering the player regaining possession, or
(b) pressing the ball holder. Thus, two constants appear to be necessary to ensure success, the defence must be in block and in numerical superiority.

4.3 Favoured zones to recover the ball
In order to get quantitative results regarding zones, the field was divided into 40 equal squares as suggested by Gréhaigne (1988).

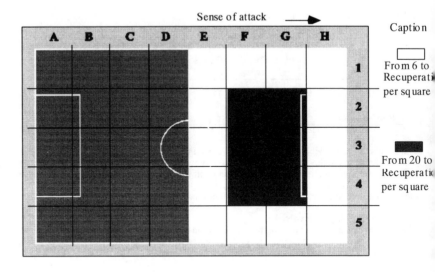

Figure 6. Zones to recover the ball (the numbers between brakes indicate the number of recuperations per zone).

Areas in which the greatest number of active recoveries occurred were located in F and G bands width-wise (81 and 82 recoveries respectively) and rows 2, 3 and 4 length-wise (54, 60, 58 recuperations respectively). Six specific zones stand out (F2, F3, F4, G2, G3, G4), considering their large number of recoveries. These six zones constitute a block at the centre of the defensive half of the pitch in the axes, and quite close to the goal. On the other hand, rows 1 and 5 do not appear to be recovery zones, especially the case for H1 and H5. Such results indicate that defence covers primarily the axes of the pitch.

Nevertheless, it is evident that the H3 area, right in front of the goal, is not an active-recovery area. In this area, rescue types of actions apply more than recoveries. It is an area where the defenders do not take risks. It is also the goalkeeper's action zone and he was not considered in this study. The G3 zone is favoured for type 3 recoveries (non-interceptable ball). One must avoid sending long passes forward in this area since heading duels are often at the defender's advantage.

Which skills do defenders perform to recover the ball actively and resume an offensive action?

4.4 Defensive skills
Over the 225 recoveries, only five types of skills have been registered: tackle; direct pass by heading the ball; direct pass by shooting the ball; taking control of the ball; taking the ball from the opponent's feet without tackling.

Taking control of the ball remains the most observed skill; that is understandable since it makes it easier to master the ball and reinitiate the play correctly (94 over 225). Types of skills are specific to each case: taking control, in cases 1 and 2, in a ground ball setting; heading, in case 3, in connection with long forward passes; tackles in case 4 in 1 vs.1 settings.

Table 4. Defensive skills according to cases.

	Case 1	Case 2	Caes 3	Case 4	Total
Tackle	–	8	–	29	37
Direct pass by heading the ball	34	–	8	–	42
Direct pass by shooting the ball	3	18	16	2	39
Taking control of the ball	10	59	23	2	94
Taking the ball from the opponent's feet without tackling him	–	–	–	13	13
Total	17	85	77	46	225

4.5 Relative speed of the players

An additional constant may be seen if one computes the players' speed. The defender who regains the ball must move faster or at least at the same speed as that of the concerned attacker who already controls the ball or is expecting it (225 over 225). The defender's speed is an essential constant for effective recovery.

Table 5. Players' relative speed.

	Case 1	Case 2	Case 3	Case 4	Total
Higher speed for the defender	3	67	37	21	129
Equal speed for players	14	17	40	25	96
Lower speed for the defender	–	–	–	–	0
Total	17	85	77	46	225

Such results may seem paradoxical since the faster the defenders move, the smaller the angle of intervention. Thus the defender needs great precision in his action.

This precision may be due to anticipation or to a faster grasp of the trajectory of the ball when all players have well interpreted the configuration of play. Indeed, the defender who is active in recovering the ball makes his move in a very limited and precise interval of time (slightly before, at the same instant, or slightly after the moment the ball has been kicked by the attacker), if he wishes to take the ball over (essentially for cases 2 and 3). In duels, higher speed and equal speed yield almost the same results. On the other hand, in all cases, a standing defender facing a moving player is one who will be bypassed and so must reposition himself.

In most cases, recovery is indeed related to a good defensive action on the defender's part but it may sometimes be due to a poor action on the attacker's part. These attacker's errors are mostly due to defensive pressure.

5 Conclusions

The main results of this study make it possible to establish some rules about the functioning of a good recovery of the ball in the defensive area:

- the defence must be in block or numerically balanced;
- when the play is in motion, a fixed defender is useless.

These rules apply with a reduced number of defensive technical skills and in specific areas of the pitch. Therefore, these characteristics of an active recovery of the ball constitute fundamentals for teachers or coaches with respect to practising defensive tactics. Indeed, with respect to defence, we know which areas must first be protected by the team on defence. From an offensive point of view, we know which types of actions or of ball movements hinder defenders. Running speed and speed of execution and of intervention remain very important elements of play.

To conclude, it would be interesting to know more specifically which collective or individual actions induce ball holders to make poor tactical choices or bad technical executions (e.g. closing the angle, defensive pressing). It would also be interesting to compare this study with the active recovery of the ball in the offensive half of the pitch in order to confirm whether this brings about more goals or not.

6 References

Bouthier, D. (1993) L'approche technologique en STAPS, représentations et actions en didactique des APS, *Habilitation à diriger les recherches,* Université Paris Sud.

Duprat, E. (1996) Technique et prise de décision: l'exemple du tacle en football, *Diplôme d'étude approfondies en didactiques des APS,* Université Paris – Orsay.

Gréhaigne, J.F. (1988) Game systems in soccer, in *Science and Football,* (eds T. Reilly, A. Lees, K. Davids, and W.J. Murphy), E. and F.N. Spon, London, pp. 316–321.

Gréhaigne, J.F. (1989) *Football de mouvement. Vers une approche systémique du jeu.* Thèse (nouveau régime). Université de Bourgogne.

Gréhaigne, J.F. (1992a) *L'organisation du jeu en football,* ACTIO: Paris.

Gréhaigne, J.F. (1992b) Modélisation pondérée de l'attaque du but en football, in *Les performances motrices: approche multidisciplinaire,* (eds M. Laurent, J.F. Marini, R. Pfister, and P. Therme), Actio/Université d'Aix – Marseille II, Paris, pp. 521–529.

Gréhaigne, J.F. and Bouthier, D. (1994) Analyse des évolutions entre deux configurations du jeu en football, *Science et Motricité,* 24, 44–52.

Gréhaigne, J.F. Bouthier, D. and David, B. (1997) Dynamic-system analysis of opponent relationships in collective actions in soccer, *Journal of Sports Sciences,* 15, 137–149.

Hughes, M. and Franks, I. (1997) *Notational Analysis of Sport.* E. and F.N. Spon, London.

Téodorescu, L. (1965) Principes pour l'étude de la tactique commune aux jeux sportifs collectifs, *Revue de la S.I.E.P.E.P.S.,* 3, 29–40.

18 AUSTRALIAN CONTRIBUTIONS TO THE ANALYSIS OF PERFORMANCE IN FOOTBALL 1963–1988

K. LYONS
Centre for Performance Analysis, University of Wales Institute, Cardiff, Wales

1 Introduction

It is common in scientific endeavour for a tradition of research or scholarly activity to be neglected or ignored. It is possible to find 'scholarly astigmatism' in most academic disciplines. To date there has been very little work in the history of ideas in sports science discourse. Hoberman (1992) has provided an excellent example of what such work encompasses. Elsewhere, I have developed a history of ideas approach to performance analysis (Lyons, 1994; 1996).

Considerable emphasis has been placed upon British and North American research traditions in the literature on the analysis of sports performance. This review focuses on a distinctive Australian contribution to the analysis of football and looks at the period 1963 to 1988 in particular. The former date marks the appearance of Nettleton and Sandstrom's (1963) report on skill and conditioning and the latter date marks the publication of the proceedings of the First World Congress of Science and Football (Reilly et al., 1988).

2 Analysing Football Performance

Australia has a rich football culture. Association football, Australian Rules, Rugby League and Rugby Union have distinctive traditions. From 1963 onwards researchers have used these codes as an empirical focus for their fundamental and applied research interests. The journal *Sports Coach*, in particular, has proved to be an excellent forum for studies in performance analysis. By the time Davis (1981) provided his overview of game analysis, the journal had published a dozen articles on performance analysis in a variety of sports (Allison, 1978; Embury, 1978; Beard, 1979; Davis, 1979; Craig et al., 1979; Davis and Fitzclarence, 1979; Hahn et al., 1979; Withers, 1979; Smith, 1980b; Barham, 1980; Evans and Davis, 1980; Walsh et al., 1980).

Enthusiasm for research into performance in football is exemplified by work in Australian Rules. One of the earliest articles to appear was Nettleton and Sandstrom's (1963) discussion of skill and conditioning. Their concern that there should be 'an interactive process between the game and training by which the events occurring in the game will affect the content of training and vice versa' became an important strand in subsequent research within and outside Australia.

They suggested that the development of an interactive process between game and training required 'an intelligent analysis of the individual performance of players during a large number of games'. They identified how an observer could chart a player's movements throughout a game and from these movements extract general patterns of effort. They also proposed a skill performance analysis and demonstrated this with data on successful and unsuccessful marks. These data collected in real-time provided opportunities for coaching interventions and player development.

Jaques and Pavia (1974) defined their study of movement patterns of players in an Australian Rules league match as an extension of Nettleton and Sandstrom's (1963) work. As consultants to a South Australian League team, the aim of their research was 'to guide the preparation of the programme for the team as a whole whilst providing information of specific workloads and patterns for individual members of the team'. They found it problematic to track a player in the way that Nettleton and Sandstrom recommended. They chose instead to film individual players with a videotape camera and used six cameras to film 20 players. Each player was filmed for at least a quarter of a game. Post-match, the movements of each player were plotted onto a scale diagram of the football oval. Four movement activities were notated: stationary, walking, jogging and running. In order to have an accurate track of each player's movements, the researchers used a lapsed-time method. That is, they played and re-played the videotape until the movement activities were appropriately encoded. From these movement plots 'the distance covered and the percentage spent on each of the activities' were calculated. They concluded that 'information far in excess of what was anticipated was gained' and suggested that their work highlighted 'the need for the development of individual training programmes to supplement the normal "team" training session'. They conjectured that it would be interesting to apply movement analysis to other sports. In subsequent years, the Australian literature on performance analysis cited their work regularly.

Pyke and Smith (1975) produced a broad-ranging book on a scientific approach to the game and thereafter a stream of articles on performance analysis flowed over the next two decades. Work on movement patterns of field and boundary umpires attracted the interest of a number of researchers (Craig et al., 1979; Smith, 1980b; Pyne and Ackerman, 1986; Ellis et al., 1989). Eagerness to apply theory to practice was evident in work in relation to: on-field training (Davis and Fitzclarence, 1979); training activities and match-play (Hahn et al., 1979); and marking skill (Walsh et al., 1980).

One chapter of Pyke and Smith's (1975) book was devoted to 'game analysis' and offered advice on basic and detailed game statistics, videotape analysis, and injury analysis. Smith (1980a) discussed the role of statistics in football. Smith, Nettleton, and Briggs (1982) further identified the role of analysis in the design of practice.

Technological developments informed the progress of research into performance in Australian Rules. Patrick (1985a, 1985b) and Patrick and McKenna (1986a, 1986b), for example, provided early accounts of the computerised analysis of performance. Patrick and McKenna (1988) and McKenna et al. (1988) also contributed communications to the inaugural World Congress of Science and Football.

3 Methodological and Empirical Significance of Australian Research

Australian Rules football as a distinct code of football offered a rich source of data for those interested in the theory and practice of the game. Australian researchers also demonstrated an interest in other forms of football (see, for example: Withers et al., 1977; Morton, 1978; Withers, 1979; Beard , 1979; Withers et al., 1982; Spink, 1988; Allen, 1989) that should have enriched a global debate about performance. These collected works addressed important methodological and empirical issues that were taken up elsewhere in the literature on performance analysis. The failure to acknowledge fully the Australian contribution to the debate about performance in football has impoverished the global research agenda. This is not to say that other research (I am thinking here of papers not written in English) has not also suffered under-reporting. My argument is that in the case of Australian research, the wider performance analysis community should have known better.

Methodologically, the time span between the reports of Nettleton and Sandstrom (1963) and Jaques and Pavia (1974) permitted the evaluation of real-time and lapsed-time notation methods. The latter's access to some of the earliest video cameras presaged debates to be found much later in other research traditions. A decade later, Patrick (1985 a,b) and his colleagues (1986 a,b; 1988) were actively exploring the use of computerised analysis methods as alternatives to hand notation and coaches' memory. Interest in movement patterns of players and officials was at least concurrent with research elsewhere and the desire to link training and competition through a detailed study of performance has established Australian research at the forefront of this kind of work. An emerging research interest in coaching behaviour (Davis, 1979; Davis and Fitzclarence, 1979) and instructional effectiveness (Tinning, 1981) also set important markers for debates that were developing elsewhere.

Whilst some of the early researchers used a limited number of games for their data, Withers et al. (1982) demonstrated that it was possible to combine empirical interest in movement patterns and training specificity with rigorous observational protocols and statistical analysis.

4 Conclusion

This review celebrates an incandescence of Australian research into performance in football. It is an incomplete account and it is acknowledged that there will be errors of omission here. In other disciplines, narrative biographies and life histories are acknowledged forms of scholarly activity. A fuller picture of the genesis of ideas requires these kinds of approaches. For at least four decades there has been a research community interested in football in Australia. An international community of scholars ought to recognise global contributions to innovative and cumulative research. The venue for the Fourth Congress in Sydney was ideal for this purpose.

5 References

Allen, G. (1989) Activity patterns and physiological responses of elite touch players during competition, *Journal of Human Movement Studies,* 17, 207–215.

Allison, B. (1978) A practical application of specificity in netball training, *Sports Coach,* 2, 9–13.

Barham, P. (1980) A systematic analysis of play in netball, *Sports Coach,* 4, 27–31.

Beard, K. (1979) A player skill analysis for soccer, *Sports Coach,* 3, 40–43.

Craig, N., Jacques, T., Pavia, G. and Squires, B. (1979) An analysis of the movement patterns of field and boundary umpires in an Australian Rules football match, *Sports Coach,* 3, 30–34.

Davis, K. (1981) Steps in comprehensive game analysis, *Sports Coach,* 5, 27–29.

Davis, K. (1979) An approach to the analysis of a coach's performance, *Sports Coach,* 3, 26–29.

Davis, K. and Fitzclarence, L. (1979) A critical analysis of on-field training of a leading VFL Team, *Sports Coach,* 3, 8–13.

Ellis, L., Gilliam, L. and Brown, P. (1989) *The Fitness Demands of VFL Umpires,* Report submitted to the Victorian Football League. Phillip Institute of Technology.

Embury, L. (1978) Analysing Netball Matches, *Sports Coach,* 2, 35–38.

Evans, J. and Davis, K. (1980) An analysis of the involvement of players in junior cricket, *Sports Coach,* 4, 26–31.

Hahn, A., Taylor, N., Woodhouse, T., Schultz, G. and Hunt, B. (1979) Physiological relationships between training activities and match play in Australian football rovers, *Sports Coach,* 3, 3–8.

Hoberman, J. (1992) *Mortal Engines,* The Free Press, New York.

Jaques, T. and Pavia, G. (1974) An analysis of the movement patterns of players in an Australian rules league football match, *Australian Journal of Sports Medicine,* 5, 10–24.

Lyons, K. (1994) Is this what current research is all about? The life and work of Lloyd Messersmith, *Paper presented at Second World Congress of Notational Analysis,* Cardiff.

Lyons, K. (1996) The long and direct road: the life and work of Charles Reep, *Proceedings of the Third World Congress of Notational Analysis,* Ankara.

McKenna, M., Patrick, J., Sandstrom, E. and Chennells, M. (1988) Computer-video analysis of activity patterns in Australian Rules football, in *Science and Football* (eds T. Reilly, A. Lees, K. Davids and W.J. Murphy), E. and F.N. Spon, London, pp. 274–281.

Morton, A. (1978) Applying physiological principles to rugby training. *Sports Coach,* 2, 4–8.

Nettleton, B. and Sandstrom, E. (1963) Skill and conditioning in Australian Rules Football, *Australian Journal of Physical Education,* November, 29, 17–30.

Patrick, J. (1985a) The CABER system. *Australian Computer Science Communication,* 7, 10-1, 10-2.

Patrick, J. (1985b) The capture and analysis of football in real time, *Australian Computer Science Bulletin,* May, 9–10.Patrick, J. and McKenna, M. (1986a) A Generalised Computer System for Sports Analysis, *Australian Journal of Science and Medicine,* 18, 19–23.

Patrick, J. and McKenna, M. (1986b) Computerised analysis of handball in Australian football, *Australian Journal of Science and Medicine in Sport*, 18, 24–26.

Patrick, J. and McKenna, M. (1988) The CABER computer system: A review of its application to the analysis of Australian Rules football, in *Science and Football* (eds T. Reilly, A. Lees, K. Davids and W.J. Murphy), E. and F.N. Spon, London, pp. 267–273.

Pyke, F. and Smith, R. (1975) *Football: The Scientific Way*, University of Western Australia Press, Nedlands.

Pyne, D. and Ackermann, K. (1986) Physiological relationships between training and match performance of Australian football field umpires, *Sports Coach*, 10, 20–24, 51.

Reilly, T., Lees, A., Davids, K. and Murphy, W. (eds.) (1988), *Science and Football*, E. and F.N. Spon, London.

Smith, R. (1980a) The Use of Statistics in Australian Football, *Towards Better Coaching: The Art and Science of Coaching* (ed F. Pyke), Canberra: Australian Government Publishing Service, pp. 276–291.

Smith, R. (1980b) An analysis of the running patterns of field umpires in Australian football, *Sports Coach*, 4, 16–18.

Smith, R., Nettleton, B. and Briggs, C. (1982) Game analysis and the design of practice, *Sports Coach*, 6, 13–20.

Spink, K. (1988) The effects of modifying football on coaching behaviours, *Sports Coach*, 11, 19–24.

Tinning, R. (1981) Improving coaches' instructional effectiveness, *Sports Coach*, 5, 37–41.

Walsh, W., Abernethy, B. and McArthur, M. (1980), Marking skill in Australian football: Front-on versus side-on approach to the ball, *Sports Coach*, 4, 37–40.

Withers, B. (1979) Specificity and the soccer coach, *Sports Coach*, 3, 16–21.

Withers, R. T., Maricic, Z. and Wasilewski, S. (1977) *A Pilot Study of the Workloads of Australian International Soccer Players*, Unpublished project, Flinders University of South Australia.

Withers, R., Maricic, Z., Wasilewski, S. and Kelly, L. (1982) Match analyses of Australian professional soccer players, *Journal of Human Movement Studies*, 8, 159–176.

19 TIME–MOTION ANALYSIS OF ELITE TOUCH PLAYERS

D. O'CONNOR
Institute of Sport and Exercise Science, James Cook University, Townsville, Australia

1 Introduction

Australia is the current 'Touch' football world champion in both the Men's and Women's divisions. Determined to stay leaders in their field, relevant and specific training programmes were to be implemented in the leadup to the third World Cup in 1999. This necessitated a study to determine if recent rule changes to Touch had brought about significant changes to the physical demands of the game. Hearsay indicates that the game is played at a faster pace with more frequent interchanges since the inception of the 'no marker' rule in 1985 and the reduction from seven to six players on the field at any one time in 1996.

Allen (1989) assessed the movement patterns of eight state open players during a 40-min game of Touch. The analysis indicated that the dominant movement pattern for both males and females was running/jogging, which accounted for 66.5% and 69.6% respectively, of total field time. To a lesser extent, walking (23.4%) and sprinting (males: 10.1%, females: 6.9%) were found to be executed. Average heart rates (HR) were recorded at 152 ± 5.5 beats·min^{-1} for males and 179 ± 4.2 beats·min^{-1} for females. This was calculated to be approximately 82% HR max for the males and 92% HR max for the females. Post-game blood lactate values for males were 3.81 ± 1.23 mmol·l^{-1} while those for females were slightly lower at 2.85 ± 1.03 mmol·l^{-1}. These readings indicated that the players were competing at below their 'anaerobic threshold'. Consequently, it was stated that the anaerobic alactic energy system was of primary importance during the high intensity activities while the aerobic system facilitated recovery from these efforts and supported the low intensity activities.

The aim of this study was to investigate the dominant movement patterns employed in elite six-aside Touch. The evaluation of energy requirements will aid in the development of effective and specific training programmes for the elite Touch player.

2 Methods

Thirty male players (19–29 years) and twenty female players (16–27 years) from the Australian squad were analysed during two men's and two women's games. Two video cameras were situated at a position 10 m from the scoreline and elevated approximately 5 m above the fields. Each game was of 30-min duration with a 5-min interval at half-time. The movement patterns of each player were analysed only when the game was in progress. Heart rate (HR) was recorded continuously using ratio telemetry (PE 3000 Sport Tester) from the start of the game until the final whistle. Blood lactate was measured during the period from 5 min of the game remaining to 5 min post-game using the ACUSPORT analyser. Warych et al. (1996) and Wigglesworth et al. (1996) have documented the high reliability and validity of this instrument.

Movement patterns were classified into the following categories (O'Connor, 1994):

- Stationary: No visible locomotor movement
- Walking: Strolling locomotor activity in either a forward, backward or sideways direction
- Jogging forward: Slow running movement, with no obvious acceleration in a forward direction
- Running backwards: Moving with purpose in a backward direction
- Sideways: A lateral movement of the body using a shuffle action of the feet
- Sprint: A fast running action with distinct elongated strides, effort and purpose
- Recovery: Time spent off the field

Intra-rater and inter-rater reliability was obtained by reanalysing one 15-min half for two players. A reliability coefficient was calculated for percentage of time, frequency and mean time for each movement classification. The correlation coefficients ranged from 0.85–0.89.

3 Results

3.1 Heart rates

Mean HR recorded in the men's and women's games were 177 ± 7 and 179 ± 7.8 beats.min^{-1} respectively. On average a player was working at 90% HR max during the entire game. During half-time the HR dropped to 60% HR max. The results showed that there were no significant differences between playing position or gender.

3.2 Blood lactate

By the conclusion of the game, the mean lactate levels of the male and female players were measured at 5.7 ± 1.4 mmol·l^{-1} and 5.9 ± 3.1 mmol·l^{-1} respectively (range 3.5-15.8 mmol·l^{-1}). When positions were taken into consideration, the outside players recorded lactate values of 7.5 mmol·l^{-1} while the middle players registered 4.6 mmol·l^{-1}.

3.3 Movement patterns

The percentage of the total game time (30 min) spent performing each activity is presented in Table 1. The two major movement categories were walking (18–19%), and jogging forward (19–20%).

Table 1. Percent of total game time for each activity.

Movement	Men	Women
Interchange	47.29 ± 6.20	47.66 ± 7.50
Jogging forward	20.04 ± 3.30	18.57 ± 2.20
Side	0.62 ± 0.80	2.31 ± 0.78
Running backward	5.57 ± 1.85	5.29 ± 1.39
Walking	17.87 ± 5.07	18.91 ± 4.01
Stationary	4.74 ± 1.40	3.46 ± 1.47
Sprinting	2.73 ± 1.05	4.19 ± 1.47

The high intensity activities of sprinting, running backward and laterally performed in either division represented just under 12% of total game time. This was interspersed among the low intensity locomotor activity that accounted for just under 42% of total game time. Interestingly, approximately 47% of a player's total game time was spent on the sideline.

Table 2 below outlines the number of repetitions completed in each movement category. For both male and female players jogging forward (67 repetitions) followed by walking (males 45 repetitions; females 59 repetitions) was the movement most frequently executed. Other than interchanges, these two activities were also executed for the longest duration before a change to another movement pattern. On average sprinting occurred on 19–22 occasions for an average duration of 2.9–3.3 s.

Table 2. Duration and Number of repetitions (reps) of each activity per game.

| Movement | Men | | Women | |
	Reps	Duration (s)	Reps	Duration (s)
Interchange	7 ± 1	111.4 ± 16.9	6 ± 1	130.35 ± 23.06
Jogging forward	66 ± 8	4.81 ± 0.7	69 ± 8	5.43 ± 0.67
Side	25 ± 6	1.88 ± 0.31	23 ± 6	2.65 ± 0.31
Running backward	30 ± 5	3.13 ± 0.55	27 ± 8	4.28 ± 0.93
Walking	45 ± 8	6.2 ± 0.97	60 ± 10	5.86 ± 0.62
Stationary	24 ± 5	4.7 ± 1.11	20 ± 8	4.52 ± 0.71
Sprinting	19 ± 4	2.88 ± 0.48	22 ± 9	3.26 ± 0.28

4 Discussion

Comparison of results between the study of Allen et al. (1989) and the present study is difficult to make as the duration of the games (40 min vs 30 min), the number of players on the field (7 vs 6), and the classification systems used to describe the movement patterns of the players in the two studies varied.

The movement patterns displayed during a competitive game will vary according to the skill of the opposition team, the closeness of the score and the environment in which the game is conducted. As all team members were part of the Australian squad, the standard of the opposition was matched as closely as possible. This resulted in final game scores within one touchdown between the teams.

4.1 Heart rates

The results of the present study demonstrate the intensity of the game of Touch. All observed HR values, other than during interchanges, were above the threshold level at which a beneficial cardiovascular effect is obtained. The high HR values suggest that Touch players require specific, high intensity training regimes in order to maximise performance. The non-significant positional finding is due to the narrow variability (± 4.69) in HR responses that was observed amongst individual players.

There appears to be little relationship between movement patterns and HR responses. However, there is some indication that those players who have a better endurance level are working at lower levels of their maximum. Consequently they are better able to cope with the physical demands of the game. In comparison to the results of Allen (1989), the HR responses for males were substantially higher during the six-a-side game, indicating the game was perhaps more aerobic in nature during the eighties. Interestingly, there is little difference in HR of females as reported in 1989 and in the present study.

4.2 Blood lactate

It has been suggested that a blood lactate level of approximately 4 mmol·l⁻¹ indicates that athletes are working around the 'ventilatory threshold' level. Players with a lower aerobic fitness (as recorded by previous fitness assessments) recorded higher lactate levels during the game. This indicated that these players spent more time using their anaerobic system to support their performance compared to other players. This trend was supported by the positional differences between outside players and 'middles'.

The lactate values were higher than those recorded by Allen (1989), indicating the increased glycolytic nature of elite Touch. In the previous study lactate concentrations remained below 4 mmol·l⁻¹, indicative of a more aerobic game. Lactate levels of 4.6–7.5 mmol·l⁻¹ reveal the need to train this energy system.

4.3 Movement patterns

The categories of stationary, walking and jogging forward were classified as low intensity activity and supported by the aerobic energy system. Sprinting, running backwards and sideways were deemed high intensity activities utilising the anaerobic energy system. Results revealed that 46–48% of total game time was spent on the sideline with interchanges occurring on five to eight occasions during a game. Players will be predominantly relying on their aerobic power to recover from the preceding work during these stints on the sideline. On-field activity revealed the intermittent nature of the game. In the men's game 11% of the total game time (or 26% of the actual on-field activity) was spent in high intensity activity. This is similar to the women's game where high intensity movements account for 12% of the total game time (or 29% of the actual on-field activity). In the men's and women's game, only 10% and 6.9% respectively, of the actual game time was spent in high intensity activity according to Allen (1989). However, there is no mention in the previous study whether any interchanges took place, and if so, how many and of what duration?

Female players walked on more occasions than male players did (60 vs 45 repetitions). This could be due to the fact that, as a group they have a lower aerobic fitness than the men while also interchanging less frequently. This pattern would contribute to the lower percentage of high intensity activity executed by the female players. More frequent and quicker interchanges may increase the amount of high intensity activity and reduce the need to walk on so many occasions. Interestingly, both genders spent 19% of total game time jogging forward and 19–22 sprints were executed in a game.

4.4 Training implications

It is evident that a strong aerobic base, a good lactate tolerance, quick acceleration and agility are advantageous to success. Due to the greater contribution from the anaerobic glycolytic energy system and the fact that movement patterns change every 2–6 s, conditioning sessions that simulate actual movements have been recommended. The basic concept of this type of workout is that the player performs ten consecutive movements with a short recovery (15–30 s). This constitutes one set. Up to five sets may be prescribed in a session, with a 2-min recovery between sets. It is imperative that each movement is executed at a high intensity rather than just 'going through the motions'. The movements can be set up using the sideline of a field and may include:

(a) 10-m sprint, diagonally backwards 7 m, sprint 10 m;
(b) 30-m sprint with the ball (zigzag every 5 m);
(c) 15-m sprint, break to right for 30 m (diagonal sprint);
(d) 10-m sprint, break to left for 30 m;
(e) 40-m sprint (scoop ball at 5 m mark);
(f) 5-m sprint, side (L) 5 m, 5 m sprint forward, side (R) 5 m, 15 m sprint;
(g) 15-m arc sprint to left;
(h) 30-m sprint (zigzag every 5 m);
(i) 5-m backwards, 5 m forwards, turn and sprint 15 m, dive on the ground, backwards 5 m;
(j) 30-m sprint (ground ball at finish).

Other examples of training include a series of short sprints with minimum recovery (e.g. 3–5 sets x 8–12 reps x 20–30 m); pyramid runs (vary movement patterns and distance – forward, backward, lateral combinations with slow jog/walk in between: 2 x 15 min). Speed and power training may include resisted (e.g., harness and parachute runs) and assisted sprints (e.g., slingshot); acceleration runs over 5–20 m; maximal speed work (30 m–70 m); and agility training.

5 References

Allen, G. (1989) Activity patterns and physiological responses of elite touch players during competition, *Journal of Human Movement Studies*, 17, 207–215.

O'Connor, D. (1994) Time motion analysis of elite touch referees, *Sports Coach*, 18, 19–21.

Warych, K., Sabina, T., Moreau, K., Fischer, K., Kaminsky, L and Whaley, M. (1996) Evaluation of the Acusport blood lactate analyser, *Medicine and Science in Sports and Exercise*, *28(Suppl.)*, *S10*.

Wigglesworth, J., LaMere, V., Rowland, N. and Miller, L (1996) Examination of the validity and reliability of a new blood lactate analyser, Medicine and Science in Sports and Exercise, 28 (Suppl.), S10.

20 APPLICATION OF AN ANALYSIS SYSTEM EVALUATING INTERMITTENT ACTIVITY DURING A SOCCER MATCH

J. OHASHI*, O. MIYAGI**, H. NAGAHAMA***, T. OGUSHI**** and
K. OHASHI*****
*   Daito Bunka University, Saitama, Japan
**   National Defense Academy, Kanagawa, Japan
***   Asia University, Tokyo, Japan
****   Sophia University, Tokyo, Japan
*****   University of Tokyo, Tokyo, Japan

1 Introduction

Soccer players move in various directions repeating intermittent activities during a match. These movements on the pitch are dependent on the players' physiological abilities and condition. To analyse such movements during a match, many researchers categorised the players' movements such as jogging and sprinting from VTR (Reilly and Thomas, 1976; Withers et al, 1982; Mayhew and Wenger, 1985). Ohashi et al. (1988) developed a new system to analyse movements of soccer players. In this method, the players' x–y co-ordinates, obtained by trigonometry, are calculated into distance and speed. This made possible the measurement of movements in more detail than in other reports. The distance and speed of movement can be analysed in time series if the movements are recorded in x–y co-ordinates in definite time sampling.

From the speed data obtained by this method and the blood lactate data in various running stages surveyed in our laboratory, it was estimated that the 10% of total movements in a match were based on the anaerobic energy supply system (Ohashi et al. 1992). However the system that we developed took too long to survey one player's movement in an actual game and to record movement in x–y c-oordinates. It was necessary to solve this difficulty. The purpose of this research was to speed up the data analysis in order to provide prompt feedback of results to coaching staff.

2 Methods

The measuring method is based on a triangular surveying technique, which calculates the point indirectly using the length of the base line and two angles. Potentiometers are attached to tripods and telescopes are attached to the tripod (Figure 1).

Surveyors follow the targeted player, keeping him in the centreline of the telescope. The signals obtained from the two potentiometers aimed at the player are digitised in 0.1 s of sampling time and data are transmitted in real time through an A–D board to a PC.

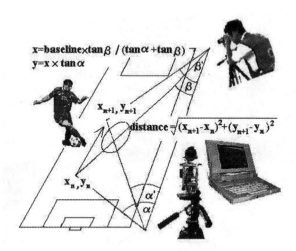

Figure 1. Measuring instruments.

Digitised data are converted to angles based on the calibration value. From the distance of two telescopes (the base line of triangle), the intersection point of two telescopes is recorded in x–y co-ordinates. The object player's movement tracks are shown in real time on a PC display by lining up the intersection points of the two telescopes (Figure 2). When the game is over, the data are reserved in the file. The data are the A–D converted value, angles and x–y c-oordinates calculated every 0.1 s from the potentiometer.

Immediately after the game, the analysis is done using the preserved data file. Errors such as accidental missing of a player by the surveyor can occur at any time. To prepare for these errors, 600 data of every 0.1s for 1 min are shown as the figures of A–D value, angle, x–y co-ordinate and distance between the x–y co-ordinates. And at the same time, changes of speed and movement tracks for every 1 min are shown so that any irregularities are detectable.

It also shows revised re-calculated movement tracks and speed immediately. Small errors are unavoidable when the surveyors track the object player during a game. This error is within 0.5 s, so small errors seen throughout the data is smoothed by a 5 point moving average. An experienced surveyor rarely loses sight of the player so the corrections are usually finalised by smoothing only. Once corrected and smoothed, the x–y co-ordinates are re-converted.

Figure 2. A player's movement track on PC display in real time.

3 Results

The distance of movement in the game is shown every 1 min. Also, the distance of 5, 15, 45 min are shown (Figure 3). From these data, it is possible to pick up a certain time in the game and check distance and speed of movement. Moreover, distributions of movement speed (m·sec⁻¹) for every minute can be shown as in figure 4. That is, for example, movement from 15 min–20 min at the speeds of 2 m·s⁻¹ and 3 m·s⁻¹ is 144.2 m and time required is 59 s, and can be evaluated as a certain percentage. This software is based on MS-Visual basic V 5.0 and can operate on both Windows 95 and 98.

Distance covered(1 min, 5min, 45min)

C:¥data¥Soccer¥Speed   45 min 5792.2m

	1	2	3	4	5	m/ 5min	m/15min
0->5	156.7	132.9	189.4	140.8	166.8	786.6	
5->10	105.2	127.0	130.4	108.9	118.8	590.3	
10->15	120.7	160.8	149.2	150.0	142.8	723.5	2100.4
15->20	42.2	168.0	150.9	141.4	83.5	586.1	
20->25	142.2	85.3	145.3	85.4	154.1	612.2	
25->30	167.7	135.1	132.9	135.5	102.3	673.6	1871.8
30->35	118.3	140.7	98.2	131.4	109.1	597.6	
35->40	82.0	162.7	111.6	154.0	134.4	644.8	
40->45	109.1	78.6	148.5	148.7	92.7	577.6	1820.0
45->50	116.4	35.0	.0	.0	.0	151.4	

Figure 3. Distance covered during a match.

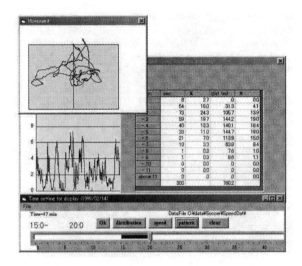

Figure 4. Change of movement speed, movement track and distribution of speed on PC display.

4 Discussion

Previous work by the authors using a PC and triangular surveying techniques enabled correct surveying of players' movements during a match but the data took too long to analyse. The current system permits real time tracking of movement and detailed analysis of the distance and speed of movement and shows the distribution of movement, distance and speed immediately after the game. This system can be used to evaluate quickly and immediately after the game the effect of conditioning and training regimes. Furthermore, this method can be used for talent identification purposes as well as for the development of new training methods that are adaptable to the actual game.

5 References

Mayhew, S.R. and Wenger, H. A. (1985) Time-motion analysis of professional soccer, *Journal of Human Movement Studies*, 11, 49–52.
Ohashi, J., Togari, H., Isokawa M., and Suzuki, S. (1988) Measuring movement speeds and distances covered during soccer match-play, *Science and Football* (eds T. Reilly, A. Lees, K. Davids and W.J. Murphy), E. and F.N. Spon, London, pp. 329–333.

Ohashi, J., Isokawa, M., Nagahama, H. and Ogushi, T. (1993) The ratio of physiological intensity of movements during soccer match-play, *Science and Football II* (eds T. Reilly, J. Clarys and A. Stibbe), E. and F.N. Spon, London, pp. 124–128.

Reilly, T. and Thomas V. (1976) A motion analysis of work-rate in different positional roles in professional football match-play, *Journal of Human Movement Studies*, 2, 87–97.

Withers, R. T., Maricic, Z., Wasilewski, S. and Kelly, L. (1982) Match analysis of Australian professional soccer players, *Journal of Human Movement Studies*, 8, 159–176.

21 AN ANALYSIS OF MOVEMENT PATTERNS AND PHYSIOLOGICAL STRAIN IN RELATION TO OPTIMAL POSITIONING OF ASSOCIATION FOOTBALL REFEREES

R.A. HARLEY, K. TOZER and J. DOUST
Chelsea School Research Centre, University of Brighton, Eastbourne, East Sussex, UK

1 Introduction

The physical fitness required of a referee to take up an optimal vantage point throughout the game depends upon the physiological demands of the task. Various researchers have analysed the physiological/work-rate requirements of association football refereeing (Asami et al., 1988; Catterall et al., 1993; Johnston and McNaughton, 1994) in order to quantify these demands. All studies published to date on work-rates and physiological strain of Association Football refereeing have been undertaken on elite level referees. The question as to whether the physiological/work-rate demands of refereeing are the same at lower levels of the officiating spectrum has not been investigated.

A minimum level of endurance fitness is required to enable referees to move around the field of play, without causing undue fatigue, in order to adopt the correct positions and to enhance their ability to make the correct decisions at the correct time. An adequate level of physical fitness is required not only for correct positioning but also to reduce fatigue and any detrimental effects this may have on the referee's decision making ability. For these reasons professional and amateur referees around the world are required to undertake a fitness test each year in order to be accepted onto the refereeing list. The most common of these tests prescribed by the Federation Internationale de Football Association (FIFA) involves referees running for 12 min, covering a minimum distance of 2600 m, with no walking. This test, however, fails to answer the question of whether referees are limited by their current levels of fitness, and whether improved fitness levels would enhance the ability of referees to position themselves appropriately during match-play. It has been suggested that understanding the physiological requirements can lead to the development of specific training programmes to prepare the soccer referee for the demands of the game (Johnston and McNaughton, 1994). However, an inherent problem with assessing the work-rate of referees during match-play is that the results indicate what the referees actually perform rather than what they should or could have been performing. This type of work-rate analysis gives no indication as to the quality of the positioning of the referee and therefore is lacking in two ways. First, no information is available as to whether the referees are unable to maintain the required work-rate due to the lack of physical fitness and second the results do not indicate whether referees are performing more physical work than is required but not necessarily taking up the optimal vantage points throughout the game. The results alone do not indicate whether improved fitness levels would lead to an improved ability of the referee to undertake more work and therefore enhance the quality of his/her positioning around the pitch.

The purpose of the present study therefore, was to assess whether there was a relationship between the physiological strain and the work-rate of County League referees and the quality of their positioning during match-play.

2 Methods

Fourteen County League referees (37.5 ± 6.7 years, 82.1 ± 9.5 kg) volunteered to be subjects in the study. All games analysed were Sussex County League Division One matches, which took place during the early part of the 1997–98 season.

2.1 Movement analysis
Each subject was video recorded during 90 min of play from the most favourable position around the touch-line or from a position in the main stand overlooking the pitch, close to the half-way line. The procedures of Catterall et al. (1993) were employed. Distances covered were categorised into four levels of intensity: walking, jogging, sprinting and backwards running. Each referee's stride length for each movement category was determined prior to the match by monitoring his movement across the centre circle, five times for each movement category. The distance divided by the average number of strides from the final three attempts was used to calculate average stride length. The distance covered was then estimated, for the first and second halves of play, by monitoring the number of strides taken under each of the four movement categories and multiplying the subsequent totals by the appropriate stride length.

2.2 Physiological analysis
Heart rate was recorded throughout each match using a heart rate monitor (Polar). Within one week of the match, referees underwent a graded exercise test on the treadmill. The starting speed used was set to elicit a heart rate of approximately 120 b·min^{-1}. Velocity was then increased 1 km.h^{-1} every 4 min with heart rate and oxygen uptake being recorded in the final minute of each

stage. Five incremental stages were completed and individual regression equations of heart rate and oxygen consumption were plotted for each subject using the least square method.

2.3 Assessment score

The quality of the referees' movements was assessed throughout the matches by three Sussex County League independent referee assessors, each evaluating three matches using a movement analysis questionnaire (see below) designed for this study.

Were the referee's movements and pacing appropriate within the first/second halves ?
Not at all   Every time 1   2   3   4   5   6   7   8   9   10
Were there any major movement discrepancies in regards to the referees diagonal system of control
Not at all   Every time 1   2   3   4   5   6   7   8   9   10
Did the referee ensure he took up the most favourable viewing position in order to see possible incidence ? E.g. The goal – Line for corner kicks
Not at all   Every time 1   2   3   4   5   6   7   8   9   10
Were the positions adopted by the referee taken up within suitable time limits ?
Not at all   Every time 1   2   3   4   5   6   7   8   9   10
Did the referee move with purpose, appropriate to the movement of play throughout the first/second halves ? Not at all   Every time 1   2   3   4   5   6   7   8   9   10

Figure 1. Movement Analysis Questionnaire.

Each referee received a mark out of 100 for the quality of his positioning in relation to match-play incidents during the game, a maximum score of 50 for each half. The total assessment score was calculated as a percentage, 0% being in the wrong place all of the time, to 100% being in the right place at the right time, all of the time.

3 Results

3.1 Movement analysis results

The mean total distances covered during 90 min of play for the 14 referees was 7496 ± 1122 m and ranged from 5760 m to 8979 m. The average distances covered consisted of walking 3137 m, jogging 3475 m, sprinting 225 m and backwards running 660 m. The total distances covered between the first (4017 ± 596 m) and second (3479 ± 574 m) halves for the 14 referees were compared using a pair t-test. Significantly ($P < 0.01$) less ground was covered during the second compared to the first halves. The distances covered within each of the movement categories for each half can be seen in Figure 2.

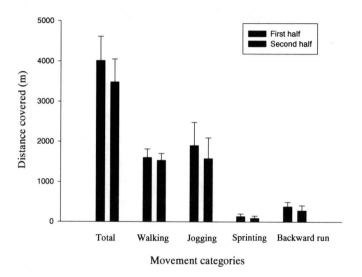

Figure 2. Distances covered within each of the movement categories for first and second halves.

3.2 Physiological strain

The mean heart-rates for the 14 subjects throughout the 90 min of play was 162 b·min^{-1}, ranging from 145 b·min^{-1} to 180 b·min^{-1}. Mean match heart rates between first (164 b·min^{-1}) and second (160 b·min^{-1}) halves were not significantly different.

The relationship between heart rate and oxygen uptake during the incremental tests was highly linear (r = 0.94 to 0.99) for all referees. The regression line for each referee was used to estimate the aerobic strain of the match by interpolation from the match heart rate data. An average aerobic strain of 80% (± 7.6) of peak VO_2 was predicted from game heart rates. A paired t-test demonstrated that average aerobic strain did not significantly alter between the first (81%) and second (79%) halves.

3.3 Assessment score

The mean assessment score for quality of positioning for the 14 referees over the whole match was 65.4%, ranging from 40% to 92%. The mean assessment scores were compared between halves of play using a paired t-test. The quality of positioning was found to be significantly reduced (P < 0.01) between the first (70%) and second (62%) halves. A significant correlation (P < 0.01, r = 0.8) was found to exist between distance covered and assessment score. This is demonstrated graphically in Figure 3.

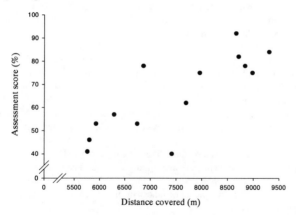

Figure 3. Relationship between total distance covered and assessment score.

4 Discussion

Referees are required to adopt certain positions on the pitch in order to uphold the laws of the game. To do this they must possess certain levels of physical fitness. It is recognised by many refereeing associations that a certain level of aerobic fitness is required. However, the question as to how physically fit referees need to be is complex and remains subject to debate. The fitness required by the referee to perform the task optimally will be influenced by many factors including the referee's positional reading of the game, the style of play adopted by the two teams, the level of performance of the two teams, the physical fitness levels of the players, and the environmental conditions. Motion analysis and physiological analysis have been used to indicate the physical requirements of a variety of sports (Reilly et al., 1990). However, in the case of football refereeing it is possible to perform the task without necessarily taking up the correct position. This can lead to mistakes by the referee and arguments from the players.

The total distances covered during games in this study of County level referees was 7496 ± 1122 m and ranged from 5760 m to 8979 m. These values are significantly lower than those reported by Asami et al. (1988) of 1118 m, Catterall et al. (1993) of 9438 ± 707 m (range 7977 to 10187 m), and Johnston and McNaughton (1994) of 9408 ± 838 m (range 7698 to 11265 m) for elite level referees. It would appear therefore that either the physical demands of the task are lower for County League level refereeing compared to elite level refereeing or that the physical fitness level of the referees is too low to cope with the demands of the task. Although the distances covered by the County level referees were significantly lower than for referees of elite level, match heart rates were similar. The average match heart rate for this study was 162 b·min^{-1} compared to 165 b·min^{-1} and 163 b·min^{-1} reported by Catterall et al. (1993) and Johnston and McNaughton (1994) respectively. The average age of the referees in this study (37.5 years) was also similar to the average age of referees (38.1 years) in the study of Johnston and McNaughton (1994). Although no ages were reported by Catterall et al. (1993), it is likely that these referees would have been in a similar age bracket since it is common for referees to take up the activity once their own playing career is finished. With age predicted maximal hearts being approximately 182 b·min^{-1}, referees are working at approximately 90% of the maximal heart rate throughout the game, equating to ~ 80% of VO$_2$ max.

The relationship between distance covered by the referees during the match and quality of positioning was investigated using a movement analysis questionnaire. The significant correlation ($P < 0.01$, $r = 0.8$) between these two variables, shown in Figure 3, demonstrates that at County League level those referees who cover the greater distances during the game attain the better assessment score for their positioning.

Referees were found to cover significantly less distance during the second compared to the first halves of play. This may well reflect the imposed lighter physical demand due to player fatigue or it may indicate that referees were not physically able to perform the required movements. Evidence gained from the assessment score would tend to indicate the latter applies, since the quality of referees' movement patterns was shown to drop significantly between the first and second halves.

To conclude, data reported by Asami et al. (1988), Catterall et al. (1993) and Johnston and McNaughton (1994) and the results of this study highlight the high levels of physiological strain placed upon Association Football referees. It would appear that at the County League level of officiating the ability of the referees to sustain the required work-rate as indicated by total distance covered is related to their ability to take up optimal vantage points throughout the game. The fact that quality of positioning and total distance both declined during the second halves of play indicates the importance of aerobic fitness for this group of referees.

5 References

Asami, T., Togari, H. and Ohashi, J. (1988) Analysis of movement patterns of referees during soccer matches, in *Science and Football* (eds T. Reilly, A. Lees, K. Davids and W.J. Murphy), E. and F.N. Spon, London, pp. 341–345.

Catterall, C., Reilly, T., Atkinson, G. and Coldwells, A. (1993) Analysis of the work rates and heart rates of association football referees, *British Journal of Sports Medicine,* 27, 193–196.

Johnson, L. and McNaughton, L. (1994) The physiological requirements of soccer refereeing, *The Australian Journal of Science and Medicine in Sport,* 6, 93–101.

Reilly, T., Secher, N., Snell, P. and Williams, C. (eds) (1990) *Physiology of Sports,* E. and F.N. Spon, London/New York.

22 PHYSIOLOGICAL ASPECTS OF SOCCER REFEREEING

S. D'OTTAVIO and C. CASTAGNA
Italian Soccer Federation (FIGC) and Italian Referees Association (AIA),
Rome, Italy

1 Introduction

In the past many researchers have addressed the metabolic demands imposed on soccer players during competitive and friendly matches (Reilly and Thomas, 1976; Van Gool et al., 1988; Bangsbo et al., 1991; Pirnay et al., 1991). Less interest have been devoted to the soccer referee's performance (Asami et al., 1988; Catterall et al., 1993) and the reports available in the international literature have been carried out employing indirect measurements (Catterall et al., 1993).

The purpose of this study was to assess the physiological demands imposed on the official during actual match-play employing direct physiological measurements. The aim was to shed light on the soccer referee's performance in order to develop specific training methodologies.

2 Methods

2.1 Subjects
Four Italian top level soccer referees (age 37.5 ± 1.2 years) were observed during the first half of friendly matches.

2.2 Match analysis
During the friendly matches (N = 4) match analysis was performed with a technology similar to that reported by Ohashi et al. (1988).

Match analyses were performed considering the following activity categories:
(1) standing;
(2) walking forward;

(3) low intensity run (less than 13 km·h^{-1});

(4) medium intensity run (from 13.1 km·h^{-1} to 18 km·h^{-1});

(5) high intensity run (from 18.1 km·h^{-1} to 24 km·h^{-1});

(6) maximal speed run (speed higher than 24 km·h^{-1});

(7) walking backwards;

(8) running backwards;

(9) running sidewards;

(10) high intensity activity (HIA, sum of the activities performed at speed faster than 18.1 km·h^{-1}

(11) unorthodox directional modes (distance in running backwards plus running sidewards, Reilly and Bowen, 1984).

2.3 Metabolic assessments

Oxygen uptake and heart rates were assessed during the matches and maximal tests with the aid of a portable lightweight device (K2, Cosmed, Rome Italy). With this device oxygen uptake and heart rates were sampled every 20 s of activity. Maximal tests were performed on separate occasions using a protocol similar to that reported by Leger and Boucher (1980).

Blood lactate analysis was performed via ear lobe blood samples at selected match time periods:

(1) after warm up;

(2) after 15 and 30 min of play;

(3) immediately at the end of the first half.

2.4 Statistics

Statistical analyses were performed using descriptive statistics. Data are presented as mean and standard deviation of the mean. Comparisons between mean were performed using the Wilcoxon matched pairs test (Vincent, 1995). Correlations between variables were performed using the Pearson coefficient of correlation. A significance level of 0.05 was chosen a priori.

3 Results

3.1 Match analysis

First half distance amounted to 6614 ± 150 m. During the first half, the referee on average walked forwards, ran at low intensity (speed slower than 13 km·h^{-1}) and ran at medium intensity (speeds ranging from 13.1 km·h^{-1} to 18 km·h^{-1}) for 7.2 ± 1.2% (480 ± 31 m), 49.8 ± 3.5% (3293 ± 48 m), and 28.8 ± 2.3% (1905 ± 27 m) respectively of the total distance (Figure 1). High intensity running and maximal speed running (speed higher than 24 km·h^{-1}) accounted for 11.7 ± 2.6% (776 ± 24 m) and 2.4 ± 0.8% (159 ± 45 m) respectively of the first half total distance covered. During the first half, the subjects never walked, ran backwards nor ran sidewards. Referees stood still for 9.6 ± 1.2% (270 ± 15 s) of the whole first-half time (Figure 2).

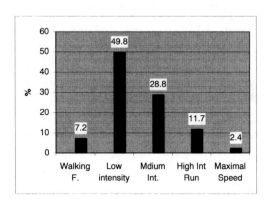

Figure 1. Percentage of total distance covered in the match categories.

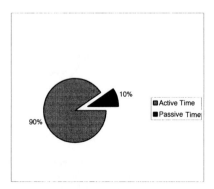

Figure 2. Percentage of match time spent with active or passive activities.

3.2 Match physiological measurements

During the first half of the match (47 ± 2.0 min) mean VO2 and heart rates attained respectively 67.6 ± 5% and 88.8 ± 6% of individual maximal values (Figures 3 and 4). Blood lactate levels after warm-up, 15, 30 min of play and immediately at the end of the match were respectively 2.1 ± 0.5, 3.5 ± 0.8, 7.3 ± 0.6 and 3.4 ± 0.9 $mmol \cdot l^{-1}$ (Figure 5). During the first half the blood lactate concentrations were revealed to be significantly higher than the post warm-up level ($P < 0.05$).

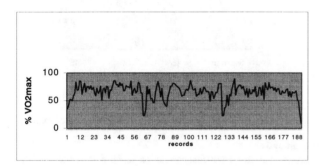

Figure 3. Percentage of VO_2 max attained by a referee during actual match play.

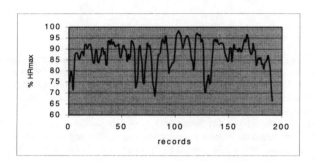

Figure 4. Percentage of the individual maximal heart rate (HR max) attained by a referee during actual match play.

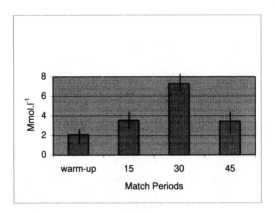

Figure 5. Lactate levels attained by the referees during the selected match periods.

3.3 Regression analysis

Regression analysis revealed a significant correlation between heart rates and VO_2 collected during the match ($r = 0.68$–0.70, $P < 0.001$).

Prediction of match VO_2 from the HR–VO_2 maximal test regression lines resulted in a $9.0 \pm 1.2\%$ overestimation ($74.3 \pm 4.2\%$ of VO_2 max; $P < 0.001$).

4 Discussion

The results of the present study were found to be similar to those reported for Italian soccer referees measured during first division matches (D'Ottavio and Castagna, 1999a,b). These referees ($N = 18$, D'Ottavio and Castagna 1999a) attained 89% of the estimated maximal heart rate and distance covered during the first half was 5854 ± 533 m ($N = 96$, D'Ottavio and Castagna, 1999b).

This comparison demonstrates that the research design adopted here was effective in simulating the competitive environment usually observed during first division matches.

In this research, mean match VO_2 attained by the referees was found to be within the ranges reported for professional soccer players in studies employing the same technology (43–69% of maximal values, Rodriguez and Iglesias, 1997). The VO_2 values found in these referees were lower than those estimated in competitive matched soccer players from HR–VO_2 regression lines (about 75%, Van Gool et al., 1988; Reilly, 1997).

Heart rates values were close to those reported for outfield soccer players during the first half (87% of HR max, Van Gool et al., 1988). Heart rates reported here are, however, lower than those estimated in top-level soccer referees by Catterall et al. (1993). Blood lactate concentrations reached levels close to that reported for soccer players (Ekblom, 1986; Bangsbo et al., 1991; Smith et al., 1993).

In this study the prediction of the involvement of the aerobic metabolism from the HR - VO_2 maximal test regression lines resulted in a 9% overestimation (see results). This difference from the values obtained with direct match measurements of the VO_2 was significant ($P < 0.001$). The difference between the two values may be due to the nature of the exercise involved during the assessment of the VO_2 which was continuous during the maximal test sessions and intermittent in match conditions. Balsom et al. (1992) reported that during intermittent high intensity short duration sprint activity, HR is higher than could be expected from VO_2 levels. It is also accepted that during match activity referees like players, have their HR raised because of emotional stress (Rhode and Espersen, 1988; Catterall et al., 1993). However, the existing significant correlation between HR and VO_2 values collected in field conditions reveals that match HR could be used, although cautiously, for the assessment of aerobic involvement during activity.

5 Conclusion

Results showed that soccer refereeing is an intermittent exercise that relies mostly on the aerobic pathway. Nevertheless, blood lactate concentrations can reach quite high levels sometimes similar to that reported for competitive soccer players.

Fitness training should focus on the development of the aerobic capacity of the referee. Nevertheless strength-endurance and speed-endurance training should also be considered in the fitness development process to allow top-level referees to cope with the high intensity bouts of activity that sometimes occur during the match. In order to assess the involvement of the aerobic metabolism during specific training drills, heart rates should be used with caution.

6 References

Asami, T., Togari, H. and Ohashi, J. (1988) Analysis of movement patterns of referees during soccer matches, in *Science and Football*, (eds T. Reilly, A. Lees, K. Davids and W.J.Murphy), E. and F N. Spon, London, pp. 341–345.

Balsom, P.D., Seger, J.Y., Sjodin, B. and Ekblom, B. (1992) Maximal-intensity intermittent exercise: Effect of recovery duration, *International Journal of Sports Medicine*, 13, 528–533.

Bangsbo, J., Nørregaard, L. and Thorsøe, F. (1991) Activity profile of competition soccer, *Canadian Journal of Sport Science,* 16, 110–116.

Catterall, C., Reilly, T., Atkinson, G. Coldwells, A. (1993) Analysis of work rate and heart rates of association football referees, *British Journal of Sports Medicine,* 27, 153–156.

D'Ottavio, S. and Castagna C. (1999a) Work rate and cardiovascular stress in top level soccer referees. Communication to the *IV World Congress of Science and Football,* Sydney, 22–26 February.

D'Ottavio, S. and Castagna C. (1999b) Activity profile of top level soccer referees during competitive matches. Communication to the *IV World Congress of Science and Football,* Sydney, 22–26 February.

Ekblom, B. (1986) Applied physiology of soccer, *Sports Medicine,* 3, 50–60.

Leger, L. and Boucher, R. (1980), An indirect continuous running multistage field test: The Université de Montréal Track Test, *Canadian Journal of Applied Sports Science,* 5, 77–84.

Ohashi, J., Togari, H., Isokawa, M. and Suzuki, S. (1988) Measuring movement speeds and distances covered during soccer match-play, in *Science and Football* (eds T. Reilly, A. Lees, K. Davids and W.J.(Murphy), E. and F.N. Spon, London/New York, pp. 329–333.

Pirnay F., Geurde, P. and Maréchal, R. (1991) Contraintes Physiologiques d'un match de football, *ADEPS, Sport,* 71–79.

Reilly, T. (1997) Energetics of high-intensity exercise soccer with particular reference to fatigue, *Journal of Sports Sciences,* 15, 257–263.

Reilly, T. and Bowen, T. (1984) Exertional costs of changes in directional modes of running, *Perceptual and Motor Skills,* 58, 149–150.

Reilly, T. and Thomas, V. (1976) A motion analysis of work-rate in different positional roles in professional football match-play, *Journal of Human Movement Studies,* 2, 87–97.

Rhode, H.C. and Espersen, T. (1988) Work intensity during soccer training and match-play, in *Science and Football* (eds T. Reilly, A. Lees, K. Davids and W.J. Murphy), E. and F.N. Spon, London/New York, pp. 68–75.

Rodriguez, F. and Iglesias, X. (1997) The energy cost of soccer: Telemetric oxygen uptake measurements versus heart rate–VO_2 estimations. Abstract presented at the *II Annual Congress of the European College of Sport Science,* Copenhagen, 20–23 August, pp. 322-323.

Smith, M., Clarke, G., Hale, T. and McMorris, T. (1993) Blood lactate levels in college soccer players during match play, in *Science and Football II* (eds T. Reilly, J. Clarys and A. Stibbe). E. and F. N. Spon, London, pp. 129–134.

Van Gool, D., Van Gerven, D. and Boutmans, J. (1988) The physiological load imposed on soccer players during real match-play, in *Science and Football* (eds T. Reilly, A. Lees, K. Davids and W.J. Murphy), E. and F. N. Spon, London, pp. 51–59.

Vincent, W. (1995) *Statistics in Kinesiology,* Human Kinetics Publishers, Champaign, IL.

23 ACTIVITY PROFILE OF TOP LEVEL SOCCER REFEREES DURING COMPETITIVE MATCHES

S. D'OTTAVIO and C. CASTAGNA
Italian Soccer Federation (FIGC) and Italian Referee Association (AIA), Rome, Italy

1 Introduction

The majority of soccer research has focused on the match and physiological analysis of outfield soccer players. The referee's activities during actual match-play has received little attention (Asami et al., 1988; Catterall et al., 1993) despite the number of referees officiating all over the world.

The aim of this study was to outline the activity profile of Italian top level soccer referees during match-play in order to gain information about the demands of refereeing and consequently to develop specific training skills.

2 Methods

2.1 Subjects

Subjects (N = 33, mean age 37.8 ± 2.1 years) were experienced top level referees enrolled in the CAN (Commissione Arbitri Nazionali) and thus officiating in the Series A and B Italian championships. The subjects were examined during the course of the 1992–1993, 1993–1994, 1994–1995 and 1995–1996 first division Italian championships (Series A).

2.2 Match analysis

Match activities were monitored with a device that utilised a technology similar to that reported by Ohashi et al. (1988).

Analysis of match activities was carried out taking into account the following categories:
(1) standing;
(2) walking forward;
(3) low intensity run (less than 13 $km\,h^{-1}$);
(4) medium intensity run (from 13.1 $km\,h^{-1}$);
(5) high intensity run (from 18.1 $km\,h^{-1}$);
(6) maximal speed run (speed higher than 24 $km\,h^{-1}$);
(7) walking backward;
(8) running backward;
(9) running sidewards;
(10) high intensity activity (HIA, sum of the activities performed at speed faster than 18.1 $km\,h^{-1}$);
(11) unorthodox directional modes (distance running backwards plus running sidewards, c/f Reilly and Bowen, 1984).

In order to detect whether or not referees modified their activity pattern at any time during the match, the first and the last 15 min of play of each half was compared. The first 15-min period of a match comes immediately after a warm-up and performance capacity should be at its highest level. This figure changes for the first 15 min of the second half which comes after a recovery period of 15 min and as for players, referees could experience lowered muscle temperature due to the lack of exercise in this intermission.

Performance in the last 15 min of each half is usually indicated as the most urgent period of a match (Reilly, 1996). Especially in the last 15 min of the second half, players tend to give out all their remaining energies in order to influence the final match outcome and this 'final rush' could also affect the referee's effort. As the first and second half each lasted more than 45 min, 0–15, 30–45, 45–60, 75–90 min period limits are indicated precisely only for the first period. The 30–45 min period means the last 15 min of play of the first half, 45–60 indicates the first 15 min of the second half, 75–90 the last 15 min of the second half.

2.3 Statistical analysis

Data are presented as means $\pm$ SD and ranges. Mean values for match activities were compared using Student's t-test. ANOVA for repeated measures was performed when comparisons between more than means were necessary. Post hoc analyses were performed using the Scheffe test. Alpha was set at 5% ($P < 0.05$) a priori.

3 Results

3.1 Whole match activity

Average total match distance amounted to 11469 ± 983 m (Figure 1, range 7818–14156 m). During an average match referees stood still 832 ± 292 (range 94–2339) s and covered running backwards and sidewards respectively 867 ± 461 (123–2758) m and 81 ± 74 (0–494) m. Walking forwards and backwards the referees covered 945 ± 161 (600–1356) m and 39 ± 22 (3–111) m respectively. Low intensity, medium intensity and high intensity runs during a match amounted to 4577 ± 561 (2906–5873) m, 2746 ± 535 (33–3962) m and 1546 ± 419 (14–2633) m respectively. Distance covered with maximal speed running amounted to 427 ± 308 (48–1441) m.

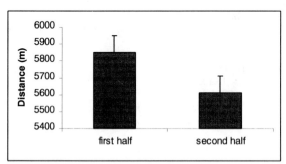

Figure 1. Distance covered during each half ($P < 0.05$).

Table 1. Distances (m) covered during the match in the categories considered for match analysis. Data are presented as means and standard deviation of the means.

Activities	Mean ± SD
Standing	832 ± 292
Forwards walk	945 ± 161
Low Int. Run	4577 ± 561
Medium Int. Run	2746 ± 535
High Int. Run	1546 ± 419
Maximal speed run	427 ± 308
Backwards walk	39 ± 22
Backwards run	867 ± 461
Sidewards run	81 ± 24
Non orthodox	948 ± 492
Total	11469 ± 983

3.2 First vs second half variations

Although no difference was observed between the duration of the two halves (P > 0.40), during the second half referees covered less distance (5612 ± 513 m vs 5854 ± 532 m, P < 0.001, Figure 2). Unorthodox directional modes distance decreased in the second half from 552 ± 280 to 433 ± 245 m (P < 0.001, Table 2). Referees stood still longer during the second half (447 ± 150 s vs 392 ± 152 s, P < 0.001). Distances covered at speeds faster than 18.1 $km·h^{-1}$ (HIA) did not show significant changes between halves (1008 ± 310 m vs 989 ± 308 m, P > 0.39, Figure 5). Nevertheless the distance covered at a speed faster than 24 $km·h^{-1}$ increased in the second half from 202 ± 164 to 225 ± 158 m (P < 0.02, Figure 2). During the second half the distance covered walking forward increased from 461 ± 82 m to 486 ± 83 m (P < 0.001, Figure 2).

3.3 Analysis of selected match periods

During respectively 0-15, 30-45,45-60,75-90 min periods the referees' total distance covered amounted to 1925 ± 211 m, 1802 ± 197, 1799 ± 195 and 1773 ± 193 respectively. The total distance covered by the referees during the 0-15 min period was superior to that achieved in other periods (P<0.001). For the four periods considered, distance run at low intensity amounted to 752 ± 106, 711 ± 123, 736 ± 115, 730 ± 107 for 0–15, 30–45, 45–60, and 75–90 min periods respectively. Differences between mean low intensity running values were not significant only between 45–60 and 75–90 min (P < 0.05) periods. Distances of medium intensity running were 502 ± 108 (0–15 min), 437 ± 98 (30–45 min), 439 ± 102 (45–60 min), 416 ± 85 m (75–90 min). In this category mean values were statistically different (P < 0.001) between 0–15 and 30–45, 0–15 and 75–90 (P < 0.001), 30–45 and 45–60 (P < 0.001) between 45–60 and 75–90 (P < 0.02) and between 30–45 and 75–90 min (P < 0.03). At high intensity (Figure 3) referees ran 257 ± 88 m, 244 ± 71 m, 236 ± 79, 243 ± 80 m respectively in the first, second, third and fourth 15-min periods considered. In that category the difference between the distances covered during the first 15 min of play of the first and the second half was significant (P < 0.02). At maximal speed running (Figure 4) referees completed 61 ± 57 m, 68 ± 62 m, 61 ± 53 m and 74 ± 59 m respectively in the course of the 0–15, 30–45, 45–60 and 75–90 min periods. Differences between periods were significant for 0–15 vs 30–45 (P < 0.04), 0–15 vs 75–90 (P < 0.01), 45–60 vs 75–90 min periods (P < 0.01). In the four periods the referees stood still for 108 ± 47, 135 ± 57, 133 ± 56 and 140 ± 55 s respectively. The time spent standing still during the first 15 min of play was significantly different from other the other selected match periods (P < 0.05). For unorthodox directional modes of motion a greater distance was covered in the first 15-min period of the match (185 ± 97) compared to the other periods taken into account (166 ± 94, 144 ± 85, 133 ± 90, P < 0.05). In this category of activity only the performance in the first 15 min of play of the second half was not significantly different from that achieved in the last 15 min of the match (P < 0.05).

Table 2. Distances (m) covered during the first and second half in the categories considered for match analysis. Data are presented as means and standard deviation of the means.

Activities	First half	Second half
Standing	392 ± 152	447 ± 150
Forwards walk	461 ± 82	486 ± 83
Low Int. Run	2295 ± 311	2282 ± 287
Medium Int. Run	1481 ± 303	1342 ± 227
High Int. Run	806 ± 200	764 ± 201
Maximal speed run	202 ± 164	225 ± 158
Backwards walk	21 ± 12	50 ± 31
Backwards run	478 ± 249	386 ± 229
Sidewards run	53 ± 6	28 ± 32
Total	5854 ± 533	5612 ± 513

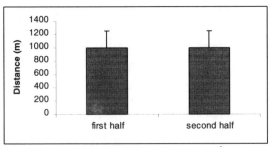

Figure 2. Distances covered at speeds faster than 18.1 km·h^{-1} during the two halves (P > 0.05).

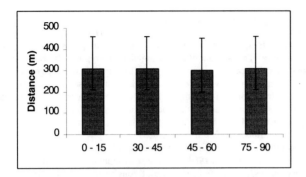

Figure 3. Distances covered at speeds faster than 18.1 km·h⁻¹ during selected match
periods (P < 0.05).

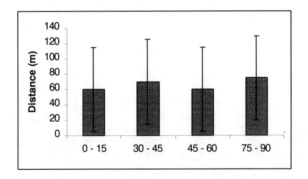

Figure 4. Distances covered at speed faster than 24 km·h⁻¹ during the selected match periods.
(for 0–15 vs 45–60 and 30–45 vs 45–60 min comparisons P > 0.05).

4. Discussion

In this study referees during an average high level soccer match covered a total distance of
11469 ± 983 m (Table 1). This distance was similar to that reported for competitive matched
midfield soccer players (Bangsbo et al., 1991) but higher than that reported for top level soccer
referees by Catterall et al. (1993) and Asami et al. (1988). The referees observed here covered
less distance during the second half ($P < 0.001$). Authors who studied the soccer referee's
performance have reported conflicting results. Caterall et al. (1993) reported a 5.49% decrease
which was significant ($P < 0.05$, $N = 13$). Conversely the research carried out by Asami et al.

(1988) showed no significant decrease in the course of the second half for total distance covered. Referees 17.20% of the whole match distance at HIA. These values seem to demonstrate that referees, at least at high level of competition, perform an important amount of high intensity activity. This amount of HIA is smaller than reported by Bangsbo et al. (1991) for competitive Danish soccer players. However, differences in the technologies employed and a different categorisation of activities obviously makes comparison between studies difficult.

The most interesting finding of this study was the conservation of the distance performed at HIA between halves (Figure 2) and in the selected match periods (Figure 3).

The finding that HIA did not change in the crucial part of the match demonstrates that the referees carefully administer their efforts in order to keep up with the game. The documented decrease of unorthodox directional modes distance, particularly demanding from the metabolic point of view (Reilly and Bowen, 1984), and the concomitant increase of time standing still demonstrate that the high level referee develops, through years of officiating, peculiar skills. In fact, referees seem to spare useless activity in order to perform at a high intensity particularly at the end of the match when game intensity often reaches its peak.

D'Ottavio and Castagna (1999) demonstrated that although there are inter-individual variations in HIA, the amount of distance performed in that category remained constant from match to match in Italian top level soccer referees. Thus the observation of just only one match can give an accurate measure of the individual ability to perform at HIA.

In this context a good level of endurance and speed endurance may prove useful in fostering the recovery of top level Italian referees after high intensity bouts of exercise and increasing the ability to perform at high intensity for longer during the match.

5 References

Asami, T., Togari, H. and Ohashi (1998) Analysis of movement patterns of referees during soccer matches, in *Science and Football* (eds T. Reilly, A. Lees, K. Davids and W.J. Murphy), E. and F.N. Spon, London, pp. 341–345.

Bangsbo, J., Norregaard, L. and Thorsoe, F. (1991) Activity profile of competition soccer, *Canadian Journal of Sport Science,* 16, 110–116.

Catterall, C., Reilly, T., Atkinson, G. and Coldwells, A. (1993) Analysis of work rate and heart rates of association football referees, *British Journal of Sports Medicine,* 27, 153–156.

D'Ottavio, S. and Castagna, C. (1999) Match to match work rate variations in top level soccer referees, *Poster presented at IV World Congress on Science and Football* 22–26 February 1999, Sydney.

Ohashi, J., Togari, H., Isokawa, M. and Suzuki, S. (1988) measuring movement speeds and distances covered during soccer match play, in *Science and Football* (eds T. Reilly, A. Lees, K. Davids and W.J. Murphy), E. and F.N. Spon, London, pp. 329–333.

Reilly, T. (1996) Motion analysis and physiological demands, in *Science and Soccer,* (ed T. Reilly), E. and F.N. Spon, London, pp. 65–81.

Reilly, T. and Bowen, T. (1984) Exertional costs of changes in directional modes of running, *Perceptual and Motor Skills,* 58, 149–150.

24 ESTIMATION OF PHYSIOLOGICAL STRAIN ON GAELIC FOOTBALL PLAYERS DURING MATCH-PLAY

T. REILLY and S. KEANE
Research Institute for Sport and Exercise Sciences, Liverpool John Moores University, Liverpool, UK

1 Introduction

The physiological strain induced in field games may vary with the standard of competition. The heart rate response to exercise provides a useful global measure of the exercise intensity. Mean heart rate during the entire game has been used as a measure of physiological strain, despite the intermittent nature of the activity (Bangsbo, 1994; Reilly, 1997). This applies to Gaelic football as well as soccer among the football codes.

The aims in this research were:
(1) to monitor heart rate during competitive Gaelic football matches;
(2) to compare heart rate responses between inter-county and senior inter-club matches;
(3) to establish the intensity of training matches compare with inter-county competitive games.

2 Methods

Twenty senior inter-county and 13 senior club-standard players acted as subjects. Each wore a short-range telemetry system (Seca Sportrance 300) during competitive matches. The duration of the games was 60 min in all cases.

Ten of the inter-county players were later examined in training sessions to establish the intensity of exercise relative to the competitive matches. The maximal heart rate of the subjects were revealed during performance of a 20-m progressive shuttle run to estimate maximal oxygen uptake (Ramsbottom et al., 1988).

Data were down-loaded following each game and analysed by means of a dedicated computer program. The mean heart rate was calculated for each half of the match. A further analysis established the times for which the heart rate lay within discrete ranges.

3 Results

The average heart rate was 9 b.min^{-1} higher for the club players than for the inter-county players, mean ± SD values being 169 ± 9 and 160 ± 6 b.min^{-1}, respectively. For the elite players the mean heart rate did not vary between the 30-min halves or between the first 10 min and the final 10 min of the game. Heart rate was more variable for the club players, tending to increase as the game progressed towards its end. Both groups attained peak heart rates of 201 ± 16 b.min^{-1} (elite) and 205 b.min^{-1} (club) during the match.

Table 1. Percent of the total time spent in various heart rate ranges by inter-county players during matches and during training.

Heart Rate Range (b.min^{-1})	Match (%)	Training (%)
< 100	1.1 ± 1.0	4.3 ± 4.1
101–130	17.8 ± 9.6	31.0 ± 7.0
131–160	28.1 ± 11.8	30.0 ± 5.4
161–180	42.5 ± 8.4	26.3 ± 9.0
> 181	10.5 ± 14.5	8.4 ± 10.8

These observations in the elite players were only periodically approached in their training regimens where the mean heart rate was 142 ± 6 b.min^{-1} for 28% of the match time and 30% for training (Table 1). The corresponding values for heart rates exceeding 180 b.min^{-1} were 11% and 8% respectively.

4 Discussion

The observations overall indicate relatively high physiological strain in Gaelic football matches, approaching 80% HR max as an average. The lower intensity (albeit longer duration) of training was due to the relatively smaller (27% versus 42%) amount of time spent with heart rates between 161-180 b.min^{-1}. It can be concluded that Gaelic football, whether inter-county or inter-club, represents strenuous exercise whereas training stress can be described currently as moderate.

The elevated values in the club players compared with the inter-county players was an unexpected observation. It contrasts with the trend of higher work-rates and therefore higher heart rates at the higher levels of competition (Reilly, 1996). The difference may be due to a lowering in exercise heart rate among the inter-county players due to superior training or a more economical use of their movement patterns during play. The better ability of the inter-county players to pace themselves throughout the game was reflected in the lower variability in heart rate compared to players participating at inter-club level. The ability to maintain performance consistently to the end of the game is important in Gaelic football, as is in the other football codes.

5 References

Bangsbo, J. (1994) The physiology of soccer: With special reference to intense physical exercise, *Acta Physiologica Scandinavica,* 150 (Suppl. 619), 1–156.

Ramsbottom, R., Brewer, J. and Williams, C. (1988) A progressive shuttle run test to estimate maximal oxygen uptake, *British Journal of Sports Medicine,* 22, 141–144.

Reilly, T. (1996) *Science and Soccer.* E. and F.N. Spon, London.

Reilly, T. (1997) Energetics of high-intensity exercise (soccer) with particular reference to fatigue, *Journal of Sports Sciences,* 15, 257–263.

25 A COMPARISON OF COMPETITION WORK RATES IN ELITE CLUB AND 'SUPER 12' RUGBY

M.U. DEUTSCH*, G.A. KEARNEY** and N.J. REHRER*
*   School of Physical Education, University of Otago, Dunedin, New Zealand
**   Department of Human Nutrition, University of Otago, Dunedin, New Zealand

1 Introduction

Specificity is a key paradigm in physical conditioning for sport, and for this purpose can be divided into two aspects. Energy specificity describes the sources from which energy is utilised in order to perform muscular work. It is thus concerned with the relative roles of the various energy pathways, and the ways in which these systems interact. Movement specificity refers to the biomechanical and motor learning aspects of sporting movements, such as co-ordination, joint angles, and so on.

As an intermittent high intensity sport, the physiological demands of Rugby Union involve a complex combination of factors that contribute to performance and fatigue. It is most likely that it is this complexity that accounts for so few scientific investigations of Rugby Union match-play. As a result, fitness testing and training methods for rugby have largely been adapted from other activities, such as track sprinting, middle distance running, bodybuilding, and Olympic weight-lifting, to name a few examples. Without a thorough understanding of the specific demands of the sport, however, it is unlikely that training and testing specificity will be achieved.

Several methods have been used to investigate the physiological demands of Rugby Union, such as heart rate (Deutsch et al., 1998), blood lactate (McLean, 1992; Deutsch et al., 1998), body temperature and fluid loss (Cohen et al., 1981), and direct muscle substrate measurements (Jardine et al., 1987). Time-motion analysis, however, remains the most commonly used measurement during match-play (Morton, 1978; Docherty et al., 1988; Treadwell, 1988; McLean, 1992; Deutsch et al., 1998).

Time-motion analysis has been used to quantify the times and/or distances in various activities during Rugby Union matches, in populations such as Australian colts (Deutsch et al., 1998), Five Nations (McLean, 1992), Scottish First Division (Docherty et al., 1988) and Welsh first division (Treadwell, 1988) players. The relevance of these investigations to current Super 12 and club rugby in New Zealand is questionable. Furthermore, only one previous investigation (Docherty et al., 1988) has sought to compare players at club and representative levels. As 'Super 12' players are drawn directly from the first grade competition in New Zealand, it is important to identify any possible differences that exist between competition levels, for developmental purposes.

2 Methods

Video data were collected from a total of 67 players over two seasons, during competition in Dunedin first grade and Otago Highlanders' Super 12 matches. The analysis of movements on individual players was based on a modification of protocols originally suggested by Reilly and Thomas (1976), and later modified by Docherty et al. (1988) and more recently by Deutsch et al. (1998). Briefly, individual players were tracked with video cameras for the entire duration of the matches investigated. Movements were later classified from video playback as either work (cruising, sprinting, tackling, jumping, rucking/mauling, and scrummaging) or rest (inactive, walking, jogging, shuffling sideways or backwards). This analysis was performed using a modified semi-professional video editing system (Pro:Log, Time Frame International, Christchurch).

Following the logging of all movements performed in each match, the editing list was processed using an in-house computer program (Deutsch et al., 1998). The program was designed to provide data such as total time and frequency of each movement, and also each instance of work, and rest. Thus the purpose was not only to investigate the specific activities performed during a match, but also to obtain information on work and rest periods, such as total work, frequency of work, mean work period, and mean rest period between work bouts.

Comparisons were made between forwards and backs at each competition level, and between competition levels. Data were first tested for the existence of normal distribution (skewness and kurtosis), and comparisons were then made using one-way ANOVA ($\alpha = 0.05$).

3 Results

Results revealed significant differences at Super 12 level between forwards and backs for total work (Figure 1), frequency of work (Figure 2), mean work duration, and mean rest duration (Figure 3). Similar differences were noted at club level between forwards and backs for total work, frequency of work, mean work duration, and mean rest duration. The only differences between club and 'Super 12' were observed among backs, who differed significantly in frequency of work ($P = 0.04$) and mean rest duration ($P = 0.004$).

It was also interesting to note that 'Super 12' backs spent significantly greater time in utility movements, when compared to the club backs (3.7 ± 2.0 and 2.25 ± 1.0, respectively; $P = 0.012$).

Table 1. Mean (SD) data for Super 12 and first division club players (* denotes significant differences (P < 0.01) between forwards and backs, $^\phi$ denotes significant differences (P < 0.05) between Super 12 and club players).

| Variable | Super 12 | | First Division Club | |
	Forwards	Backs	Forwards	Backs
Total Work (min)	10.2 (1.8)*	3.6 (0.6)*	10.2 (1.8)*	3.9 (0.6)*
Work Frequency (n/game)	121.9 (18.4)*	46.9 (9.0)*$^\phi$	118 (23.4)*	54.8 (10.8) *$^\phi$
Mean Work Period (s)	5.1 (0.5)*	4.7 (0.6)*	5.2 (0.6)*	4.3 (1.1)*
Mean Rest Period (s)	37.1 (7.0)*	99.5 (3.5) *$^\phi$	33.5 (1.6)*	78 (3.5) *$^\phi$

Figure 4 illustrates the relative (%) contribution of each of the high intensity modes to the total amount of work performed. Data are presented as means only, and do not indicate statistical comparisons.

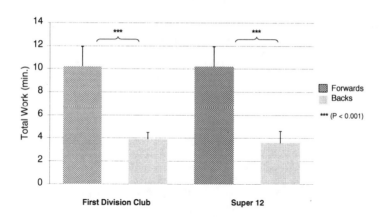

Figure 1. Total work performed (min) by forwards and backs at Super 12 and first division club level.

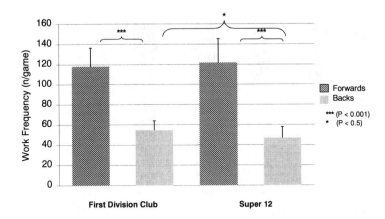

Figure 2. Frequency of work (n/game) by forwards and backs at Super 12 and first division club level.

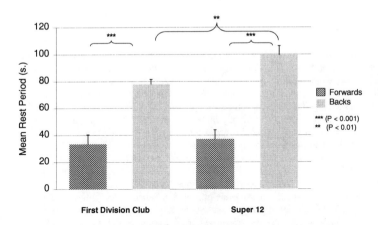

Figure 3. Mean rest period (s) between work bouts for forwards and backs at Super 12 and first division club level.

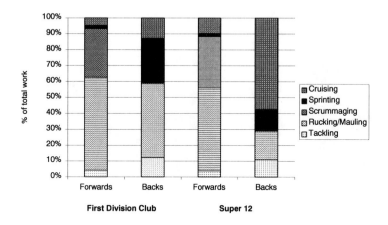

Figure 4. Relative contribution of various high intensity movements to total work performed by forwards and backs at Super 12 and first division club level.

4 Discussion

As in previous investigations, forwards were found to perform approximately 2.5 times as much more high intensity work than backs. Thus while forwards spent 12% of total match time in a state of high intensity work, the corresponding value for backs was only 4.5%. The present data indicate a slightly lower work-rate than data from 70-min Australian Colts matches, where forwards and backs averaged relative work times of 16% (~ 11.2 min) and 5.8% (~ 3.5 min), respectively (Deutsch et al., 1998). The slightly higher amount of total work performed in Colts rugby, particularly by forwards, is especially surprising, given the 70-min match duration (Deutsch et al., 1998). The data from this previous study also reflected a slightly higher work-rate when compared to other studies of senior level rugby (Morton, 1978; Docherty et al., 1988; McLean, 1992). Possible reasons for this difference include a higher level of exertion when working during senior rugby, and/or a less structured, more intermittent style of rugby in Colts players. In the present study, the lower work rate observed in Super 12 backs may reflect a more structured and higher quality of play (particularly by forwards) in professional Rugby Union, even when compared to elite club play. The current finding of a greater use of utility movements by Super 12 backs may further indicate more effective defensive and offensive patterns during the higher level of rugby competition.

Figure 4 demonstrates that approximately 90% of high intensity work performed by forwards was in rucking/mauling, scrummaging, and tackling. As these activities are highly specific to rugby, their development may have been overlooked. Such activities involve the production of power and force in a horizontal direction, and therefore the relative merits of vertical power development techniques used in Olympic power lifting and athletics warrant questioning. Developing training techniques that improve power, strength, and endurance in these specific activities should be sought as alternatives.

Forwards were found to perform approximately 120 instances of work in the present study, with a mean duration of approximately 5 s, separated by rest periods of ~35 s. As the creatine phosphate pathway is unlikely to be significantly replenished during such short rest periods (Balsom et al., 1992), much of the anaerobic energy for forwards must come from anaerobic glycolysis. Previous investigations of blood lactate concentration have shown only moderate accumulation during Rugby Union. Reported values ranged from 1-3 mmol·l^{-1} (Docherty et al., 1988) to approximately 6-9 mmol·l^{-1} (McLean, 1992; Deutsch et al., 1998). The validity of these measurements during intermittent high intensity exercise is questionable (Deutsch et al., 1998)

Training the peak power (ATP supply rate) of the anaerobic glycolytic system over a short duration (~5 s) may be more valuable than training to tolerate excessive lactate levels following longer, more exhausting intervals (~30 to 60 s). In cyclists, repeated 30-s maximal efforts (4-min recovery) have been shown to result in the inhibition of PFK (the rate-limiting enzyme in anaerobic glycolysis) after the first sprint. During a subsequent sprint, a further decrease in pH results in the inhibition of phosphorylase (the rate limiting enzyme in glycolysis), causing an almost complete inhibition of anaerobic glycolysis (McCartney et al., 1986). Perhaps anaerobic training for forwards should, therefore, focus on the development of short duration glycolytic power, rather than the tolerance of high lactate concentrations.

Current and previous data (Docherty et al., 1988; Deutsch et al., 1998) suggest that backs rely predominantly on energy from the creatine phosphate pathway during high intensity activities. In backs, unlike forwards, the mean rest period of ~90 s shown for backs would be sufficient to allow an almost complete recovery of creatine phosphate stores between work bouts, (Balsom et al., 1992). When fully replenished, creatine phosphate has been shown to provide almost all of the energy for ATP replenishment at the beginning of the work bout, and approximately 50% after 5 s of sprinting (Hultman et al., 1990), with most of the remainder coming from anaerobic glycolysis. As anaerobic glycolysis is an important source of energy after only 1.2 s of exercise (Hultman et al., 1990), developing the peak ATP turnover rate of this system should also be beneficial for back-line players.

The importance of aerobic fitness for Rugby Union players in all positions has previously been highlighted (Morton, 1978; Docherty et al., 1988; McLean, 1992; Deutsch et al., 1998), due to its role in recovery from high intensity periods of play, and in providing energy for the lower intensity elements of the game. Aerobic conditioning should, however, remain specific to the highly intermittent nature of Rugby Union. The development of running economy for rugby players is questionable, as frequent changes in direction and body position are more common in the game situation.

The present data indicate that Rugby Union requires fitness in many aspects of physiological function. As the interaction of these elements is unique to the game of rugby, physical conditioning should be aimed at developing these elements (and their interactions) in ways that remain physiologically and/or biomechanically specific as often as possible.

5 Acknowledgement

Sport Science New Zealand supported this study.

6 References

Balsom, P.D., Seger, J.Y., Sjödin, B. and Ekblom, B. (1992) Maximal intensity intermittent exercise: Effect of recovery duration, *International Journal of Sports Medicine,* 13, 528-533.

Cohen, I., Mitchell, D., Seider, R., Kahn, A. and Phillips, F. (1981) Effect of water deficit on body temperature during rugby, *South African Medical Journal,* 60, 11-14.

Deutsch, M. U., Maw, G.J., Jenkins, D. and Reaburn, P. (1998) Heart rate, blood lactate, and kinematic data of elite under 19 (Colts) match play, *Journal of Sports Sciences,* 16, 561-570.

Docherty, D., Wenger, H. A. and Neary, P. (1988) Time-motion analysis related to the physiological demands of rugby, *Journal of Human Movement Studies,* 14, 269-277.

Hultman, E., Bergstrom, M., Spriet, L.L. and Soderlund, K. (1990) Energy metabolism and fatigue, in *Biochemistry of Exercise VII* (eds A.W. Taylor, P.D., Gollnick, P.D. et al.), Human Kinetics, Champaign, IL, pp. 73-92.

Jardine, M.A., Wiggins, T.M., Myburgh, K.H. and Noakes, T. D. (1988) Physiological characteristics of rugby players including muscle glycogen content and muscle fibre composition, *South African Medical Journal,* 73, 529-532.

McCartney, N., Spriet, L.L., Heigenhauser, G.J.F., Kowalchuk, J.M., Sutton, J.R. and Jones, N.L. (1986) Muscle power and metabolism in maximal intermittent exercise, *Journal of Applied Physiology,* 60, 1164-1169.

McLean, D.A. (1992) Analysis of the physical demands of international rugby union, *Journal of Sports Sciences,* 10, 285-296.

Morton, A.R. (1978) Applying physiological principles to rugby training, *Sports Coach,* 2, 4-9.

Reilly, T. and Thomas, V. (1976) A motion analysis of work rate in different positional roles in professional football match-play, *Journal of Human Movement Studies,* 2, 87-97.

Treadwell, P.J. (1988) Computer-aided match analysis of selected ball games (soccer and rugby union), in *Science and Football* (eds T. Reilly, A. Lees, K. Davids and W. J. Murphy), E. and F. N. Spon, London, pp. 282-287.

26 CONTRIBUTING FACTORS TO SUCCESSIVE ATTACKS IN RUGBY FOOTBALL GAMES

K. SASAKI, J. MURAKAMI, H. SHIMOZONO, T. FURUKAWA, T. KATUTA and I. KONO
Research Centre of Health, Physical Fitness and Sports, Nagoya University,Chikusa,Nagoya, Japan

1 Introduction

Successive attack is one of the game skills in rugby football. Successive attack may be defined as the gain which shows more than one 'Maul' or 'Ruck' and more than a 5 m-gain. The purpose of this study was to investigate the factors contributing to successive attacking scenes in Rugby football games.

2 Methods

Three plays were selected as factors contributing to so-called 'second' attacks. These were individual gain (running, kicking), pass and finally making ground (Maul, Ruck).

Multiple regression analysis was used for the examination of the effective plays. The dependent variable was total gain (TG) of successive attacks in a game (m). Five selected independent variables were: the sum of 'individual gain' (SuIG) in successive attacks in a game (number of times), sum of running gain (RunG) in 'individual gains' (m), sum of gain from kicking (KcKG) in 'individual gains' (m), sum of all passes (PASS) in successive attacks (number of times), sum of points scored (MaPo) in successive attacks (number of times).

Nine games of the Japanese representative team in 1998 were studied. They included the games of the Asian elimination round of the World Cup. The technical staff of the Japan Rugby Union filmed all games. The total numbers of players analysed were over 270 persons.

3 Results and Discussion

The correlation coefficient between the balance of gains (m) of successive attack in the games and the balance of each team's points in the games was 0.91 which represents a high correlation value. The regression equation between 'Total gain' and the contributing factors is shown in Table 1. A high multiple regression coefficient (R square = 0.994, s = 15.01, F = 275.39, $P < 0.01$) was obtained in this analysis. The findings presented in Table 1 indicate that five independent variables explain the contributing factors for successive attacking. Furthermore, a statistically significant correlation was found between the independent variable RunG and the index TG. The value of KcKG also showed a relatively high correlation, which means a contribution to successive attacks similar to RunG.

Table 1. Regression equation between 'Total Gain' and the factors contributing to successive attacks.

TG= □1.54*SulG□□4.55*PASS □□2.52*MaPo □0.92*RunG□□0.93*KckG□□98.1					
S→ (3.20)	(3.93)	(1.54)	(0.13)	(0.36)	(28.0)
t→ (-0.48)	(-1.16)	(1.64)	(7.20)*	(2.62)	(3.51)
r^2=0.994 s = 15.01		F=275.39*		*p<0.01	

The successive attacks in game-time sequence for both the 1st half and 2nd half were expressed as 10 time-parts, each 4 minutes. This division can help to understand each team's attacking frequency. The gain image in a real field of each team can be displayed in this way to include the time of successive attacks, (Figure 1). The findings in the analysis showed that the quantity of individual gains contributes to successive attacking in Rugby Football. The next step was to clarify the quantity of individual gain for concrete playing skill analysis. In this step the contents of running gain were "an open run (inside direction)", "an open run (straight direction)", an "open run (outside direction)", "blind side run" and "point (Scrum, Maul, Ruck) side run". The contents of a kicking gain were "short punt", "long punt", "blind punt" and "drop goal". Each player's details may be presented both for those play patterns and the gain in metres (see Table 2). The data clarify who contributes to successive attacks and by what kind of play. It can be applied as an index for assessing player performance and potentially also for player selection.

Table 2. Details of plays contributing to successive attack in a specific game.

		Sum of IG(times)		Pass	Po Making		Try	Contents of Run Gain(m)					Contents of Punt Gain(m)			
		Run (m)	Kick (m)	(times)	Main (times)	Sub (times)	(times)	Open (inside)	Open (straight)	Open (Out)	Blind	Side Attack	Short Punt	Long Punt	Blind Punt	Dro Goa
1	Prop	1	5		3		5			5							
2	Hooker	1	5		7		6			5							
3	Prop						4										
4	Lock						9										
5	Lock	2	20		2	3	4			10			10				
6	Flanker	6	40		2	4	5	2			0		5,5,10,10,5,5				
7	Flanker	7	45		4	9	14		10	5, 5, 5	0	10	5,5				
8	No,8	5	35		8	6	7			5, 10	5, 5	10					
9	SH	5	10	65	34	1	1						5, 5	30		5, 30	
10	SO	13	20	235	12	4	1			5, 5, 5 5				30,10	50,35,20,15	20	20,
11	Wing	5	45		2	6	5			15	15		5, 5, 5				
12	Centre	9	45	85	7	7	4			10,15,5,5	5		5	25	50, 10		
13	Centre	4	25	40	3	5	10	1		5, 15			5		40		
14	Wing	2	60		2	2	1	1		20	40						
15	FB	4	70		3	3	6			20,10,30	10						
Total(m)		425	425					10	195	95	25	95	95	220	55	
	(times)	64 49	16	89	50	82	4	1	19	9	3	17	4	7	3	

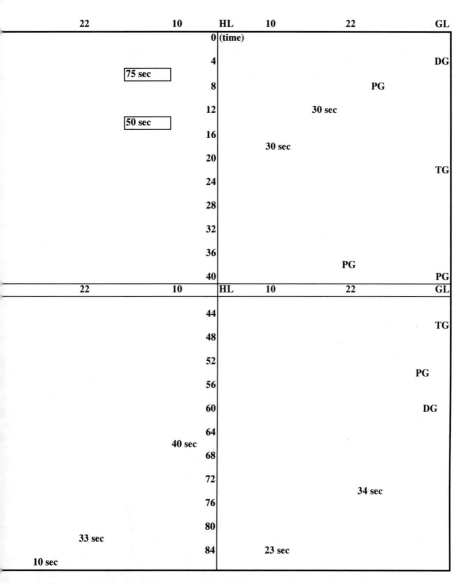

Figure 1. Gain image of each team (J→ A←) in a real pitch on a test match.

4 Conclusion

It could be concluded that the contributing factors to successive attacks were the 'Individual Gain (N of times)', 'Pass (N of times)', 'Making Ground (N of times)', 'Individual Running Gain (m)' and 'Kicking Gain (m)'. If the team's quantity and quality of running gain skill improves, the successive attacking pattern can be developed.

PART FOUR

Medical Aspects of Football

27 FURTHER INJURY IN A COHORT OF RUGBY UNION PLAYERS: PRELIMINARY REPORT

Y.N. BIRD, S.M. WILLIAMS, D.J. CHALMERS, B.D.WILSON and D.F. GERRARD
University of Otago Medical School, Dunedin, New Zealand

1 Introduction

A history of previous injury is reported to increase an athlete's risk of having a future sports injury (Ekstrand and Gillquist, 1983; Jones et al., 1993; Requa et al., 1993; Korkia et al., 1994; van Mechelen et al., 1996). In research on rugby union injuries, further injuries have tended to be reported as recurrent, with no further definition given. Between 15% and 18% of injuries to rugby players have been reported as recurrent (Dalley, Laing et al., 1992; Hughes and Fricker 1994; Garraway and MacLeod, 1995). In general, little research has been presented on previous injury as a risk factor for further injury. Much of the research on recurrent injury is related to specific injuries and their treatment (Stuart, 1994; Gross, Clemence et al., 1997; Inoue and Tamura, 1998).

The purpose of this research was to investigate the issue of previous injury as a risk factor for further injury within rugby union football. The hypotheses were that a higher risk of further injury is associated with the time elapsed after the first injury before returning to play in competition games; a previous history of injury; first injuries to joints, first injuries being dislocations or sprains/strains; the first injury being more severe; and players reporting high alcohol use scores.

2 Methods

Data from the Rugby Injury and Performance Project (RIPP) were used. The design of the RIPP has been described in detail elsewhere (Waller et al., 1994). In summary, a cohort of 356 male and female rugby players was followed throughout the 1993 competitive club season. Players in the cohort came from the following grades: First XV Schoolgirls and Schoolboys (Under 18/Under 19 grades), Colts (Under 21), Senior A Men and Senior Women.

At the start of the season, information on potential risk and protective factors was obtained by means of a self-administered questionnaire and an anthropometric and physical fitness assessment. Players were telephoned weekly throughout the club season to obtain information on rugby exposure and injury experience.

Detailed information was collected for all injury events that caused the player to seek medical attention *or* to miss at least one scheduled game or team practice. This included information on the site and type of injury, and playing position. Information was collected for new injuries only. Further injury was defined as: any second injury to a player who had been previously injured once within the same rugby season, regardless of the site or type of injury.

This paper reports on injuries occurring in scheduled competition games. Players were assigned to a single positional category based on the position in which they played most often during the season. Over 90% of participants played in the same position for the whole season. Team positions were grouped into two categories, forwards and backs. The forwards consisted of front row (props, hooker), locks, flankers and number eight. The backs consisted of half-back, first-five-eighth, second-five-eighth, centre, wings and full-back.

Three measures of severity were used: i) whether the player continued to play once injured, ii) whether the injury interfered with what the player had planned to do the next day (this measure relates to the player's activities of daily living), and iii) injuries were coded according to the Abbreviated Injury Scale (Association for the Advancement of Automotive Medicine, 1990).

The variables included in the univariate analyses were selected from information collected when the first injury was reported and from the baseline assessment. The variables relating to the first injury were time elapsed before returning to play after the first injury, site of the first injury, type of the first injury, and severity of the first injury (as described above). Return to play was recorded in weeks and was taken from when the first injury occurred to when the next competition game was played. Variables included from the baseline assessment were alcohol use (Quarrie et al., 1996) and previous injury experience.

Of the 356 players recruited into the cohort, eight were not followed because of incomplete consent, not having a telephone or living/travelling out of Dunedin. A further three players did not participate in any games or practices during the season and were also excluded. This left a total of 345 players (258 males and 87 females) who participated in rugby and were followed throughout the season. Given the small number of females in the cohort, meaningful analyses could not be performed on the data for these players. The analyses were therefore restricted to male players who had at least one musculoskeletal injury during the season (n = 185; concussive injuries were excluded from the analysis). Musculoskeletal injury was defined as: any injury involving soft tissues and/or bones, to include muscle contusions, ligamentous strains, joint dislocations or fractures. A Cox's proportional hazards analysis was undertaken. The outcome measure was number of games to second injury. Ideally, practice exposure would have been included in the outcome measure. Injured players could have attended practice as an observer, and as participation and observation could not be differentiated, there would have been an unknown bias in the outcome measure. Univariate analyses were performed initially, with adjustment for grade and position, as these two variables have been shown to be directly related to rugby injury (Seward et al., 1993; Garraway and MacLeod, 1995; Chalmers et al., 1996; Bird et al., 1998). The criterion for inclusion in the multivariate model was a value of P < 0.25 at the univariate stage. Grade and player positions were automatically included in the preliminary multivariate model.

3 Results

One hundred and eighty five male players received at least one injury in a rugby game during the season, with 91% of them receiving some form of medical attention. Ninety-eight of the 185 players (53%) received two or more injuries. Only 11 players (6% of the 185) had further injuries to the same site and of the same type as the first injury.

From the univariate analyses, time elapsed before return to play was found to be associated with the risk of further injury, the longer the elapsed time being associated with a lower risk of further injury (see Table 1).

Table 1. Modelling: Relative risks and 95% confidence intervals, adjusted for grade and position, for the association between the independent variables and rugby games played between first and second injury.

	N	%	RR	95%CI		p value
Age (years)						
17 and under	26	14.3	1.00			
18-19	52	28.6	1.11	0.31	3.90	0.870
20-22	61	33.5	0.78	0.19	3.16	0.732
23 and over	43	23.6	0.97	0.22	4.19	0.972
Position (Adjusted for grade only)						
Forwards	104	56.2	1.00			
Backs	81	43.8	1.29	0.83	1.99	0.242
Grade (Adjusted for position only)						
Senior A	77	42.3	1.00			
Senior B	27	14.8	0.90	0.44	1.84	0.777
Colts	44	24.2	0.79	0.45	1.39	0.427
Schoolboys	34	18.7	0.56	0.29	1.11	0.097
Previous Injury						
No previous injury	12	6.7	1.00			
Injury in last 12 mths	105	58.3	0.67	0.26	1.74	0.419
Injury at pre-season	63	35.0	0.92	0.33	2.53	0.876
Site of first injury						
Head	16	8.6	1.00			
Neck	9	4.9	1.42	0.45	4.48	0.541
Trunk	7	3.8	1.02	0.18	5.78	0.973
Upper Limb	47	25.4	0.89	0.37	2.13	0.795
Lower Limb	106	57.3	0.94	0.43	2.09	0.895
Type of first injury						
Sprain/strain	100	55.6	1.00			
Fracture	9	5.0	1.33	0.28	6.17	0.710
Haematoma	50	27.8	1.19	0.71	2.01	0.493
Laceration	14	7.8	1.34	0.62	2.89	0.450
Dislocation	7	3.9	3.12	0.70	13.84	0.133

	N	%	RR	95%CI		p value
Time away from play after injury						
Did not return	17	9.2	1.00			
Less than one week	96	51.9	0.65	0.38	1.10	0.110
Two-three weeks	44	23.8	0.36	0.18	0.71	0.003
More than three weeks	28	16.7				
AIS - The Abbreviated Injury Scale						
Minor	136	76.0	1.00			
Moderate/Serious	43	24.0	0.96	0.55	1.67	0.900
Continued playing after injury						
Did not continue	56	31.3	1.00			
Continued playing	123	68.7	1.10	0.67	1.81	0.688
Injury interfered with next days activities						
Did not interfere	103	56.0	1.00			
Interfered with next day	81	44.0	0.73	0.46	1.15	0.181
Alcohol AUDIT						
<8	37	20.4	1.00			
8-10	40	22.1	0.59	0.30	1.18	0.139
11-12	30	16.6	0.74	0.36	1.50	0.411
13-16	36	19.9	0.42	0.20	0.86	0.018
17-30	38	21.0	0.49	0.23	1.02	0.058

Return to play time, type of first injury, alcohol use, grade of play and player position, were all included in the multivariate model. In this model the strongest predictor of further injury in the season was return to play time (P = 0.002, Table 1). The longer the player was away from play after the first injury, the more games they played until they were injured a second time.

4 Discussion

The hypotheses that a higher risk of injury is associated with first injuries to joints, a previous history of injury and high alcohol use were not supported by the results of this preliminary regression analysis.

A higher risk of further injury was evident for those players with a shorter elapsed time before returning to competition games. Return to play time may be related to a number of other factors including severity of first injury (van Mechelen et al., 1992) and the rehabilitation process (Garrick, 1981; Reid, 1983). Players may not be giving their bodies enough time to recover from their injuries as the players with rapid return times had fewer games until their second injury occurred. It may be that players with injuries require some time away from playing rugby to let the injury heal. Missing one rugby game a week after injury, then playing two weeks post-injury would place that player at a lower risk of further injury, this may be beneficial for the team in the long term.

Healing time varies according to the injury type. In general it is considered that there are three phases in the healing of body tissue. The first of these is the acute phase lasting approximately 72 hours after the injury occurs. Phase two, the matrix and cellular proliferation phase, follows for a period of 72 hours to six weeks. Finally, phase three, the remodelling and maturation phase of healing can last from six weeks to several months post-injury (Oakes, 1992). The time required for initial healing and recovery of different tissue types can also be estimated:

muscles – 6 weeks, tendons and ligaments – 12 weeks, bones and joints – 6-12 weeks (Purdam et al., 1992). In general, most players in this cohort would have returned to play within the second phase of healing and well within the specific tissue healing time estimates. A high percentage of the cohort received medical attention for their injury. However, there was limited information collected about the rehabilitation phase of injury in this study. Accordingly, specific comments about the rehabilitation process for the players in this study cannot be made here.

5 References

Association for the Advancement of Automotive Medicine. (1990) *The Abbreviated Injury Scale,* Des Plaines, Association for the Advancement of Automotive Medicine.

Bird, Y.N., Waller, A.E., Marshall, S.W., Alsop, J.C., Chalmers, D. J. and Gerrard, D.F. (1998) The Rugby Injury and Performance Project: V. Descriptive epidemiology of a season of rugby injury, *British Journal of Sports Medicine,* 32, 319-325.

Chalmers, D.J., Sharples, K. et al. (1996) *A Model of Risk and Protective Factors for Rugby Injury: A Report Prepared for the Accident Rehabilitation and Compensation Insurance Corporatio,* Injury Prevention Research Unit and the Human Performance Centre, Dunedin.

Dalley, D.R., Laing, D.R., et al. (1992) Injuries in rugby football, Christchurch 1989, *New Zealand Journal of Sports Medicine,* 20, 2-5.

Ekstrand, J. and Gillquist J. (1983) Soccer injuries and their mechanisms: a prospective study, *Medicine and Science in Sports and Exercise,* 15, 267-270.

Garraway, M. and MacLeod D. (1995) Epidemiology of rugby football injuries, *The Lancet,* 345, 1485-1487.

Garrick, J.G. (1981) 'When can I...?' A practical approach to rehabilitation illustrated by treatment of an ankle injury, *American Journal of Sports Medicine,* 9, 67-68.

Gross, M.T., Clemence, L.M., et al. (1997) Effect of ankle orthoses on functional performance for individuals with recurrent lateral ankle sprains, *The Journal of Orthopaedic and Sports Physical Therapy*, 25, 245-252.

Hughes, D.C. and Fricker, P.A. (1994) A prospective survey of injuries to first-grade rugby union players, *Clinical Journal of Sports Medicine*, 4, 249-256.

Inoue, G. and Tamura Y. (1998) Recurrent dislocation of the extensor carpi ulnaris tendon, *British Journal of Sports Medicine*, 32, 172-174.

Jones, B.H., Cowan, D.N., Tomlinson, J.P., Robinson, J.R., Polly, D.W. and Frykman, P. N. (1993) Epidemiology of injuries associated with physical training among young men in the army, *Medicine and Science in Sports and Exercise*, 25, 197-203.

Korkia, P.K., Tunstall-Pedoe, D.S. and Maffuli, N. (1994) An epidemiological investigation of training and injury patterns in British triathletes, *British Journal of Sports Medicine*, 28, 191-196.

Oakes, B.W. (1992) The classification of injuries and mechanisms of injury, repair and healing, *Textbook of Science and Medicine in Sport* (eds J. Bloomfield, P. A. Fricker and K.D. Fitch), Blackwell Scientific Publications, Carlton.

Purdam, C.R., Fricker, P.A. et al. (1992) Principles of treatment and rehabilitation, *in Textbook of Science and Medicine in Sport* (eds J. Bloomfield, P.A. Fricker and K.D. Fitch), Blackwell Scientific Publications, Carlton.

Quarrie, K.L., Feehan, M. et al. (1996) The New Zealand Rugby Injury and Performance Project: Alcohol use patterns within a cohort of rugby players, *Addiction*, 91, 1865-1868.

Reid, D.C. (1983) Sports Medicine: Functional assessment and return to game fitness, in *Physiotherapy in Sport* (eds M.L. Howell and M.I. Bullock), University of Queensland, St Lucia, pp. 3-7.

Requa, R.K., DeAvilla, N. and Garrick, J.G. (1993). Injuries in recreational adult fitness activities, *American Journal of Sports Medicine*, 21, 461-467.

Seward, H., Orchard, J. et al. (1993) Football injuries in Australia at the elite level, *The Medical Journal of Australia*, 159, 298-301.

Stuart, M.J. (1994) Treatment of chronic chondral injuries, *Sports Medicine and Arthroscopy Review*, 2, 50-58.

van Mechelen, W., Hlobil, H. and Kemper, H. C. G. (1992) Incidence, severity, aetiology and prevention of sports injuries: A review of concepts, *Sports Medicine*, 14, 92-99.

van Mechelen, W., Twisk, J., Molendijk, A., Blom, B., Snel, J. and Kemper, H. C. G. (1996) Subject-related risk factors for sports injuries: a 1-yr prospective study in young adults, *Medicine and Science in Sports and Exercise*, 28, 1171-1179.

Waller, A.E., Feehan, M., Marshall, S.W. and Chalmers, D.J. (1994) The New Zealand Rugby Injury and Performance Project: I. Design and methodology of a prospective follow-up study, *British Journal of Sports Medicine*, 28, 223-228.

28 AN INJURY PREVENTION STRATEGY FOR RUGBY UNION

H. BROUGHTON, F. PHILLIPS AND K. MCKINNEY
Ponsonby District Rugby Football Club (Inc), Western Springs, Auckland, New Zealand

1 Introduction

Rugby Union injuries continue to be of concern to the player, coach and the administration of this code. Whilst various researchers have identified a number of factors which give rise to these injuries (Bird et al., 1995; Watson, 1998; Wetzel et al., 1998), it is the nature of the injury process within a game situation which requires more attention. In this study a simple method is implemented to collect injury data as the event occurs. From a rugby training point of view a knowledge of game situations in which injuries occur is important. The tackle has been identified as a high-risk event for which there is concern as to the frequency and nature of the injuries. In the scrum situation there has been a reduction in the number of acute injury episodes but chronic injuries are likely to require further inquiries. By far the most common types of injuries are soft tissue in nature affecting ligaments, the capsule, tendons, muscle and subcutaneous tissue.

Such injuries result either from direct forces or excessive stress along different planes of tissue resulting in a strain or tear, with bone such excesses result in a fracture. The high frequency of soft tissue injuries suggests that conditioning may be an important factor and that the nature of tackle injuries not only can be avoided by, mastering appropriate techniques but also requires the player being tackled to 'think more with his feet'.

2 Methods

A questionnaire was designed in which the manager of each team was required to complete following each match and training session throughout the 1998 season.

The information on the questionnaire sought details from the player including the grade played and the ground conditions played on. Player number, initials and position were designated, then details on injury site, injury type, quarter of the event, mechanism of injury and treatment given. At the bottom of the sheet, coded items were referred to when completing the form for each player. At the end of each month all questionnaires we collected for analysis.

3 Results

Altogether 227 injuries were recorded for 133 events which had included 10 training sessions where injuries had occurred.

The more common injury types recorded were *strains, bruising* and *sprains*. These made up 77.5% of the injuries recorded. The more common injury sites were the *upper leg* (28%), *head* and *face* (20%) and *lower leg* (13%). The three main actions that resulted in injuries were '*being tackled*', '*running*' and '*tackling*'. These three actions accounted for 70% of all injuries.

'*Running*' mainly caused injuries to the legs and '*being tackled*' mainly caused injuries to the head and face, the shoulders and the legs. A total of 94% of the 227 injured players wore mouth-guards, as required by the New Zealand Rugby Football Union. The player positions with the greater recorded injuries were (in order) *wing three quarters, props, flankers* and *half-backs*. These four positions accounted for 50% of all the player injuries. About 15% of the injured players had applied strapping even if the site of injury was unrelated to the injured site. Other data collected showed high strapping to the shoulder (acj), wrist, ankle and thumb.

4 Discussion

This was a descriptive study where frequency data had portrayed the injury events. Similar results have been found from other epidemiological studies by Garraway and McLeod (1995) (Scotland), Waller et al. (1994) (RIPP, Otago), notably ~75% of soft tissue injuries, ~48%, tackle injuries ~5%, concussion. However, this study has highlighted '*running*' as a means of being injured, 47 cases (28%) were reported during this period.

The upper part of the leg (including the knee) was affected more so than the lower part of the leg. Strains or tears to this region contributed to these injuries. Initial body position and explosive movements by the props, flankers and inside backs appear to be a factor with respect to hamstring strains. High speed, acceleration–deceleration contact in a tackle situation experienced by wing three-quarters might possibly explain by this position and the defensive player. Other factors such as pelvic functional stability and running technique must be considered as associated factors.

In the tackle situation there is a need for ball carriers always to anticipate that they could be tackled from a defensive player coming from anywhere. In a more recent study by Garraway and McLeod (1995) where 71 matched sets of tackle injuries occurred, 52% were from behind the tackled player. They also found that the tacklers came off worse. Some 85% of tackling players who were injured were wing three-quarters.

The need for the ball carrier either to retract and/or move to the side from the defensive player should be encouraged.

Foot drills utilising the dribble method may assist with the need to be more agile. Wing three-quarters need to be agile but also to have developed a running technique in which speed (acceleration) would require skills of judgement in respect of maintaining space.

The relatively high occurrence of soft tissue injuries (75%) suggests that conditioning may be a factor in prevention of such injuries. More specifically, conditioning protocols should include motor and propioceptive training. However, hand-grip and forearm functional strength, pelvic stability and lower limb co-ordination are the important regions which need attention in respect of conditioning.

Given that well designed 'fitness and conditioning' programmes for the rugby player are followed during the off-season, such a player will be well prepared for formal training during the in-season period. Much of rugby training employs explosive short distance activities. The activities which need more attention are in movement skills, especially with running and agility.

Whilst the data in respect of strapping, incorporating 15% of the injured players, were limited, they nevertheless raise the question of the need for strapping. A planned conditioning programme might well be a better option. Strapping, however, has a role when rehabilitation is a part of management for these affected parts. The wearing of mouth-guards certainly would have minimised dental injuries.

5 Recommendations

- This questionnaire should be employed as a means of documenting injuries as they occur both in training and match events.
- Methods of assessing fitness and conditioning of players should be reviewed in terms of relevance to rugby practice and training.
- More emphasis must be placed on conditioning techniques especially to the hands, pelvic region and feet.
- Running training should be an integral part to rugby practice and training.
- Technical efficiency should be practised in all facets of training in rugby.
- Dribbling skills should be a regular feature to rugby practice and training.

Table 1. Action of players when injured.

Action	Number of injuries	Proportion (%)
Being tackled	68	29.6
Tackling	44	19.1
Running	47	20.4
Maul	20	8.7
Scrum	13	5.6
Ruck	3	1.3
Other	3	14.3

Table 2. Injury site by action.

Injury site	Being tackled	Tackling	Running	Scrum	Maul
Head & face	17	7	2	-	11
Neck	5	5	2	3	3
Shoulders	11	11	1	1	1
Torso	7	2	5	5	1
Arms	4	9	-	1	3
Upper leg	14	5	27	2	1
Lower leg	10	2	12	1	-

6 Acknowledgements

The following are acknowledged for their assistance in the undertaking of this study:
The Accident Rehabilitation and Insurance Corporation, The Minstry of Maori Development (Te Puni Kokiri), The New Zealand Herald, The Australian Rugby Review, The Ponsonby District Rugby Football Club (Inc), Fiona Phillips and Kathryn McKinney.

7 References

Bird, Y.N., Waller, A., Marshall, S.W., Alsop, J.C., Chalmers, D.J. and Gerrard, D.F. (1995) The rugby injury and performance project playing experience and demographic characterisitics, *Journal of Physical Education New Zealand,* 28, 12-16.

Garraway, M. and McLeod, D. (1995) Epidemiology of rugby football injuries, *Lancet,* 345, 1485-1487.

Waller, A.E., Feehan, M., Marshall, S.W. and Chalmers, D.J. (1994) The New Zealand rugby injury and performance project: I. Design and methodology of a prospective follow-up study, *British Journal of Sports Medicine,* 28, 223-228.

Watson, A.W.S. (1998) Sports injuries related to intrinsic factors, *New Zealand Journal of Sports Medicine,* 26, 18-19.

Wetzler, M.J., Akpata, T. et al. (1998) Occurrence of cervical spine injuries during the rugby scrum, *American Journal of Sports Medicine,* 26, 177-180.

29 RISK ASSESSMENT OF HAMSTRING INJURY IN RUGBY UNION PLACE KICKING

P. GRAHAM-SMITH and A. LEES
Research Institute for Sport and Exercise Sciences, Liverpool John Moores University, Liverpool, UK

1 Introduction

Many authors have assessed pre-season muscle strength and flexibility and investigated their relation to injury during the competitive season. While the results of such studies are often conflicting, it is generally accepted that low Hamstrings:Quadriceps (H:Q) ratios over a range of movement speeds and poor flexibility are high risk factors in hamstring injury (Jonhagen et al., 1994). In Australian Rules Football for example, Orchard et al. (1997) found that players who had a low isokinetic H:Q ratio of around 0.55 at $60°·s^{-1}$ (1.05 rad·s^{-1}) were more likely to suffer an injury than those who had a ratio of around 0.66. Generally, a ratio of around 0.60 is thought to be sufficient to reduce the risk of injury (Orchard et al., 1997). Rugby Union players have also been reported to have a high incidence of hamstring strains. Upton et al. (1996) observed that 80% of match and training time lost due to injury was a direct result of hamstring strain and reported a rate of 57 injuries per 1000 h of play. The causes of high levels of hamstring strains in Australian Rules Football were suggested to be due to repetitive punt kicking and repeated sprinting efforts during a game. Considering the similar nature of the game, it is likely that such actions are also major causes of hamstring strain in Rugby Union.

While screening studies are important throughout the season, it becomes apparent that there is also a need to assess the risk of injury during sport-specific actions. In the place kicking action, with its characteristic high straight leg follow through, it would be useful to know how flexible a player needs to be to perform the action safely. It would also be useful to determine whether the hamstrings are lengthening or shortening throughout the action and to evaluate the importance of eccentric strength of the hamstrings (Aagard et al., 1998). The relevance of the traditional concentric hamstrings:concentric quadriceps ratio for a specific action can also be assessed.

The aims of this study were to devise a method for assessing the risk of hamstring injury to players during sport-specific actions, and to apply this method to the place kicking action in Rugby Union.

2 Methods

2.1 Subjects

Eight place kickers of University standard took part in the study. Subject characteristics are presented in Table 1. Prior to any strenuous exercise each player completed a 15-min warm-up which consisted of jogging on a treadmill and stretching exercises usually performed prior to a game. Following this, static and dynamic flexibility and isokinetic muscle strength were assessed.

Table 1. Mean (± SD) age, height and mass of the eight subjects.

	Age (years)	Height (m)	Mass (kg)
Mean	22	1.81	82.6
SD	2.6	0.9	10.0

2.2 Assessment of static and dynamic flexibility

Static and dynamic flexibility were assessed using a maximal assisted hamstring stretch and during the performance of a maximal place kick, respectively (Figure 1). While it is possible to assess static flexibility using a goniometer or the 'sit-and-reach' test, such techniques cannot be adopted when a player is performing a movement. In this respect a video-based technique was used where static and dynamic actions were recorded in the sagittal plane at 50 Hz. Each player performed three maximal assisted stretches and three maximal place kicks.

A fluid-filled goniometer (Biokinetic Inc., England) was also used in the static stretches to identify which trial produced the greatest stretch, and the best place kick was identified through maximal ball velocity as measured by a radar gun (JUGS Inc., Tualitan, Oregon). The best trials of static and dynamic conditions were then digitised for each player to determine hip and knee flexion-extension angles. The conventions were such that a fully extended knee joint represented zero degrees knee flexion and zero degrees hip flexion occurred when the trunk and thigh were in alignment (Figure 1). Data were smoothed using a Butterworth 4th order zero lack filter with padded end points and a cut-off frequency of 7 Hz.

Having determined the hip and knee flexion-extension angles, the length of the hamstrings (biceps femoris muscle) was estimated using the equations presented by Visser et al. (1990). As the biceps femoris muscle spans both the hip and knee joints, the change in length is firstly determined for changes in both joints independently. The total length change is then calculated as the sum of the independent changes. Such a method provides the link between static and dynamic conditions. The rationale is that if during the dynamic condition, the length of the biceps femoris exceeds that of the static condition, then there is an increased risk of straining the hamstrings.

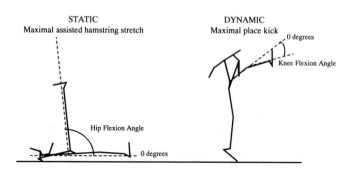

Figure 1. Assessment of static and dynamic flexibility and conventions used to determine hip and knee flexion-extension angles.

2.3 Assessment of isokinetic muscle strength

Hamstrings and quadriceps muscle strength was assessed using an isokinetic dynamometer (LIDO Active, Loredan, Davis, CA). Four gravity-corrected tests were conducted on the kicking leg to determine peak concentric torque values at slow (1.05 rad·s^{-1}), intermediate (2.09 rad·s^{-1}) and fast (5.24 rad·s^{-1}) angular velocities and peak eccentric torque values at the intermediate speed. The subjects were in a seated position. The H:Q ratios were then calculated using two methods;

i) the traditional ratio $\dfrac{\textit{Concentric Hamstrings}}{\textit{Concentric Quadriceps}}$ at 1.05, 2.09 and 5.24 rad·s^{-1}, and

ii) the 'Dynamic Control Ratio' (DCR) $\dfrac{\textit{Eccentric Hamstrings}}{\textit{Concentric Quadriceps}}$ at 2.09 rad·s^{-1}.

The latter is more 'functional' and is becoming more widely used in the literature (Aagard et al., 1998). Ideally the hamstrings should be able to resist as much force as the quadriceps can produce, which equates to a DCR of 1.0. However, due to characteristics of the force-velocity curve it is unlikely that this value will be attainable for all movement speeds, but it is important that a value of 1.0 is attained at movement speeds typical of place kicking. Estimates of hip flexion and knee extension velocities from Figure 2 are both greater than 600°·s^{-1} (10.47 rad·s^{-1}). Currently such movement speeds are not measurable on isokinetic equipment and so a convenient angular velocity of 2.09 rad·s^{-1} was chosen for the test. The aim was to establish a reasonable value of the DCR at this angular velocity.

3 Results

Figure 2 shows the angles of hip and knee flexion-extension during the place kick for one player. The typical movement pattern is characterised by hip hyperextension (negative) as the leg is taken backwards in the withdrawal phase. This is followed by hip flexion as the leg is brought through towards the ball and further flexion occurs after impact in the follow through phase. The knee joint can be seen to flex through around 100° in the early part of the kick. The knee begins to extend as impact approaches and this continues into the follow through phase where the knee reaches full extension. In the middle of the follow-through phase, the knee begins to flex through approximately 20°.

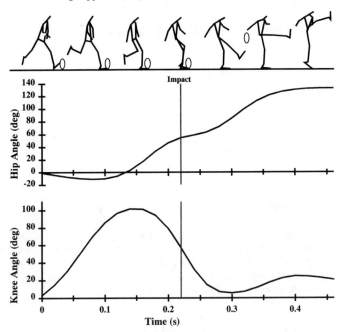

Figure 2. Characteristic hip and knee flexion-extension graphs during the place kicking action (Subject 4).

The combined effect of hip and knee flexion-extension is represented by the change in length of the biceps femoris muscle (Visser et al., 1990) in Figure 3. The change in length of the biceps femoris muscle is expressed as a percentage of segment length. The graph shows that the biceps femoris shortens in length in the withdrawal phase due to hip hyperextension and knee flexion and then lengthens as the foot approaches the ball due to hip flexion and knee extension. Muscle length continues to increase in the follow-through phase even though the knee shows some degree of flexion. Flexion of the hip joint therefore has the greatest influence on the muscle length.

By plotting a line to represent the muscle length in the static condition, it can be seen that the muscle length exceeds that of the maximal assisted stretch in the mid-part of the follow-through phase. Interestingly, the point of intersection occurs around the time that the knee begins to flex. Knee flexion is undoubtedly a mechanism to relieve some of the strain in the muscle, but as can be seen, this has a minimal effect.

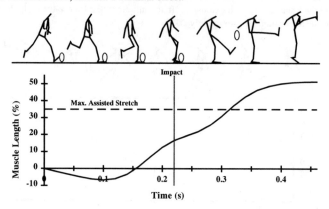

Figure 3. Characteristic graph of the change in length of the biceps femoris muscle during the place kick, including maximal assisted stretch value (Subject 4).

Figure 4 shows the change in muscle length in static and dynamic conditions for all subjects. It can seen that all players exhibited greater increases in length during the place kick (mean 45.2% ± 5.3) than in the static stretch (mean 34.3% ± 6.1). The mean difference was 10.9% of segment length, which for a typical player with a thigh segment length of approximately 40 cm, relates to an additional stretch of 4.5 cm.

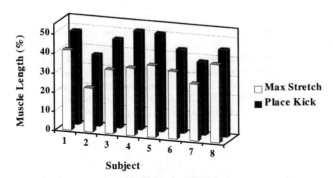

Figure 4. Change in length of the biceps femoris muscle during a maximal assisted stretch and during a place kick for all subjects

Figure 5 shows the mean muscle strength profile from the isokinetic tests. The mean peak torque values can be seen to follow the typical force (torque)-velocity relationship where muscle strength decreases with increasing speed of movement. The quadriceps muscle group showed a 46% decrease from slow speed (246 Nm ± 38) to fast speed (133 Nm ± 21), compared to only 23% decrease for the hamstrings (149 Nm ± 26 down to 114 Nm ± 17). Eccentric quadriceps strength (241 Nm ± 79) was found to be 1.20 times greater than concentric strength (201 Nm ± 27) at an angular velocity of 2.09 rad·s^{-1}. For the hamstrings, eccentric strength was found to be 1.09 times greater than concentric strength (152 Nm ± 31 and 139 Nm ± 20, respectively).

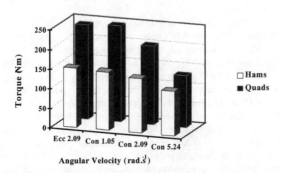

Figure 5. Mean peak torque values for the quadriceps and hamstrings in eccentric and concentric modes of contraction and at various angular velocities.

The traditional H:Q ratio was found to increase from $0.61 \pm .07$ to 0.69 ± 0.07 to 0.86 ± 0.09 as the speed of movement increased from 1.05 to 2.09 to 5.24 rad·s^{-1}. The DCR at 2.09 rad·s^{-1} revealed a mean value of 0.75 ± 0.08, which is somewhat lower than the suggested ideal ratio of 1.0. However, the DCR gets progressively closer to 1.0 as movement speed increases – concentric strength decreases with speed of shortening while eccentric strength remains fairly constant with increasing speeds of lengthening. This observation is supported by Aagard et al. (1998) who found the DCR to increase from 0.61 to 1.01 as movement speed increased from 0.52 to 4.19 rad·s^{-1}. A DCR value of 0.75 would therefore appear to be sufficient for place kickers at a movement speed of 2.09 rad·s^{-1}.

4 Discussion

The results showed that a place kick imposed an additional stretch of 10.9% in the biceps femoris muscle than in a maximal assisted stretch. Based on the rationale of this study, the place kicker would be at a greater risk of injury. It is acknowledged that this does not necessarily mean that injury will occur, but through repeated kicking and fatigue towards the end of the game the player would be at a greater risk. To provide an answer as to the level of flexibility a player requires to be able to perform place kicking safely the calculation procedure can be reversed. A mean stretch of 45% of segment length in the place kick equates to a hip range of motion of 118°, providing the knee remains extended. This is approximately 25° further than that found in the maximal assisted stretch, 93°. Such a deficit indicates that advanced stretching techniques (eg., Proprioceptive Neuromuscular Facilitation) may be beneficial to the place kicker prior to a game or as part of a structured flexibility training programme. However, this might still be unachievable and so would need to be supplemented by a change in technique, i.e., greater knee flexion and pelvis backward rotation (not measured) in the follow-through phase.

The observation that the biceps femoris increased in length throughout the follow-through phase indicates that the DCR is probably a better indicator of injury risk to place kickers than the traditional H:Q ratio. It is acknowledged that the quadriceps are unlikely to be actively working against the hamstrings at the end of the follow-through due to the ballistic nature of the kick. However, the eccentric force required by the hamstrings will be related to the amount of force produced by the quadriceps prior to ball contact, and as such the DCR is a relevant measure. Although a DCR of 0.75 was less than the desired value, it is likely that a ratio of around 1.0 would be attained when eccentric hamstring strength is assessed at a faster speed of movement, typically greater than 4.19 rad·s^{-1} (Aagard et al., 1998). However, the results of this study indicate that a DCR of 0.75 at 2.09 rad·s^{-1} is sufficient for place kickers and values less than this would indicate an increased injury risk.

The use of muscle length observations has been shown to provide valuable information on the risk of hamstring injury. Such a technique provides a link between static stretch assessment and dynamic performance, giving the player, coach or physiotherapist an indication as to how flexible a player needs to be to perform a skill safely. It also has the ability to identify whether the muscle is lengthening or shortening and this can be assumed to relate to eccentric or concentric muscle activity. By incorporating an isokinetic evaluation in concentric and eccentric modes of contraction, this method can be applied to other sports-specific actions to identify the potential risk of hamstring injury.

5 Acknowledgements

The authors would like to acknowledge Neil Callis, Iain Fowler, Ed Hannon, Clare Todd and Christine Yeates for their contribution to this study.

6 References

Aagard, P., Simonsen, E.B., Magnusson, S.P., Larsson, B. and Dyhre-Poulsen, P. (1998) A new concept for isokinetic Hamstring:Quadriceps muscle strength ratio, *American Journal of Sports Medicine*, 26, 231-237.

Jonhagen, S., Nemeth, G. and Eriksson, E. (1994) Hamstring injuries in sprinters. The role of concentric and eccentric hamstring muscle strength and flexibility, *American Journal of Sports Medicine*, 22, 262-266.

Orchard, J., Marsden, J., Lord, S. and Garlick, D. (1997) Preseason hamstring muscle weakness associated with hamstring muscle injury in Australian footballers, *American Journal of Sports Medicine*, 25, 81-85.

Upton, P.A.H., Noakes, T.D. and Juritz, J.M. (1996) Thermal pants may reduce the risk of recurrent hamstring injuries in rugby players, *British Journal of Sports Medicine*, 30, 57-60.

Visser, J.J., Hoogkamer, J.E., Bobbert, M.F. and Huijing, P.A. (1990) Length and moment arm of human leg muscles as a function of knee and hip-joint angles, *European Journal of Applied Physiology*, 61, 453-460.

30 VALIDATION OF AN INSTRUMENT FOR DATA COLLECTION IN RUGBY UNION

A. MCMANUS
Curtin University, Perth, Western Australia

1 Introduction

Australia has over one million sporting injuries annually costing almost one billion dollars (Egger, 1990). While injury rates are relatively high in the rugby codes in comparison to other sports (Nicholl, Coleman and Williams, 1995), a meta-analysis on rugby injuries conducted from 1974-1994 concluded no study met all the desirable criteria necessary to ascertain the extent and nature of injury (Lower, 1995).

Significant injury rates in Rugby Union are of national and international concern. Egger (1990) found at least 50% of all Australian Rugby Union players could expect at least one injury per season, with the major cause of injury being collision. O'Brien (1992) reported a high incidence of injury in Irish Rugby Union citing speed and contact as major risk factors, a finding also supported by Roux et al. (1987) in a study of 3,350 rugby matches in South Africa. While Nicholl, Coleman and Williams (1995) reported the risk of injury playing soccer was high, they found the risk of substantive injury in British Rugby to be three times higher. New Zealand established the Rugby Injury and Performance Project in an endeavour to reduce the high incidence, severity and consequences of injury in that country (Chalmers, 1994).

2 Literature Review

To reduce the high incidence of injury in Rugby Union scientifically based research must be undertaken. MacLeod et al. (1993) stated 'Prevention of sport injuries must be based on accurate, reproducible data which can be readily interpreted,' There is limited research available that provides accurate, reproducible data which can be readily interpreted in Rugby Union and Rugby League (Chalmers, 1994; Gibbs, 1993).

Development of a standardised data collection instrument to assess incidence rates, is a critical measure of methodologically sound, injury research (Van Mechelen, Hlobil and Kemper, 1992).

An extensive literature review failed to locate a valid instrument for injury data collection in Rugby Union. Several researchers in this area did, however, suggest necessary criteria for future research. Garraway, MacLeod and Sharp (1991) suggested injury data collection forms should be user friendly to non-medical personnel and include an acceptable definition of injuries. Garraway in MacLeod (1993) noted incidence rates based on exposure of all players was essential and Jones, Cowan and Knapik (1994) suggested injury data from both games and training should be recorded, as all injury impacts upon performance.

There is a paucity of data regarding valid and reliable instruments for injury collection in Rugby Union. Of those studies that did report this information about its instruments, little was written regarding the processes used. The majority of articles reviewed concur with the findings of Finch, Ozanne-Smith and Williams (1995) and Lower (1995) who examined data collection methods of sports-related injuries. These findings indicated there was a lack of valid and reliable measures in sports injury studies and that there was a need for injury data collection forms to be simple, standardised and concise. These instruments should be sports specific, however, they should incorporate a standardised injury coding system to allow for comparison of findings between and within sports.

3 Methods

The aim of this research was to provide the basis from which Rugby Union injury data can be collected using a validated injury report form. This aim was achieved using a seven-stage process determined from the literature reviewed. The first stage involved the design of the Rugby Union Injury Report Form – For Games and Training (the Form). The Form was then assessed for face, content and criterion validity in stages two to four. A 22-member panel plus four sporting bodies assessed the Form for face validity and an expert panel assessed the Form for content and criterion validity. Panel members were consulted until consensus was reached.

The gold standard, developed in stage five, was used to assess the reliability of the Form in stage six of the process. A further 40 triallists completed the Form 'in situ' (stage seven) during four games to further assess inter-rater reliability.

4 Results

4.1 The Rugby Union Injury Report Form – For Games and Training
The design and content of the draft Rugby Union Injury Report Form was devised from recommendations for future research elicited from the literature reviewed and a review of existing data collection instruments.

The following information is collected using the Form: environmental conditions that impact upon injury; mechanism of injury; phase of play or aspect of training; if play was legal or illegal; position played specifically and in general; relationship of ball and injured player; severity of injury; time of game injury occurred; and where the injury occurred (game or training).

The design of the front of the Form (Table 1) comprises closed-ended questions and instructions indicate the need to merely circle an option for each question.

Table 1. The Rugby Union Injury Report Form - For Games and Training

ID# ☐☐☐☐☐1-5

	(Office use only)	
Name _____	**1.** Grade _____	☐☐6-7
2. Date ☐☐dd☐☐mm☐☐yy 8-13 **3.** Age (in Years)____		☐☐14-15

	(Office use only)
4. Site of Injury (circle) Head, Face, Neck, Shoulder, UArm, LArm, Wrist, Hand, Fingers, Thumb, Chest, Abdomen, Spine, Back, Pelvis, ULeg, LLeg, Knee, Ankle, Foot, Toes, Other _____	☐☐16-17
5. Severity of injury (circle) minor / mild / moderate / severe	☐☐18-19
6. Mechanism of injury (circle) extrinsic / intrinsic	☐☐20-21
7. Where (circle) game / training	☐☐22-23
8. Phase of play or aspect of training (circle) Scrum, Lineout, Ruck, Maul, Tackle, Kicking, Pileup, Collision, Other _____	☐☐24-25
9. If terrain a factor of injury (circle) hard / soft / muddy / other _____	☐☐26-27
10. If weather a factor of injury (circle) hot / cold / wet / other _____	☐☐28-29

If injured in game continue. At training, go to Question 16. on the reverse.

	(Office use only)
11. Time of game (circle) 1st half / 2nd half / time on	☐☐30-31
12. Relationship of ball and injured player (circle) near ball / behind play	☐☐32-33
13. Play (circle) legal / illegal	☐☐34-35
14. Position played (circle) LHP H THP LL RL LF RF 8 HB 5/8 LW IC OC RW FB	☐☐36-37
15. Back or forward (circle) B / F	☐☐38-39

NOTE: SEVERITY OF INJURY - required treatment but:
MINOR - if able to return to game/training in which injury occurred   **MILD -** if missed one week
MODERATE - if missed two weeks   **SEVERE -** if missed more than two weeks

POSITION PLAYED - 1. **LHP** - Loosehead prop, 2. **H** - Hooker, 3. **THP** - Tighthead prop, 4. **LL** - Left lock, 5. **RL** - Right lock, 6. **LF** - Left flanker, 7. **RF** - Right flanker, 8. No 8, 9. **HB** - Half back, 10. **5/8** - Five eight, 11. **LW** - Left wing, 12. **IC** - Inside centre, 13. **OC** - Outside centre, 14. **RW** - Right wing, 15. **FB** - Full back.

The reverse of the Form (Table 2) allows space for a written record of assessment, treatment and management of injury. A separate Form is completed for each injury.

Table 2. The Rugby Union Injury Report Form – For Games and Training (**reverse**)

16. Classification of injury ☐☐☐40-42

Assessment

Treatment

Instruction to player/carer

Referred to

Other information

Medic/Sports Trainer's Name _____ Signature _____

The reverse side of the Form incorporates the Orchard Sports Injury Classification System (1997) (OSICS) to streamline data input. The OSICS is currently used by the Australian Institute of Sport and the Australian Rules Football Commission, and is endorsed by the Australian Sports Medicine Federation (Orchard, 1995).

A key to coding the Form was also developed to assist data input and analysis using a statistical computer program. The key was subjected to review by gold standard panel members and representatives from various sporting bodies.

4.2 Face, content and criterion validity

A 22-member panel plus four sporting bodies assessed the Form for face validity. During this process the only alteration related to the division of questions into three separate sections. A seven member expert panel assessed the Form for content validity using the Delphi Technique (Reid in Williams and Webb, 1994). As no gold standard was available, the expert panel also assessed the Form for criterion validity (Goodman, 1987). A four round Delphi Technique was conducted before consensus on content and criterion validity was reached.

4.3 Development of a gold standard

Streiner and Norman (1989) described a gold standard as a measure or scale used as a yardstick or standard against which new measures are correlated. Development of a gold standard to validate an injury report form provides an additional means of increasing study rigour. The literature review failed to locate a gold standard for data collection in Rugby Union.

Seven methodological steps used to devise and validate Clinical Decision Rules (gold standards) in clinical medicine lend themselves to the development of a gold standard for injury data collection. The seven steps used in this study were: review of literature; development of a desirable criteria for inclusion; development of a standardised data collection form including standardised descriptions; development of an information/instruction sheet; validation of the gold standard by an expert panel; trial of the gold standard by trained raters on a predetermined set of injuries; and trial of the gold standard 'in situ' (Seaberg and Jackson, 1994; Bauer et al., 1995; Stiell et al., 1995; and Stiell et al., 1996). A gold standard panel of experts in the field of study developed a gold standard for the Form using the Delphi Technique on a predetermined set of five videotaped injuries. Panel members were consulted until consensus was reached.

4.4 Intra-rater and inter-rater reliability

Ten independent raters were asked to complete separate Forms for the five videotaped injuries three times over a five-week period. Completed Forms were collected after each viewing. Forms from the first viewing were used to assess inter-rater reliability with results indicating a 98% agreement. Intra-rater reliability agreement of raters assessed against the gold standard devised by the gold standard panel indicated a 98% agreement.

4.5 Inter-rater reliability of the Form 'in situ'

Forty raters were randomly selected from spectators present at four games (10 per game) to trail the Form 'in situ'. All raters were screened prior to selection to ensure all had a fundamental knowledge of sport injury by using only people who understood the terms intrinsic and extrinsic in relation to mechanism of injury. Hennekens and Buring (1987) stated the maintenance of blindness was the most important way to minimise bias. Raters were asked to remain in one position during the game under review thus blinding them to other raters results. An inter-rater reliability agreement of 98% was achieved for the nine injuries sustained during these four games.

5 Conclusions

This study has validated a measurement instrument for injury collection in Rugby Union, thus providing injury researchers with a basis for future studies in this area, as well as a procedure to develop future instruments.

The following are recommendations for use of the Rugby Union Injury Report Form – For Games and Training:

Widespread adoption of the Rugby Union Injury Report Form: The Form could be used for injury data collection at all levels of Rugby Union, from elite to social. The Form could also be used at all games and training. This will ensure all injuries sustained by players are accurately reported, as no matter how minor, all injuries affect performance.

Compatibility of results: As the Form includes the Orchard Sports Injury Classification System (OSICS), it will enable comparison of results between similar studies. Furthermore, the Form can be used as a basis for comparison of other sports injuries that use the OSICS.

Assessing injury trends: The Form could be used to collect data in longitudinal studies to identify injury trends over time. The Form facilitates this through the incorporation of a computer coding system for data analysis, and collection of injury data through all levels of Rugby Union.

Calculation of incidence rates: The Form could be used in conjunction with a player log or diary of game and training hours, to calculate incidence rates based on exposure time.

6 References

Bauer, S.J., Hollander, J.E., Fuchs, S.H. and Thode, H.C. (1995) A clinical decision rule in the evaluation of acute knee injuries, *Journal of Emergency Medicine.* 13, 611-615.

Chalmers, D.J. (1994) New Zealand's injury prevention research unit: reducing sport and recreational injury, *British Journal of Sports Medicine.* 28, 221-222.

Egger, G. (1990) Sports injuries in Australia: causes, costs and prevention, *A Report to the National Better Health Program October 1990.* Centre for Health Promotion and Research, Sydney.

Finch, C., Ozanne-Smith, J. and Williams, F. (1995) The feasibility of improved data collection methodologies for sports injuries, *Murdoch University Accident and Research Centre.* Report No 69, January 1995.

Garraway, W.M., MacLeod, D.A.D. and Sharp, J.C.M. (1991) Rugby injuries: the need for case registers, *British Medical Journal,* 303, 1082-1083.

Gibbs, N. (1993) Injuries in professional rugby league: a three-year prospective study of the South Sydney professional rugby league football club, *American Journal of Sports Medicine,* 21, 696-700.

Goodman, C.M. (1987) The delphi technique: a critique, *Journal of Advanced Nursing,* 12, 729-734.

Hennekens, C.H. and Buring, J.E. (1987) *Epidemiology in Medicine.* Little, Brown and Company, Boston.

Jones, B.H., Cowan, D.N. and Knapik, J.J. (1994) Exercise, training and injuries, *Sports Medicine,* 18, 202-214.

Lower, T. (1995) Injury data collection in the rugby codes, *The Australian Journal of Science and Medicine in Sport,* 27, 38-42.

MacLeod, D.A.D., Moughan, R.J., Williams, C., Madeley, C.R., Sharp, J.C. and Nutton, R.W. (1993). *Intermittent High Intensity Exercise Preparation, Stresses and Damage Limitation.* E. and F.N. Spon, London.

Nicholl, J.P., Coleman, P. and Williams, B.T. (1995) The epidemiology of sports and exercise related injury in the United Kingdom, *British Journal of Sports Medicine,* 29, 232-238.

O'Brien, C. (1992) Retrospective survey of rugby injuries in the Leinster province of Ireland 1987-1989, *British Journal of Sports Medicine.* 26, 243-244.

Orchard Sports Injury Classification System. (1997) Internet address http://www.med.unsw.edu.au/physio...ernal/sportmed/Orchard/index1.htm

Orchard, J. (1995) Orchard Sports Injury Classification System (OSICS), *Sport Health,* 11, 39-41.

Roux, C.E., Goedeke, R., Visser, G.R., Van Zyl, W.A. and Noakes, T.D. (1987) The epidemiology of schoolboy rugby injuries, *South African Medical Journal,* 71, 307-313.

Seaberg, D.C. and Jackson, R. (1994) Clinical decision rule for knee radiographs, *American Journal of Emergency Medicine,* 12, 541-543.

Stiell, I., Wells, G., Laupacis, A., Brison, R., Verbeek, R., Vandemheen, K. and Naylor, C.D. (1995) Multicentre trial of introduce the Ottawa ankle rules for use of radiography in acute ankle injuries, *British Medical Journal,* 311, 594-597.

Stiell, I.G., Greenberg, G.H., Wells, G.A., McDowell, I., Cwinn, A.A., Smoth, N.A., Cacciotti, T.F. and Sivilotti, L.A. (1996) Prospective validation of a decision rule for the use of radiography in acute knee injuries, *Journal of the American Medical Association,* 275, 611-615.

Streiner, D.L. and Norman, G.R. (1989) *Health Measurement Scales: A Practical Guide to their Development and Use.* Oxford University Press, Oxford.

Van Mechelen, W., Hlobil, H. and Kemper, H.C.G. (1992) Incidence, severity, aetiology and prevention of sports injuries, *Sports Medicine,* 14, 82-99.

Williams, P.L. and Webb, C. (1994) The delphi technique: a methodological discussion, *Journal of Advanced Nursing,* Feb, 18-23.

31 HAMSTRING REHABILITATION FOR FOOTBALL PLAYERS

A. MURPHY*, M. ANDREWS** and W. SPINKS*
*   Human Movement Department, University of Technology, Sydney, Australia
**   Sydney Academy of Sport, Sydney, Australia

1 Introduction

Individuals playing sports that combine a significant volume of running and kicking, for example Australian Rules Football, appear to have a high risk of injury to the hamstring musculature. For example, the 1997 Australian Football League (AFL) Injury Report stated that 'Hamstring injury is the most common significant injury and responsible for more missed playing time than any other injury' (Orchard, 1998). In such sports, hamstring injuries have the potential to affect a team's performance both acutely and also over a whole season. This is not only due to the fact that the injury occurs, but also due to possibility that the athlete will be re-injured either during rehabilitation or on returning to competition. While research efforts have predominantly been directed at determining physiological factors that predispose an individual to hamstring impairment, the results of this work remain equivocal. In practice, injury prevention strategies typically focus on the strength and flexibility of the hamstrings. Indeed Orchard et al. (1997) reported that AFL players who had relatively low hamstring strength had a higher risk of injury. However, Gleim and McHugh (1997), after reviewing the literature relating to flexibility and its effects on sports performance, concluded that 'There is no scientifically based prescription for flexibility training and no conclusive statements can be made about the relationship of flexibility to athletic injury.' Similarly, Knapik et al. (1992) noted that 'assumptions regarding strength, flexibility and injuries have little or no data to support them'.

If it is accepted that to some extent hamstring injuries are inevitable in sports which involve repeated maximal kicking and sprinting actions, it is surprising to find that very little work has been specifically directed at developing a detailed and valid rehabilitation programme for muscle injuries. Such a programme would encapsulate medical, sports science and various other management concerns related to the rehabilitation process and provide objective information on potential re-injury risk and suitability for return to the game. It would also have the flexibility to cope with an injury that takes longer than anticipated to heal and alternatively, an injury which heals relatively quickly. Finally, as support staff numbers in professional sporting teams increase, a prevailing problem in many professional sports teams is poor communication between the coaches, players, medical and conditioning personnel during the rehabilitation process. For these reasons, and given the significance of hamstring injuries in AFL, it is apparent that the process of injury rehabilitation is gaining new status as a major area within the sport and as such warrants the development of objective and valid rehabilitation programmes.

Therefore, the purpose of the current research was to construct a hamstring rehabilitation model specifically for use with an AFL team. The model was researched and developed with the following specific outcomes in mind:

- Minimise the number of games lost after a hamstring injury has been incurred by providing a framework, which permits, but safely regulates, an aggressive rehabilitation programme.
- Minimise the risk of hamstring re-injury both during and after the rehabilitation process through injury management and a progressive training programme.
- Provide a co-ordinated framework through which doctors, physiotherapists, sport scientists, coaches and players can communicate regarding injury and rehabilitation.

2 Development of the Model

The model that was developed is an attempt to provide sports medicine and rehabilitation practitioners with a valid framework with which to evaluate and progress a hamstring rehabilitation programme. It is a holistic model, which seeks to incorporate all aspects of rehabilitation including the initial and continuing treatment of the injury, facilitation of the healing process, skill maintenance, aspects of conditioning and musculo-skeletal assessment.

The model includes four phases of progression;(1) Acute treatment, (2) Remodelling, (3) Reconditioning and, (4) Integration (see Table 1). Movement from one phase to the next is based on a number of set criteria, which must be completed satisfactorily. Within each phase a player is allocated activities from various progressive programmes covering all aspects of the rehabilitation process including injury management, running, kicking, strengthening/stretching and fitness.

Table 1. A model for the rehabilitation of hamstring injury.

Phase	Acute	Remodelling	Reconditioning	Integration
Running	None	Stages 1-4	Stages 5-6	Stage 7 + Training
Strength/Stretching	None	Stages 1-2		Stage 3
Kicking	None	Stages 1-2	Stages 3-4	Stages 5-6

Of these activities, the running, kicking and stretching/strengthening programmes have been divided into stages of progressing intensity. For example, the running programme is divided

into seven stages. Beginning at Stage 1, which involves low intensity running for limited time periods, an individual in rehabilitation would, over the course of several weeks, progress through to
Stage 7. This stage involves running in straight lines and changing direction while moving at maximal velocity. The strengthening programme begins with isometric exercises in Stage 1 and progresses to include both concentric and eccentric muscle actions at increasing velocity by the final stage (Stage 3). The kicking programme initially (Stages 1-3) focuses on execution of the standing drop punt over increasing distances (10-30 m). Following these stages, individuals begin drop punt kicking while running and by the end of the programme are kicking maximally and three-dimensionally in a variety of game-like scenarios. As mentioned previously, Table 1 shows how each of the programmes progresses through each phase of the model. The following is an outline of each phase in more detail.

3. Phases of the Model

3.1 Acute treatment
The primary concern in the first phase is to ensure the correct treatment of the injury and therefore physical or skill training is not allowed. Standard treatment involves application of the 'Rest/Ice/Compression/Elevation' principles to limit swelling and bleeding of the injured tissue. Rest is required for a minimum period of 48 h from the time of injury. Ice is applied to the injured site on and off every 20 min for the first 6 h, disrupting sleep if necessary. Following this, the injury is iced every 20 min for every waking hour (i.e., approximately 12-15 times per day). Finally a compression bandage is used at all times when not icing and elevation of the injured limb is encouraged whenever possible during the first 48 h. After the 48-h rest period, crutches are utilised if necessary for locomotion, in order to encourage normal gait patterns and limit the development of secondary injuries. Light stretching of the injured muscle also begins under medical supervision only. Research has shown that when mobilisation occurs after a period of short immobilisation, a better penetration of muscle fibres through connective tissue is found and there is better alignment with the non-injured muscle fibres (Järvinen and Lehto, 1993).

To assist with injury diagnosis a magnetic resonance image (MRI) is taken after initial swelling has been controlled and while there is still some pain on movement. In addition, anti-inflammatory medication is prescribed as required. The criterion for passing from the acute injury phase to the remodelling phase is that 48 h of rest be undertaken and that the individual can walk pain free.

3.2 Remodelling
The progressions in the remodelling phase are guided by the fundamental understanding that the structural and functional integrity of the injured muscle is not compromised. As such, a pain-free philosophy overrides all activity in this phase of the programme. This is important because any member of the conditioning/medical team can take a session and determine what

level of effort is appropriate. Ongoing injury treatment revolves around icing the site regularly, particularly after exercise, and massage of the injured tissue, which continues throughout the phase.

The running, strengthening and kicking programmes begin in this phase (Table1). Each programme has a number of stages, which increase in both volume and intensity as the individual progresses through the programme. In the running programme, the emphasis is on maintaining the appropriate stride pattern while still progressively overloading the injured muscle. In order to give the injured tissue adequate recovery time, running is not undertaken on successive days. On these 'off' days, cardiovascular conditioning not involving the injured limb and the strengthening and kicking exercises are performed. By the end of this phase, individuals would normally be running 100 m at approximately 80% maximum. Strengthening and stretching begin with low level isometric contractions increasing to maximum as tolerated. Towards the end of the remodelling phase, low level concentric and eccentric exercises are introduced and are progressed in both force and velocity.

The primary criterion for transition from the remodelling to the reconditioning phase is that the individual successfully completes Stage 4 of the running programme. Because running is such a fundamental activity, the running programme now becomes the primary measure by which an individual's progress through the hamstring rehabilitation model is evaluated. In addition, the athlete should be able to contract the injured hamstring isometrically at a level that approximates 95-100% pain-free maximum. Once both of these objectives have been achieved, the player shifts to the reconditioning phase.

3.3 Reconditioning

In this phase, the focus changes from one of injury management to reconditioning the injured site such that it is able to withstand the forces and speeds to which it is exposed in a training and/or game situation. This philosophy underpins the shift to a programme with increased intensity and volume. Consequently, in all of the supplementary programmes there is a shift from linear, two-dimensional movements, to three-dimensional actions, which exert forces on the muscle cells at many angles and speeds. Together with the increase in training intensity there is a shift in the treatment of the injured area with the focus being on deep massage of the injured site to break up remaining scar tissue. The increase in intensity of eccentric stretching (Stage 3 in the Strength/Stretch programme) also helps achieve this goal.

The goal at the end of this phase is to have the player ready to participate safely in ball work training. In the running, kicking and strength programmes that align with this phase (Table 1), the injured muscle has been exposed to various stresses above that which will be encountered in training and games. For example, to complete Stage 6 of the running programme successfully, the player must run 12 x 100 m at top speed as well as maximally performing 800 m of agility and shuttle type running. This volume and intensity of sprinting would never occur in a game and thus the assumption is that if the individual completes the stage, then he is ready to participate safely in team training.

In order pass from the reconditioning phase to the integration phase an individual must as a minimum, complete Stage 6 of the running programme pain free as outlined above. Other assessable criteria include the absence of palpable soreness over the injured site and absence of pain during an eccentric stretch of moderate force. On successful fulfilment of these criteria the player shifts to the integration phase.

3.4 Integration

Entry to this phase is a clear indication that the player will soon be ready to be integrated back into team training with a view to playing again in the near future. Having previously completed the phases of the programme as outlined above, combined with what needs to be done in this phase, entry into the team training environment is considered low risk. The last stage of the running programme (Stage 7) shifts further in emphasis from straight line running to running involving change of direction with maximal acceleration and deceleration. After this has been completed, the player then resumes normal team training. Two full team training sessions must be completed before the player is eligible to play.

At this point in the model, and in addition to the successful completion of two team training sessions, other criteria that are considered prior to a return to play include a follow-up MRI and an isokinetic evaluation of the injured limb. Preliminary experience using the MRI data has shown oedema still present several weeks after the initial hamstring injury in spite of the player having full functional capacity. As such, the value of MRI scanning as an indicator of repair is limited. However, what is clear is that at this stage of the model, the follow up MRI should show a clear improvement (i.e., reduced oedema) when compared to the MRI taken in the acute injury phase. The isokinetic force data are evaluated with respect to strength comparisons between opposite limbs and agonist/antagonist ratios and compared to pre-injury measurement. If either of these data fall outside predefined ranges then a decision will be taken to delay return to play as the risk of re-injury is deemed too high.

4 Discussion and Conclusions

As the hamstring model is relatively new, no objective data are yet available to analyse its efficacy. Objective analysis of the model's success will include such factors as player compliance, re-injury rates (both during the programme and on return to the game) and assessment of first game performance. However, three players with Grade 2 hamstring tears have completed the model as outlined in this paper. The players took three to five weeks to complete the programmes and none re-injured the hamstring during any phase or on return to game play. The model was flexible in allowing one player to return after missing only three games while limiting the others until the injury had healed to an appropriate level. One other factor that clearly stood out in the short period we have been using the model is the running programme. It became the backbone of the whole model and was the number one indicator of progression through the rehabilitation programme.

The players responded well to clear guidelines and did not develop false hopes of an early return to play. Specifically, the defined benchmarks that were set (model phases and programme stages) appeared to help the players focus on the rehabilitation process and gave them a mechanism by which to evaluate their own progress. This would then seem to be the reason why compliance among the players was perceived to be very high among the medical and conditioning staff. Communication with the coaching staff was also improved using the model. 'Guesstimates' of injury progress were replaced with phases and stage information (specifically the running programme stages) giving the medical staff a more objective means of assessing progress and possible return-to-play dates. As such, our preliminary assessment of the efficacy of this hamstring programme is very good and we are confident that this model can easily be extended to other common soft tissue injuries.

5 Acknowledgements

The authors would like to acknowledge Matthew Cameron, Paul Hedger, Tom Cross and Nathan Gibbs for their contributions to this research.

6 References

Gleim, G.W. and McHugh, M.P. (1997) Flexibility and its effects on sports injury and performance, *Sports Medicine,* 24, 289-299.

Järvinen, M.J and Lehto, M.U. (1993) The effects of early mobilisation and immobilisation on the healing process following muscle injuries, *Sports Medicine,* 15, 78-89.

Knapik, J.J., Jones, B.H., Bauman, C.L. and McA. Harris, J. (1992) Strength, flexibility and athletic injuries, *Sports Medicine,* 14, 277-288.

Orchard, J. (1998) *AFL Injury Report 1997,* 10.

Orchard, J., Marsden, J., Lord, S. and Garlick, D. (1997) Preseason hamstring muscle weakness associated with hamstring muscle injury in Australian footballers, *American Journal of Sports Medicine,* 25, 81-85.

32 SPORTS MEDICINE CHANGES THE RULES

H. SEWARD
Corio Bay Sports Medicine Centre, Geelong, Australia

1 Introduction

Footballe, wherein is nothing but beastlie furie and extreme
violence,wherefore it is to be put in perpetual silence.
An English edict in the Middle Ages

Concern about injuries in football is not a new phenomenon, even in football's embryonic
stages injured players caused consternation amongst the authorities (Dunning, 1994). When
football began in the Middle Ages as a European village game it was seen as a distraction from
the important activities such as archery and other military skills. Injuries would reduce the
fitness of the population for the purposes of war.

During the 19th Century when the different codes of football were in their formative years,
prevention of the potential for injury often influenced the development of their rules.

Today the identification of mechanisms that may give rise to injury is an important focus of
sports medicine and has become the single greatest influence on rule changes. Many of these
changes have improved the safety for players, but also in some cases, improved the spectacle of
the game (Noakes and Jakoet, 1995).

2 History

Australian Rules Football is the major football code in all Australian states, except New South
Wales and Queensland where Rugby League and Rugby Union are the more popular. Soccer is
becoming increasingly popular and is played throughout Australia.

Australian Football developed in the 19th Century when Australian settlers from England
brought their various football codes with them. The formation of this code demonstrates the
influence of injury concern in rule development. The founders of the Australian game wanted a
code that could not only be compared to the English public school games of 'Rugby' or 'Eton'

but would also 'combine their merits while excluding the vices of both', including some of their more brutal elements (Thompson, 1860). At Rugby, Shrewsbury and Eton, for example, players were allowed to hack (or kick) their opponents' shin, a practice that many early colonial players opposed. As a result 'hacking' was banned. Soon after the Australian game was founded tripping was prohibited as well.

Despite some of these early concerns about injury prevention, the game of Australian Football, like its Rugby cousins, was always associated with a hard body collision style. One of the founding fathers of the code, H.C.A. Harrison, was a recognised tough player.

In 1870 amidst criticism for his rough style of play, he defended himself in the press claiming that 'football is essentially a rough game all the world over, and is not suitable for men - poodles or milk sops' (*Australasian* newspaper, 1870).

3 Rule Changes to Prevent Serious Injury

All football codes have undergone gradual change in style since their inception. Rules were usually changes in response to a desire to improve the spectacle of the game, for the enjoyment of players and supporters alike. However, in the last 20 years an increasing concern about severe injuries has led to significant rule modification.

American football has led the way in modern times with injury research leading to rule changes. First, helmets were introduced to reduce severe brain injury and death in that code, second were the changes to reduce the rising incidence of cervical spinal cord injuries, following the identification of the mechanisms of that injury, such as spear tackling. More recently Rugby Union has altered scrum and ruck rules in a successful effort to reduce the incidence of severe cervical spinal cord injuries (Noakes and Jakoet, 1995). Rugby League also has rules to control tackling that minimise injury to the head and neck. American football has introduced tackling rules that attempt to reduce the incidence of knee injuries, while in soccer restrictions on tackling that can cause severe leg injuries have been introduced. Most codes at the elite level conduct video surveillance to identify illegal violence and engage tribunals to administer penalties.

While many of the serious injuries have been addressed, many players remain at risk, particularly when the rules are not appropriately administered. The rule interpretation and vigilance of referees and umpires play a vital role in the protection of players from unnecessary injury.

4 Injury Prevention by Other Methods

Beyond rule changes there are two broad areas where sports medicine offers footballers further potential for injury prevention. First, extrinsic factors such as equipment and playing surfaces are the subject of research to determine optimum safety levels. American football incorporates extensive protective equipment, but this itself can be a causative factor for injury. By contrast

Australian footballers rarely use any protective equipment. Recent research focusing on ground surfaces has identified harder grounds as an influencing factor for injury (Orchard et al., in press).

Second, training programmes have been subjected to extensive changes over the last 20 years as sports medicine knowledge increases. Training programmes are now designed to produce maximum skill and fitness required for the game, while producing minimal injury. Further, rehabilitation programmes are carefully designed to return the injured player to their optimum performance as quickly as possible.

5 The Future: Injury surveillance and research

Demands on footballers are constantly increasing due to the increase in size and speed of players (Norton et al., 1998), together with the influence of the growing professionalism in football. Sports medicine now faces the challenge to move ahead of these changes and maintain the previous gains in injury prevention.

The key to the future is to undertake accurate injury surveillance to identify injury patterns. When injury surveillance identifies common injuries or changing trends, further research is required to identify causative factors and develop preventative programmes, if simple rule changes are not sufficient. Commencing with a comparative study for elite Australian Football, Rugby League and Rugby Union in 1992 (Seward et al., 1993), the AFL Medical Officers Injury Survey is an excellent example of effective injury surveillance. Research is the next stage in this phase and is just beginning

The critical role that sports medicine and science has begun to play in football, means that these professions join the traditional coach–administrator duopoly as the guardians of their football codes: overseeing the welfare of players and nurturing the continued evolution of their games. To facilitate this change the administrators must now recognise the essential role of sports medicine and science and their ability to enhance players' participation and performance. Changes to rules or training techniques that come from this source must be based on high quality scientific research, not just ad hoc decisions based on impressions or anecdotal evidence. Future research must be well funded and supported by the senior administrative bodies of all football codes.

6 References

Australasian Newspaper 1870: quoted from Hess, R. and Stewart, B. (1998) *More Than A Game,* Melbourne University Press, Australia.

Dunning, E. (1994) The History of Football, in *Football (Soccer)* (ed B. Ekblom), Blackwell Scientific Publication, London, pp. 1-19.

Noakes, T. and Jakoet, I. (1995) Spinal cord injuries in Rugby Union Players, *British Medical Journal,* 310, 1345-1346.

Norton, K., Craig, N. and Olds T. (1998) Evolution of Australian Football: the impact on injury potential, *Proceedings: Football Australasia. Reducing the Injury List,* Royal Australian College of Surgeons.

Orchard, J., Seward, H., McGivern, J. and Hood, S. (in press) Soft grounds reduce the risk of non contact anterior cruciate ligament injury in the Australian Football League, *Medical Journal of Australia.*

Seward H., Orchard J., Hazard H. and Collinson, D. (1993) Football injuries in Australia at the elite level, *Medical Journal of Australia,* 159, 298-301.

Thompson, J.B. (1860) Victorian Cricketers Guide, quoted from Hess, R and Stewart, B. (1998) *More Than A Game,* Melbourne University Press, Australia.

33 PUBIC SYMPHYSIS STRESS TESTS AND REHABILITATION OF OSTEITIS PUBIS

A. HOGAN and G. LOVELL
SA Sports Medicine Centre, Adelaide, Australia.

1 Introduction

Sports medicine practitioners have taken a great interest in the diagnosis and management of chronic groin pain in the football codes which involve running, kicking and explosive side-stepping manoeuvres to gain possession of the ball or to avoid being tackled.

A number of conditions have been associated with chronic groin pain including osteitis pubis, pubic instability, sportsman's hernia, musculotendinous lesions of the adductor muscle group and nerve entrapments (Fricker, 1997). There is a variety of combinations of one or more of these diagnoses which can make it difficult for the clinician to isolate osteitis pubis as the singular condition responsible for chronic groin pain in a footballer. The clinical presentation of the appropriate signs and symptoms and confirmatory investigations should allow the clinician to confidently state whether osteitis pubis is a component of the chronic groin pain experienced by the footballer (Fricker et al., 1991; Fricker, 1997; Verrall, 1998).

The treatment of choice for osteitis pubis is conservative management (Fricker et al., 1991), and the design of a successful rehabilitation programme will need to address the primary pathology, secondary adaptations and other predisposing and contributing factors within the neuromusculoskeletal system. Fricker et al. (1991) reported the average recovery time for osteitis pubis for soccer (N = 8) was 7.5 months, rugby (N = 6) was 17.7 months and Australian Rules Footballers (N = 2) was 15 months. This length of lay-off represents a very high financial and professional cost to the footballer, team and administration.

Preliminary evidence from the retrospective review of 20 footballers and the rehabilitation of a further 62 with osteitis pubis as a component of their presentation has identified that optimal rehabilitation requires correct timing in the progression of exercise and avoidance of exacerbations of pain. A successful rehabilitation programme can return a compliant professional footballer to competition within a four to five month timeframe (i.e., off-season) which compares favourably with the data published by Fricker et al. (1991).

2 Preliminary Study

A group of footballers (N = 20) aged 18-33 years was included in a preliminary study of the relationship between reassessment criteria and successful return to running if they had :
1. persistent groin pain for longer than eight weeks;
2. a diagnosis of osteitis pubis as a component of their presentation made independently by a sports physician and a physiotherapist;
3. unable to run because of the groin pain they described;
4. did not have a lumbar spine dysfunction being currently treated;
5. any other lower limb injury.

Each footballer underwent a subjective and physical examination, a test run and players were classified 'Unable to Run' if they could not run because of their groin pain. All were placed on the same pain free exercise programme and were encouraged not to perform any activity that reproduced their groin pain. Each subject was assessed on a fortnightly basis and was classified 'Able to Run' once the test run was completed without any groin pain. Providing they were able to complete the test run on three occasions on alternate days without pain, subjects retained that classification. Three subjects were later excluded from the group as they underwent groin surgery after they been classified in the Able to Run group, leaving a group of 17 footballers for analysis.

Footballers were assessed using 24 criteria which included the description of pain and stiffness, pelvic and hip joint mobility and stability tests, muscle length and strength tests (for muscles that attach to the pelvis) and palpation. The prediction for the study was that one or more assessment criteria would be significantly different between the first assessment (Unable to Run) and the second assessment (Able to Run).

After evaluation, the following five assessment criteria were statistically significant $(P < 0.05)$:
1. no groin pain during cycling/swimming and muscle strengthening/stretching exercises;
2. no groin pain and/or stiffness the next morning after these cycling/swimming etc.;
3. no groin pain during isometric hip adduction (in 60, 40, 20 and 0° hip flexion);
4. normal Bent Knee Fall-Out test;
5. no pain during Modified Thomas Test.

3 Discussion

Fricker et al. (1991) suggested that 'Activity should be encouraged as much as pain allows' but more recently pain free activity has been advocated (Fricker, 1997). It is the experience of the authors that pain-free exercise (cardiovascular, muscle strengthening and stretching) is an essential pre-requisite for the successful progression to pain-free running. Morning pain and stiffness may indicate the degree of inflammatory response to the exercise performed the day before.

Young et al. (1998) reported that the Isometric Hip Adduction test in 30° of hip flexion was the gold standard in the diagnosis of osteitis pubis based on a strong correlation with abnormal findings on computerised tomography scan and surgery. This preliminary study supports the inclusion of the four test positions (60, 40, 20 and 0° hip flexion) rather than a single test position.

3.1 Pubic symphysis stress test

The Modified Thomas test is used by clinicians to assess the length of hip flexor muscles and adductor muscles; however, muscle length was not found to be a useful assessment criterion. Of great interest to the authors was the reproduction of ispilateral and/or contralateral groin pain during the muscle length tests whilst the clinician performed passive hip extension and hip abduction. It is proposed that the position of the lower limbs during the Modified Thomas test produced a biomechanical stress that exceeded the pain threshold of the injured tissues causing the groin pain.

To avoid confusion, the Modified Thomas test was renamed as the Pubic Symphysis (PS) Stress Test to emphasise the pain provocation component (Hogan, 1998). Isometric Hip Adduction (0 and 20° of hip abduction) was added to enhance the isometric hip adduction test assessed in 0, 20, 40 and 60° of hip flexion. Hopping on one leg was added as a PS Stress Test to stress the pubic symphysis vertically, consistent with the radiological method used for pubic instability (Fricker et al., 1991). Table 1 compares the direction of pubic symphysis movement identified by Walheim et al. (1984) using an electromechanical method and the proposed direction of movement which occurs during each of the PS Stress Tests.

Table 1. Comparison of pubic symphysis movement (Walheim et al., 1984) and the proposed direction of movement during PS Stress Tests.

Walheim et al. (1984)	PS Stress Tests (Hogan, 1998)
Sagittal Rotation (yz Plane)	PS Stress Test (Passive Hip Extension)
Transverse (x translation)	PS Stress Test (Passive Hip Abduction)
	PS Stress Test (RSC Hip Adduction)
Vertical (y translation)	PS Stress Test (Hopping)
Sagittal (z translation)	Antero-posterior glide on pubic bone
	PS Stress Test (RSC Hip Flexion)

3.2 Modification of the PS Stress Tests

The abdominal muscles and adductor muscles, which attach to the pubic bones and anterior fascia as it crosses the midline of the pubic symphysis, provide the primary myofascial stabilization of the pubic symphysis. If the PS Stress Tests provoked pain, a contraction of the abdominal muscles was used to facilitate stabilization and the effect on pain response was noted.

3.3 PS Stress Test examination procedure

Step 1: Starting position: check pain free.
Step 2: PS Stress Test: pain response noted (abdominal/groin pain described by athlete).
Step 3: Return to Starting position: check pain free.
Step 4: Low abdominal contraction.
Step 5: Repeat PS Stress Test (compare pain response noted with Step 2).

If the abdominal/groin pain associated with the osteitis pubis was decreased or abolished, it is proposed that facilitation of the myofascial stabilization of the pubic symphysis provided increased resistance to the stresses placed on the pubic symphysis by the PS Stress Tests.

3.4 Reliability or the PS Stress Tests

The reproducibility of the Pubic Stress Tests was analysed in a group of subjects with groin pain (N = 10) with the order of testing randomised and found to have a high degree of reliability except for the PS Stress Test (Hopping) which demonstrated weaker predictability (Table 2). Due to the lack of clinical significance ($r = 0.58$), the PS Stress Test (Hopping) has not been retained in the assessment to commence a footballer running.

Table 2. Reliability of PS Stress Tests (N = 10)

PS Stress (Passive Hip Extension)	r=0.97
Abdominal contraction + Passive Extension	r=0.91
PS Stress (Passive Hip Abduction)	r=0.89
Abdominal contraction + Passive Abduction	r=0.93
PS Stress (RSC Hip Adduction)	r=0.90
Abdominal contraction + RSC Adduction	r=0.87
PS Stress (Hopping)	r=0.58
Abdominal contraction + Hopping	r=0.54

3.5 PS Stress Test 'Crossover Sign' and successful return to running

To date 82 footballers (including the 20 involved in the preliminary study) with an osteitis pubis component to their groin pain have been rehabilitated to pain-free running. As the condition settles and the isometric tests become pain free, the PS Stress tests are the most useful indicator of the capacity to commence running ($r = 0.94$, N = 82 subjects).

The report by the footballer of contralateral suprapubic abdominal pain or groin pain during the PS Stress Tests has been descriptively labelled the 'crossover sign' by the authors. The crossover sign was present in 18 footballers who ran when recommended not to run and six footballers who were instructed to run as a last resort after more than 20 weeks of exercise rehabilitation. None of these footballers was successful in the return to running. All of the 18 footballers made a successful return to running once their PS Stress Tests were negative and the crossover sign resolved and MRI investigations of the six recalcitrant footballers demonstrated severe pathology.

The crossover sign is a common presentation in footballers (54 of 82). The authors feel that it indicates significant pathology in the early weeks and delayed healing of a lesion if it persists longer than 20 weeks. The "crossover sign" is now the subject of further investigation.

3.6 Rehabilitation of professional footballers to competition

It is always difficult to get a homogeneous group of footballers suitable for analysis as it is impossible to control the multitude of factors and issues that arise during a long rehabilitation programme. A group of 11 professional footballers (soccer N = 2, rugby N = 3, Australian Rules N = 6) has undergone the rehabilitation described above using the 5 key assessment criteria to determine the correct timing for exercise progression. This group demonstrated that it is possible to resume training in a pain free state between 12-20 weeks and resume normal competition after 17–25 weeks. It is strongly recommended that footballers returning from osteitis pubis should be prepared for the early weeks of the main season and not the pre-season competition.

4 References

Fricker, P. (1997) Osteitis pubis, *Sports Medicine and Arthroscopy Review,* 5, 302–312.

Fricker, P., Taunton, J. and Ammann, W. (1991) Osteitis pubis in athletes: Infection, inflammation or injury? *Sports Medicine,* 12, 266–279.

Hogan, A. (1998) A rehabilitation model for pubic symphysis injuries, *Australian Conference on Science and Medicine in Sports Book of Abstracts.*

Verrall, G. (1998) Osteitis pubis in Australian Rules footballers: Clinical aspects and guidelines for investigations and management, *Australian Conference on Science and Medicine in Sports Book of Abstracts.*

Young D., Zimmerman, G. and Toomey, M. (1998) Osteitis pubis, *Proceedings Football Australasia Conference,* Melbourne, Australia.

34 OSTEITIS PUBIS: A STRESS INJURY TO THE PUBIC BONE

G. VERRALL, J. SLAVOTINEK and G. FON
SPORTSMED•SA Sports Medicine Clinic, Adelaide, Australia

1 Introduction

Australian Rules Football (AFL) is a high energy, collision contact ball sport with speed, strength and agility demands similar to Rugby or American football. It differs from these football codes with more kicking, more evasion and higher aerobic demands due to the continuous nature of play and in this manner is more similar to soccer.

The diagnosis of chronic groin pain in sportsmen is difficult. Athletes who participate in sports with repetitive kicking, side-to-side movement and twisting such as AFL and soccer seem to be most at risk. Potential diagnoses include osteitis pubis, adductor muscle/tendon dysfunction and posterior abdominal wall weakness/conjoint tendon dysfunction, i.e., Sportsman's hernia. The MRI scan has been shown to differentiate between the causes of chronic groin pain but to date this technique has not been used extensively in the investigation of athletic groin pain.

2 Methods

Altogether, 116 MRI scans were performed in this study. The subjects included 89 AFL players recruited at the end of pre-season, after at least six weeks of intensive training along with 17 running athletes from the umpires training for the AFL competitions as active control subjects and 10 sedentary males as inactive control subjects. The AFL players from each team (N = 3) included the following: (1) Athletes with current symptomatic groin pain, (2) Athletes with previous groin pain and (3) Athletes who had never had any previous episode of groin pain.

The clinical history was not known to the examiner (GV). The subjects underwent examination of the groin region followed by having their history taken by direct interview that included current groin pain symptoms and details of any previous groin pain episodes.

Subjects with pain occurring in the pre-season period that was during and/or after exercise located in the groin region (pubic symphysis and adjacent pubic bone region, adductor area and/or lower abdominal area) with tenderness of their pubic symphysis and/or adjacent superior pubic rami were considered currently symptomatic (CS).

Following the examination and history a MRI scan was performed (time after examination 0-12 days, mean 4 days) using T1 and T2 Fat Suppression sequences (1.5T GE Signa, 1.0T Siemens Impact). The scans were obtained in the coronal and axial planes with respect to the pubic body (3-4 mm slices). The images were independently reviewed without clinical information by two musculoskeletal radiologists.

Scans were graded for (1) signal intensity (Grade 0-I, II, III) and (2) size of signal change (< 2 cm, ≥ 2 cm) and area of signal change (pubic body and adjacent pubic ramus, i.e. subchondral area).

3 Results

The AFL players (N = 89) were classified into Currently Symptomatic CS n = 47 (53%) Currently Asymptomatic CA n = 42 (47%). Of the umpires (N = 17), classifications were CS n = 5 (29%) and CA n = 12 (71%).

The MRI signal intensity changes can be seen in Table 1. The statistical analyses can be seen in Table 2.

Maximal signal intensity was subjectively graded according to signal intensity and was correlated with current clinical symptom category (Table 1). There was a statistically significant finding of being currently symptomatic CS and having Grade III signal intensity changes on the MRI scan (P < 0.05, phi 0.25). As the size of change became larger (> 2 cm) and the region affected involved the superior pubic ramus and was not confined to the pubic body (subchondral), the number of CS athletes in these groups becomes proportionately larger (Table 1). As this occurs the statistical significance increases. Subjects with Grade II signal intensity changes also showed a similar pattern and when the Grade II and III changes were combined an even stronger association, best demonstrated by following the phi coefficient (Table 2), can be shown. This table depicts that a very strong association exists between current symptomatic groin pain in AFL players and the MRI scan findings of increased signal intensity. In these analyses only the AFL population has been used. If the other two groups (umpires and sedentary) are added then the significance and phi coefficient become more pronounced.

Table 1. MRI signal intensity – grade, size, region and subjects by clinical symptom category.

Grade, Size and Region	AFL 89	CS 47	CA 42	UMP 17	CS 5	CA 12	SED 10
III	33	22	11	3	3	0	0
II	35	20	15	5	2	3	0
0, I	21	5	16	9	0	7	10
III, > 2 cm	22	17	5	3	3	0	0
II, > 2 cm	19	15	4	0	0	0	0
III, > 2 cm, subchondral	16	14	2	3	3	0	0
II, > 2 cm, subchondral	14	12	2	0	0	0	0

Table 2. Correlation between current symptomatic groin pain in AFL players and the MRI scan findings of increased signal intensity.

Grade, size, region	P <	phi
III	0.05	0.245
III, > 2 cm	0.007	0.320
III, > 2 cm, subchondral	0.002	0.400
II & III	0.001	0.400
II & III, > 2 cm	0.001	0.448
II & III, > 2 cm, subchondral	0.001	0.549

4 Conclusions

The increased signal intensity of the pubic bone marrow in this study is consistent with bone marrow oedema, a non-specific finding that may be seen with stress injury. The differential diagnosis of marrow oedema, includes bone bruising due to direct trauma, osteomyelitis and infiltrative neoplasm but these are most unlikely in the described clinical situation. A tension stress injury seems the likely aetiology of MRI scan appearances seen in this study.

PART FIVE

Metabolism and Nutrition

35 METABOLIC AND PHYSIOLOGICAL RESPONSES TO A LABORATORY BASED SOCCER-SPECIFIC INTERMITTENT PROTOCOL ON A NON-MOTORISED TREADMILL

B. DRUST*, T. REILLY** and N.T. CABLE**
*   Centre for Sport, Health and Exercise Performance, University of Durham, Stockton-on-Tees, UK
**   Research Institute for Sport and Exercise Sciences, Liverpool John Moores University, Liverpool, UK

1 Introduction

Few researchers have attempted to devise soccer-specific laboratory based protocols that seek to replicate the exercise patterns observed during match-play. Laboratory based protocols present an opportunity to utilise apparatus that increases both the detail and accuracy of the data generated as well as the controlled conditions that are associated with experimental investigations.

The development of non-motorised running treadmills has helped to overcome some of the problems associated with devising laboratory based intermittent exercise protocols. As well as facilitating the attainment of near maximal sprinting speeds (Lakomy, 1987), the non-motorised treadmill also allows almost instantaneous changes in the intensity of exercise. Such abrupt changes in the intensity of action are characteristic of soccer match-play.

Previous studies have utilised a non-motorised treadmill to re-create activity patterns representative of soccer (Fallowfield et al., 1997; Wilkinson et al., 1997). These re-creations have, however, been concentrated on the effects of fluid supplementation on repeated sprint performance as opposed to the physiological and metabolic responses that accompany soccer-specific intermittent exercise performance. Few data also are available on the physiological responses to international level match-play. Changes in work-rate profile, as a result of differences in the level of competition (Reilly, 1994), may have consequences for the physiological demands imposed on players. The aim of this study was to determine the physiological and metabolic responses to soccer-specific intermittent exercise protocols representative of work-rates observed during international match-play on a non-motorised treadmill.

2 Methods

Seventeen full-time professional soccer players were filmed during international match-play for the determination of work-rate profiles. The activities of one player during 90 min of match-play were recorded using a video camera (Sony TR 75E, Taiwan). Videotapes were replayed on a television monitor and video (Sony Trinitron, Taiwan; Panasonic NV-F77, Japan) and analysed using a personal computer (Amstrad PC7486, Japan). Activities were separated into four activity categories based on the intensity of action. These were walking (forward and backward movements), jogging (forward, backward and sideways), cruising and sprinting. Static pauses were also recorded. Analysis was linked to a time base to enable the percentage of the total time for each movement classification to be determined.

The data obtained from the work-rate analysis were then used to form the basis of the soccer-specific intermittent exercise protocol conducted on a modified non-motorised treadmill (Woodway, Vor Dem, Auf Schrauben, Germany). Four movement categories were incorporated: walking, jogging, cruising and sprinting. Static periods were also included in the protocol in which the subject was stationary on the treadmill. Utility movements (backwards and sideways) were not included in the protocol because of the technical limitations of the equipment used. The percentage time spent in these two activity categories during match-play was divided between the walking and jogging categories. Table 1 shows the percentage of the total time spent in each activity for the soccer-specific intermittent protocol.

Table 1. Percentage of total time spent in each activity in the soccer-specific
intermittent exercise protocol.

Activity	% time
Static	15
Walking	50
Jogging	30
Cruising	4
Sprinting	1

Treadmill speeds were assigned to each of the four types of activity incorporated in the protocol to ensure that the absolute physiological load remained constant across subjects. Treadmill speeds for each activity were based on the data of Van Gool et al. (1988) and corrected for the intrinsic resistance associated with non-motorised treadmill running. The respective speeds chosen for each activity pattern were :- walking 4 km·h^{-1}, jogging 8 km·h^{-1} and cruising 12 km·h^{-1}. No speed restrictions were placed on the sprinting category as subjects were instructed to produce a maximum effort.

The protocol was arranged around a 15-min activity cycle (see Figure 1). This was performed six times in all to make up the 90-min protocol duration (2 x 45 min separated by a 15-min half-time period). This 15-min cycle was further sub-divided into three separate 5-min sub-cycles. The order of presentation of the activities was determined by the researcher. High-intensity bouts were separated by low intensity recovery periods and static pauses in an attempt to replicate the acyclical nature of the exercise pattern in soccer. The duration of each bout of activity was determined by dividing the percentage time of each activity category observed during match-play by the time for one activity cycle in the intermittent protocol (i.e. 5 min). Once the total time for each activity category was established, it was possible to determine the time for each discrete bout of activity within each category by dividing the total time for each category by the number of bouts required. Each 5-min cycle consisted of 11 bouts of activity. The 11 bouts of activity were comprised of 3 static pauses, 3 walking bouts, 3 jogging bouts, 1 cruise and 1 sprint.

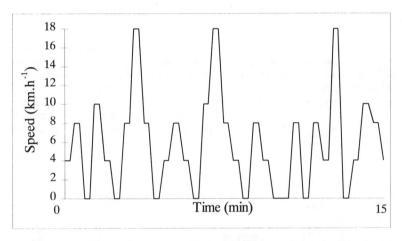

Figure 1. Representation of the soccer-specific intermittent protocol developed for the non-motorised treadmill. The 15-min cycle was repeated in accordance with 90 min of match-play.

Six male university soccer players completed the soccer-specific intermittent exercise protocol (mean ± SD age 27 ± 2 years; height 1.77 ± 0.03 m; mass 72.2 ± 1.5 kg; VO_2 max 58.9 ± 3.5 ml·kg^{-1}·min^{-1}). Expired air was collected and analysed continuously throughout the exercise period for the determination of the oxygen consumption associated with the exercise trial (Medgraphics Cardi0$_2$ system, Minnesota, USA). Heart rate was continuously monitored using short-range radio telemetry (Sports Tester, Polar Electro, Kempele, Finland) during exercise and half-time recovery periods (15-s intervals) to provide an indication of circulatory strain.

Venous blood samples (<10 ml) were taken from an antecubital vein by an experienced phlebotomist for the analysis of blood-borne substrates. Blood samples were taken approximately 5 min prior to exercise, during the half-time recovery period and upon cessation of exercise. Blood samples were immediately centrifuged (ALC PM180R, high-speed refrigerated centrifuge, Camlab, Cambridge, UK) and the plasma removed and frozen at -83° C for analysis. Plasma samples were analysed for glucose, free fatty acids, glycerol and lactate using a centrifugal analyser (Monarch Chemistry System, Instrumentation Laboratory, Lexington, USA).

A paired sample *t*-test was used to compare differences in oxygen consumption, heart rate and the total distance covered between each of the two halves. Differences in plasma free fatty acids, plasma lactate and plasma glucose concentrations prior to exercise, following 45 min (half-time) and post-exercise were examined using one-way ANOVA. The probability level for statistical significance was $P < 0.05$.

3 Results

The mean ± SD total distance covered during the soccer-specific intermittent exercise protocol was 9500 ± 400 m. Figure 2 shows the mean ± SD total distance covered during the first and second halves. There was no significant difference in the total distance covered between the first and second half of the protocol.

Mean ± SD oxygen consumption was 2.5 ± 0.3 l·min^{-1}. This value corresponds to approximately 65–70 % VO$_2$ max. Figure 3 shows the mean ± SD heart rate response to the soccer-specific intermittent protocol. There was no significant difference in the oxygen consumption and heart rate between the first and second half.

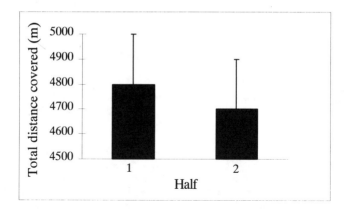

Figure 2. Mean ± SD total distance covered during each half during the soccer-specific intermittent exercise protocol.

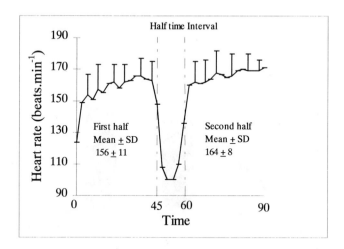

Figure 3. Mean ± SD heart rate for each half during the soccer-specific intermittent exercise protocol.

Figures 4, 5 and 6 show the mean ± SD plasma lactate, glucose and free fatty acid response to the soccer-specific intermittent protocol. Plasma lactate did not significantly increase with time at any stage of the protocol. Plasma glucose concentrations were elevated significantly, $(F_{2, 10}) = 5.46$, $P < 0.05$) after 45 min and remained elevated for the remainder of the protocol. Free fatty acids were also elevated after 45 min over resting values though this difference was not significant. Significant differences were found between the concentration at the end of the protocol and both pre-exercise and at half-time ($F_{2, 10} = 24.18$, $P < 0.05$).

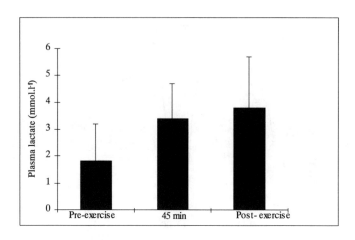

Figure 4. Mean ± SD plasma lactate response to the soccer-specific intermittent exercise protocol.

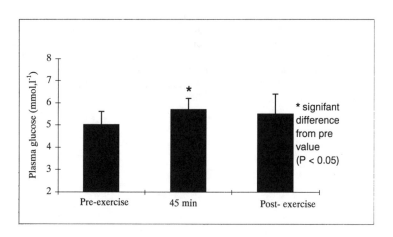

Figure 5. Mean ± SD plasma glucose response to the soccer-specific intermittent exercise protocol.

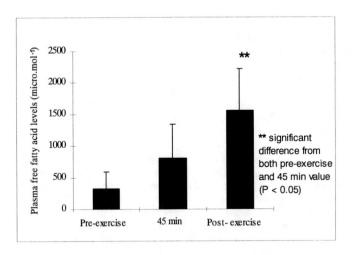

Figure 6. Mean ± S.D plasma free fatty acid response to the soccer-specific intermittent exercise protocol.

4 Discussion

The aim of the present investigation was to determine the physiological and metabolic responses to a soccer-specific intermittent protocol completed on a non-motorised treadmill. The total distance covered during the soccer-specific intermittent protocol was 9500 ± 400 m. This distance is similar to the total distance of 8638 ± 1158 m observed for South American international players (Drust et al., 1998). It is also in the order of the 10 km total distance covered by Danish players according to Bangsbo et al. (1991). Some differences, with respect to the activity profile, exist between match-play and the soccer-specific intermittent protocol in the current investigation. The replication of utility movements (i.e. backing and sideways movements) is impractical and the total number of activity changes in match-play cannot be modelled adequately due to their frequency and irregular nature. Performance of game skills is also neglected in the laboratory simulation.

The physiological and metabolic responses to the soccer-specific intermittent protocol on the non-motorised treadmill nevertheless approximated the physiological strain associated with match-play despite these shortcomings. The oxygen consumption associated with the soccer-specific intermittent protocol can be expressed as a percentage of maximum to enable comparisons to be drawn with values estimated during match-play. The oxygen consumption during the intermittent protocol was calculated to be approximately 65–70 % VO_2 max. Such values are similar to the mean relative oxygen consumption associated with soccer match-play (Bangsbo, 1994) and indicate a high aerobic contribution to energy provision.

Mean heart rate was observed to be 160 ± 6 b·min^{-1} for the soccer-specific intermittent protocol. This compares well with the 157 b·min^{-1} obtained by Reilly (1986) for soccer match-play. Other authors have noted a slightly higher heart rate (approximately 10 b·min^{-1}) during games (Van Gool et al., 1988; Bangsbo, 1994). These results indicate that the soccer-specific intermittent protocol in the current investigation provokes a reasonable representation of the cardiovascular demands of match-play. The slightly lower heart rates are probably the result of the omission of game skills and a decrease in the number of activity changes in the protocol.

The blood lactate concentration can be used as an indicator of the anaerobic energy provision. The data suggest that there is an anaerobic component to the energy demands of the intermittent protocol, as indicated by the increases over resting levels in plasma lactate levels during the first and second halves. Plasma lactate concentrations during the soccer-specific intermittent protocol were lower than values at half and full time reported for soccer match-play (Ekblom, 1986; Rohde and Espersen, 1988). These differences could again reflect for example, the exclusion of activities that increase the physiological demands on players (i.e. match skills).

The dominant substrates utilized in soccer are carbohydrate and fat stored within the exercising muscle or delivered via the blood stream (Bangsbo, 1994). Plasma glucose levels were significantly elevated upon completion of the first half of the soccer-specific intermittent protocol. Blood glucose levels are similarly elevated over resting values during competitive matches with mean values of between 3.2 to 4.5 mmol.l^{-1} being observed for elite Swedish and Danish players during match-play (Ekblom, 1986; Bangsbo, 1994). Bangsbo (1994) noted increases in free fatty acid concentration in the blood during a competitive soccer match, with the increase being more pronounced in the second half of the match. This trend is identical to that found for the simulation in the current investigation. The increase in free fatty acid concentrations is probably the result of elevated blood flow to the adipose tissue and changes in catecholamine concentrations, thus promoting a high free fatty acid release in the second half.

The data provided by means of the soccer-specific intermittent protocol in the current investigation seem to indicate that the protocol replicates the overall energy demands of match-play as well as the physiological responses observed during the game. Such similarities make the current soccer-specific intermittent protocol suitable for the investigation of the metabolic and physiological responses to intermittent exercise.

5 References

Bangsbo, J. (1994) Physiology of soccer – with special reference to intense intermittent exercise, *Acta Physiologica Scandinavica*, 151, Suppl., 619.

Bangsbo, J., Norregaard, L. and Thorsoe, F. (1991) Activity profile of competition soccer, *Canadian Journal of Sports Science*, 16, 110–116.

Drust B., Reilly, T. and Rienzi, E. (1998) A motion-analysis of work-rate profiles of elite international soccer players, *Journal of Sports Sciences*, 16, 260.

Ekblom, B. (1986) Applied physiology of soccer, *Sports Medicine*, 3, 50–60.

Fallowfield, J.L., Jackson, A.G., Wilkinson, D.M. and Harrison, J.J.H. (1997) The influence of water ingestion on repeated sprint performance during a simulated soccer match, in *Science and Football III* (eds T. Reilly, J. Bangsbo and M. Hughes), E. and F.N. Spon, London, pp.61–65.

Lakomy, H.K.A. (1987) The use of a non-motorized treadmill for analysing sprint performance, *Ergonomics*, 30, 627–637.

Reilly, T. (1986) Fundamental studies in soccer, in *Sportswissenshraft und Sportspraxis* (ed. R. Andresen), Ingrid Czwalina Verlag, Hamburg, pp. 114–121.

Reilly, T. (1994) Motion characteristics, in *Football (Soccer)* (ed. B. Ekblom), Blackwell Scientific Publications, Oxford, pp. 31–43.

Rohde, H.C. and Espersen, T. (1988) Work intensity during soccer training and match-play, in *Science and Football* (eds T. Reilly, A. Lees, K. Davids and W.J. Murphy), E. and F.N. Spon, London, pp. 68–75.

Van Gool, D., Van Gervan, D. and Boutmans, J. (1988) The physiological load imposed on soccer players during real match-play, in *Science and Football* (eds T. Reilly, A. Lees, K. Davids and W.J. Murphy), E. and F.N. Spon, London, pp. 51–59.

Wilkinson, D.M., Garner, J., Fallowfield, J.L. and Harrison, J.J.H. (1997) The influence of carbohydrate ingestion on repeated sprint performance during a simulated soccer match, in *Science and Football III* (eds T. Reilly, J. Bangsbo and M. Hughes), E. and F.N. Spon, London, pp. 67–73.

36 EFFECT OF AN ACTIVE WARM-DOWN FOLLOWING COMPETITIVE SOCCER

T. REILLY and M. RIGBY
Research Institute for Sport and Exercise Sciences, Liverpool John Moores University, Liverpool, UK

1 Introduction

The majority of the body's physiological systems are stressed during the course of a soccer game and often also by a strenuous training programme. These include metabolic energy systems (Saltin, 1973), the musculoskeletal system and perhaps also the nervous and immune systems (Reilly, 1999). Following training or competitive matches there is a need to switch attention to regenerative processes so that the player regains capabilities in readiness for the next engagement. This is particularly applicable to professionals who may be obliged to play more than one full game within the same week.

Warm-down has been a tradition in track and field sports but not in field games. Active recovery removes lactate from the blood more quickly than does a passive recovery. Clearance of lactate from the blood is related directly to the exercise intensity up to about 50% of the maximal oxygen uptake (VO_2 max). Body temperature continues to rise for some minutes on cessation of exercise, due to the recirculation of the blood from warmed muscles (Reilly and Brooks, 1982). Consequently, exercise at a light intensity at this time should effect a smoother decline in body temperature and blood flow than if exercise is terminated abruptly.

There are other possible benefits of an active warm-down. It may help to dampen activity in the nervous system, promoting sleep the night after a game which otherwise might be adversely affected by elevated levels of stress hormones. The immune system may be depressed for a window of some hours after competition and it is thought that a warm-down can reduce vulnerability to minor illnesses such as upper respiratory tract infections. Stretch-shortening cycles of muscle actions occur repeatedly during games play, normally inducing so-called 'delayed onset muscle soreness'. Active warm-down may be of benefit in reducing the damage associated with this phenomenon.

The aim of the present investigation was to establish the effects of active warm-down on the recovery process following a soccer match. A subsidiary aim was to examine the influence of the warm-down on muscle soreness after a match.

2 Methods

Fourteen males, aged 20.9 ± 1.5 years, participated in the study. Their mean height (± SD) was 174.9 (± 5.7) cm, body mass 77.5 (± 5.6) kg, mean percent body fat determined by bioelectric impedance (Bodystat, Isle of Man) 13.2 (± 2.7) %. All participants were students and were active members of their University's first soccer team squad.

The subjects were divided into two equal groups in order to balance the order of administering the experimental warm-down. One group did the active warm-down first, and a controlled recovery a week following, this order being reversed in the remaining seven players.

The warm-down lasted 12 min and consisted of three phases: (i) jogging (5 min); (ii) stretching (5 min); (iii) lying prone with legs raised and 'shaken down' by another player (Football Association, Coaching Certificate notes). In the corresponding 'controlled' recovery, the subjects returned to the changing rooms and rested (seated) for 12 min.

Following the recovery (12 min) period, the subjects completed a battery of performance tests. These included a standing broad jump, vertical jump, three trials of 30-m sprints and a sprint-fatigue test (Williams et al., 1997). This consisted of seven successive 30-m sprints (a 10-m deceleration zone followed the 30-m marker) with 20 s in between the sprints. Mean time over the seven efforts was calculated by comparing the slowest and the fastest times. The running tests were conducted in football boots on a good quality grass surface.

Altogether the tests were implemented on four occasions on each of two weeks. Pre-match observations were made after competition on Wednesdays. Tests were repeated at the same time of day on the two subsequent days. On each of the occasions muscle soreness was rated by each player using a visual analogue scale.

Data were analysed using either multiple t-tests or where appropriate Mann–Whitney tests. A P-value of 0.05 was taken to indicate statistical significance.

3 Results

The results for the broad jump and the vertical jump showed a similar pattern. Performances were impaired following the game in both groups (P < 0.01). The mean reductions were 20 cm and 9.5 cm for the experimental group and 21 cm and 14 cm for the control group, for broad jump and vertical jump, respectively. The group doing the active warm-down had improved performance 24 h later (by 9 cm and 2.5 cm), whereas the control group showed a further deterioration (of 7 cm and 1 cm). Significant differences between the groups remained 48 h following the game, although the experimental group had not even then recovered performances fully.

Performance in the 30-m sprint was also impaired throughout the observations and had not recovered to baseline levels 48 h after the matches. The largest difference between the two groups (averaging 0.22 s) was 24 h following the game where sprint performance was almost 5% slower in the control group. A mean difference of 0.6 s remained one day later.

The sprint-fatigue test was impaired throughout the observations after the game, but not in the group who had done the active warm-down. By 48 h following the game, the mean of the seven times was not different from baseline measures in the experimental group. A more striking difference between the groups was evident in the fatigue index. The decrement was almost 50% greater in the control group immediately following the game, but the difference between the groups narrowed with the subsequent two days.

Ratings of muscle soreness were significantly different between the groups on the three days of measurement following playing ($P < 0.05$). The mean difference immediately after the warm-down was 1.4 units in a 5-point scale. The muscle soreness increased further 24 h after the game (mean difference between the groups was 2.6 units) and were still above the post-game value another 24 h later. In contrast, muscle soreness had effectively disappeared in the experimental groups two days after finishing the game.

4 Discussion

The present observations provide evidence of the benefits of active warm-down on the process of recovery from competitive soccer play. Differences between the groups were evident in the tests following the match. This effect alone should raise questions about the consequence of an inactive half-time period where teams may have cooled down rather than warmed down. An alteration in the incidence of goals scored in the 3-min period following resumption of play at the start of the second half has been reported for World Cup matches (Grant et al., 1998). It may be that preserving 'warmth' and maintaining arousal are linked, both during the game and in the short-term following its completion.

More evident were the faster recoveries of performance in the group doing active warm-down in the two days following the match. This was a general observation, consistent across all the performance measures. The trends in the data would indicate that the group warming down after the mid-week match was adequately recovered for a further weekend match. This did not extend to the control group who would have undertaken a Saturday game with performance profiles not completely restored to their true values.

It appears that active warm-down also attenuated the induction of muscle soreness. Delayed onset muscle soreness is subject to a 'repeated bouts effect' but otherwise appears resistant to quick cures (Cleak and Eston, 1992). Whether the benefits of the experimental warm-down regimen were attributable to the activity (jogging), stretching or facilitation of blood flow from the legs is not possible to conclude from their observations.

Professional soccer teams should pay attention to warming down following matches in order to facilitate the recovery process. Further modifications of habitual regimens could include use of deep-water running (Dowzer et al., 1998), flexibility exercises and conditioned training drills. The immediate need is to change the tradition of inactivity once a game is over and this requirement applies to other football codes in addition to soccer. Further research is recommended in order to identify the mechanisms implicated in the delayed recovery process in the absence of a warm-down.

5 References

Cleak, M.J. and Eston, R.G. (1992) Delayed onset muscle soreness: Mechanisms and management, *Journal of Sports Sciences*, 10, 325–341.

Dowzer, C.N., Reilly, T. and Cable, N.T. (1998) Effects of deep and shallow water running on spinal shrinkage, *British Journal of Sports Medicine*, 32, 44–48.

Grant, A., Reilly, T., Williams, M. and Borrie, A. (1998) Analysis of goals scored in the 1998 World Cup, *Insight*, 2, (1), 18–20.

Reilly, T. (1999) Recovery from strenuous training and matches, *Sports Exercise and Injury*, 4, 156–158.

Reilly, T. and Brooks, G.A. (1982) Investigation of circadian rhythms in metabolic responses to exercise, *Ergonomics*, 25, 1093–1108.

Saltin, B. (1973) Metabolic fundamentals in exercise, *Medicine and Science in Sports*, 5, 137–146.

Williams, M., Borrie, A., Cable, T., Gilbourne, D., Lees, A., MacLaren, D. and Reilly, T. (1997) *Umbro Conditioning for Football*, Ebury Press, London.

37 DIETARY ANALYSIS OF A GROUP OF ENGLISH FIRST DIVISION PLAYERS

R. CRAVEN, M. BUTLER, L. DICKINSON, R. KINCH and R. RAMSBOTTOM
School of Biological and Molecular Sciences, Oxford Brookes University, Headington, Oxford, UK

1 Introduction

Soccer matches last for a minimum of 90 min during which players cover between 8 and 12 km, of which sprinting accounts for 11% of the total activity (Reilly 1996). Many of the critical phases of play are performed at high intensity and so it is very important that players are able to attain consistently high levels of exercise intensity at intervals throughout the match. The ability to maintain sprinting speed during repeated efforts seems to be strongly correlated with intramuscular glycogen stores. Players with low initial glycogen concentrations in the leg muscles cover less distance and sprint less often compared to players with higher initial glycogen stores (Saltin, 1973). Muscle glycogen levels are characteristically depleted by at least 50% during a soccer match (Shephard and Leatt, 1987) and so the initial levels of muscle glycogen must influence a player's performance towards the end of the match (Saltin, 1973). Consequently, although a diet which replenishes and maintains adequate muscle glycogen levels is crucial to professional footballers, their diets are reported to be sub-optimal both in terms of total energy intake and energy balance, with too little energy consumed as carbohydrate and too much as fat (Bangsbo, 1994). The present study examined both total energy intake (MJ) and the percentage contribution from carbohydrates, fat and protein, in a group of senior professional players from an English First Division club.

2 Methods

Fourteen players with a median age of 23 (range 18-30) years, body mass 80.1 ± 9.2 (mean ± SD) kg and height 1.81 ± 0.06 (mean ± SD) m, volunteered for the study. Verbal and written instructions about completing a 7-day food diary were given to the players. Each player was given a list of the weights of medium-sized portions of commonly consumed foods, an example of a completed food record for 1 day, together with approximately 12 food record sheets. Five additional sheets contained, in total, 10 photographs of a selection of medium-sized portions;

each photograph was accompanied by a written description of the food items. The players were asked to choose a week in which they had no mid-week match and to begin the food diary on a Wednesday and continue until the following Tuesday, creating a complete weekly record which included a match weekend. The daily diets were then analysed using 'Dietplan 5' dietary analysis software.

3 Results

Table 1. The mean (± SEM) total daily energy intake of the players calculated from their food diaries, compared with the estimated average requirement (EAR) for energy for an adult male and a professional footballer.

	Mean actual daily energy intake (± SEM) (MJ)	EAR for average male (MJ)	EAR for a professional footballer (MJ)
Wednesday	12.3 ± 1.4	10.5	14.9
Thursday	12.0 ± 1.3	10.5	14.9
Friday	13.1 ± 0.9	10.5	14.9
SATURDAY	14.1 ± 1.1	10.5	14.9
Sunday	12.8 ± 1.0	10.5	14.9
Monday	13.0 ± 1.2	10.5	14.9
Tuesday	13.0 ± 0.9	10.5	14.9

The basal metabolic rate (BMR) for the players in this study was estimated using the following equation:

$$BMR (MJ \cdot day^{-1}) = 0.064 \text{ Body Mass (kg)} + 2.84. \text{ (WHO, 1985)}$$

The EAR for a professional footballer is estimated using the following equation:

$$EAR (MJ \cdot day^{-1}) = (BMR \times 1.7) + 10\% \text{ for TEF (thermic effect of food intake).}$$
(Ekblom 1994).

The mean daily energy intake for the group of 12.9 ± 2.8 MJ was significantly ($P<0.05$) lower than the EAR of 14.9 ± 1.1 MJ for a professional footballer of similar weight.

3.1 Pattern of food consumption

Analysis of the pattern of food consumption showed that the mean interval between breakfast and the next meal was 5.27 ± 0.97 h (mean ± SD). There was very little 'snacking' in the interval between breakfast and the next meal; most snacks consumed were potato crisps and/or chocolate bars. For the majority of players the mid-day meal was a sandwich. Alcohol consumption was restricted almost exclusively to Saturday, when it accounted for between 0% and 40% of the day's energy intake. This was at the expense of carbohydrate intake which fell by between 10% and 25% compared to the previous day.

Table 2. The mean (± SEM) percentage contribution to total daily energy intake made by carbohydrate, fat, protein and alcohol compared to recommended values (Bangsbo, 1994; Reilly, 1996).

	Total Fat (%)	Total CHO (%)	Protein (%)	Alcohol (%)
Wednesday	36.7 ± 1.2	47.0 ± 2.0	15.6 ± 0.9	0.7 ± 0.5
Thursday	32.5 ± 2.8	49.7 ± 1.6	15.3 ± 1.0	0.0 ± 0.0
Friday	29.1 ± 2.3	52.2 ± 2.1	17.9 ± 1.5	1.1 ± 0.9
SATURDAY	28.4 ± 2.1	43.2 ± 2.4	14.3 ± 1.8	14.0 ± 3.9
Sunday	34.2 ± 1.6	44.8 ± 2.1	16.9 ± 1.1	4.0 ± 2.5
Monday	30.5 ± 2.1	52.6 ± 2.0	16.7 ± 0.9	0.2 ± 0.2
Tuesday	30.5 ± 1.7	52.3 ± 2.1	17.0 ± 1.3	0.2 ± 0.2
Recommended	21.5	61.5	14.0	

4 Conclusions

The players were shown to be consuming significantly less energy than the EAR for footballers of similar weight, although they exceeded the EAR for the general population. The deficit in actual daily energy intake between the players and the EAR for professional footballers may be explained by an over-estimate of the energy requirements combined with under-reporting of food intake. The players involved in this study showed no weight loss over the course of the season.

The percentage of daily energy obtained from carbohydrate (48.8 ± 3.8%) was too low and that from fat (31.8 ± 3.0%) too high.

The 5-h interval between breakfast and the midday meal, which was often only a sandwich, was too long. An inadequate number of snacks were being consumed. 'Snacking ' is essential if the players are to consume an adequate carbohydrate intake.

For some players, alcohol consumption contributed excessively to the daily energy intake on the Saturday. It is probable that, in addition to substituting for carbohydrate, the alcohol prevents adequate rehydration.

Remedial Action:

1 Encourage the consumption of more carbohydrate containing foods by recommending that:
(a) the players eat appropriate snacks regularly during the day;
(b) the players eat more of the high carbohydrate foods which they already enjoy.
2 Encourage the players to ensure adequate rehydration and carbohydrate replenishment following training and matches, using appropriate carbohydrate containing drinks.

5 Acknowledgements

Dr Craven was supported by a grant from the Wellcome Trust.

6 References

Bangsbo J. (1994) *Fitness Training in Football - A Scientific Approach*, HO+Storm, Bagsvaerd.

Ekblom B. (1994) *Handbook of Sports Medicine and Science: Football (Soccer)*, Blackwell Scientific Publications, Oxford, UK, p. 146.

Reilly T. (Ed) (1996) *Science and Soccer*, E. and F.N. Spon, London.

Saltin B. (1973) Metabolic fundamentals in exercise, *Medicine and Science in Sport,* 5, 137–146.

Shephard R.J. and Leatt P. (1987) Carbohydrate and fluid needs of the soccer player, *Sports Medicine*, 4, 164–176.

WHO (1985) *Energy and protein requirement*, Technical Report Series 724. World Health Organisation, Geneva, p. 71.

38 THE EFFECT OF CARBOHYDRATE SUPPLEMENTATION ON THE WORK-RATE OF GAELIC FOOTBALL PLAYERS

T. REILLY and S. KEANE
Research Institute for Sport & Exercise Sciences, Liverpool John Moores University, Liverpool, UK

1 Introduction

Gaelic football is played at an intensity similar to that reported for soccer. This correspondence is borne out from measurements of work-rates in both games expressed as distance covered each minute of play (Keane et al., 1993). It is further corroborated in measurements of heart rate responses which average 158 and 161 $b·min^{-1}$ in Gaelic football and soccer, respectively (Florida-James and Reilly, 1995).

The fall in muscle glycogen levels towards the end of endurance activities, including soccer match-play, is linked with a fall in work-rate (Saltin, 1973). Carbohydrate supplementation by means of glucose polymer solutions administered pre-game and at half-time has proved beneficial to performance of players towards the end of a soccer game (Kirkendall, 1993). Leatt and Jacobs (1989) employed a similar nutritional strategy and showed that ingesting a glucose polymer drink decreased the net muscle glycogen utilisation during a soccer game.

Inter-county Gaelic football championship matches are 70 min in duration rather than the 90 min time for competitive soccer matches. The corresponding time for inter-county league and for inter-club matches is 60 min. The shorter duration of Gaelic football may allow players to sustain activity levels more easily than in soccer and it is not known whether carbohydrate supplementation can be of benefit to the work-rate of Gaelic footballers. The aim of this study was to monitor the effect of ingesting a glucose solution on the performance of Gaelic football players, including any influence on fatigue as reflected in a fall in work-rate towards the end of a game.

2 Methods

A cross-sectional design was employed in this research. Eight inter-county players were subjected to the experimental treatment and 56 other inter-county players were used as a reference group.

The study was replicated in senior inter-club players. Eight players constituted the experimental group whilst 14 players not administered a glucose drink acted as a reference group.

The work-rates of all the players were analysed by means of video-recordings of players' movements during an entire game. The camera followed one player at a time from an elevated position 20–25 m above the playing pitch and for an entire game. The stride length corresponding to a range of velocities of movement was determined for each player as described by Keane et al. (1993). Motion categories included walking, jogging, striding, sprinting, moving with the ball, moving sideways, walking backwards, jogging backwards. Time 'static' represented periods >1 s when the subject was still and pauses for injury were recorded separately. Playing activities such as jumping, tackling, frees conceded, frees won and pick-ups were also monitored. Ball contacts were divided into movements with the ball and single contacts (one-touch). The profiles were divided into 10-min blocks: the profile for the final 10 min of the second half was compared with the first 10 min for evidence of a 'fatigue factor'.

A 5% glucose solution (150 ml) was ingested by the experimental subjects 5 min prior to the start of the game and again at half-time. The subjects undertook this procedure twice and their work-rate profiles were compared to those players of the same standard who were not given the solution. No attempt was made to change the practices of members of the reference group, who generally did not utilize ergogenic drinks.

Data were analysed using analysis of variance. When significant differences were detected, the least significant difference test was used to isolate the experimental effect.

3 Results

The work-rate profiles are presented for the experimental group in Table 1 and for the reference group in Table 2. The total distance covered for the group receiving the CHO supplement was 7898 (± SD = 1177) m compared with 7575 (± 1207) m in the reference group. Overall, work-rate indices did not differ between the groups except for a higher total frequency of activities and a shorter mean distance walking for the group ingesting the drink supplement ($P < 0.05$).

The club players covered significantly less distance than the inter-county players ($P < 0.05$). The reduction in distance covered averaged 1854 m for the experimental groups and 526 m for the groups not given the carbohydrate solution.

Comparison of profiles between the opening 10 min of the game and the final 10 min displayed evidence of fatigue occurring. This fatigue effect was not offset by ingesting the carbohydrate drink and was evident also among the club players.

Table 1. Work-rate and team-motion profile of inter-county players (n = 8) following ingestion of a carbohydrate supplement.

Activity	Frequency	Mean Distance (m)	Total Distance (m)	% Total Distance	Mean Time (s)	Total Time (s)	% Total Time
Walking	277 ± 45	10 ± 1.6	2774 ± 550	35.4 ± 6.9	7 ± 0.8	1894 ± 354	46.0 ± 6.8
Jogging	187 ± 38	17 ± 2.6	3144 ± 814	39.3 ±7.2	5 ± 0.7	924 ± 250	22.7 ± 66
Striding	43 ± 6	15 ± 2.4	634 ± 160	8.1 ± 2.6	3 ± 0.4	134 ± 31	3.3 ± 0.9
Sprinting	15 ± 3	15 ± 2.1	233 ± 57	3.1 ± 1.0	3 ± 0.4	42 ± 12	1.0 ± 0.3
Sideways	37 ± 8	3 ± 0.5	113 ± 34	1.7 ± 0.6	2 ± 0.4	59 ± 19	1.5 ± 0.5
Walk Back	158 ± 42	5 ± 2.3	716 ± 202	8.7 ± 4.6	3 ± 0.9	529 ± 189	15.9 ± 4.1
Jog Back	35 ± 7	5 ± 1.4	173 ± 68	2.2 ± 0.7	2 ± 0.4	73 ± 24	1.8 ± 0.6
With/Ball	7 ± 4	15 ± 4.6	111 ± 59	1.4 ± 0.7	3 ± 0.7	20 ± 10	0.5 ± 0.2
Static	78 ± 29	-	-	-	4 ± 0.8	300 ± 151	7.3 ± 3.5

Table 2. Work-rate and time-motion profile of inter-county players (n = 56) not administered a carbohydrate supplement.

Activity	Frequency	Mean Distance (m)	Total Distance (m)	% Total Distance	Mean Time (s)	Total Time (s)	% Total Time
Walking	241 ± 44	12 ± 3.0	2700 ± 551	36.9 ± 7.0	8 ± 2.0	1778 ± 501	48.5 ± 6.1
Jogging	189 ± 35	17 ± 2.8	3145 ± 823	40.0 ± 6.0	15 ± 2.1	932 ± 246	24.0 ± 6.1
Striding	47 ± 11	14 ± 2.3	636 ± 167	8.5 ± 2.4	3 ± 0.5	134 ± 32	4.0 ± 1.0
Sprinting	15 ± 4	14 ± 3.3	210 ± 73	3.0 ± 1.0	3 ± 0.7	40 ± 11	0.7 ± 0.3
Sideways	39 ± 16	4 ± 1.6	141 ± 60	1.8 ± 0.8	2 ± 0.9	58 ± 24	1.5 ± 0.7
Walk Back	131 ± 47	4 ± 1.5	540 ± 322	6.9 ± 3.2	3 ± 0.8	452 ± 174	11.7 ± 4.3
Jog Back	30 ± 12	4 ± 1.3	131 ± 78	1.7 ± 0.9	2 ± 0.4	57 ± 28	1.5 ± 0.3
With/Ball	9 ± 4	14 ± 4.1	121 ± 60	1.6 ± 0.8	3 ± 0.8	25 ± 12	0.7 ± 0.3
Static	63 ± 27	-	-	-	4 ± 1.1	258 ± 169	6.5 ± 4.3

Table 3. Work-rate and time-motion profile of senior club players (n = 8) who ingested a carbohydrate supplement before and during a competitive match.

Activity	Frequency	Mean Distance (m)	Total Distance (m)	% Total Distance	Mean Time (s)	Total Time (s)	% Total Time
Walking	207 ± 12	11 ± 1.1	2277 ± 126	37.9 ± 9.5	9 ± 1.7	1746 ± 266	45.5 ± 6.8
Jogging	158 ± 48	16 ± 1.4	2528 ± 929	39.8 ± 7.6	5 ± 0.6	829 ± 318	21.7 ± 8.4
Striding	14 ± 4	12 ± 3.0	254 ± 52	4.2 ± 0.9	3 ± 0.4	38 ± 9	1.0 ± 0.3
Sprinting	12 ± 1	11 ± 2.7	131 ± 48	2.0 ± 0.8	2 ± 0.4	27 ± 5	0.7 ± 0.1
Sideways	59 ± 8	3 ± 0.5	177 ± 34	3.9 ± 0.6	2 ± 0.4	55 ± 19	1.5 ± 0.5
Walk Back	96 ± 19	4 ± 1.4	428 ± 215	6.7 ± 2.6	4 ± 0.8	415 ± 150	10.8 ± 3.9
Jog Back	23 ± 12	6 ± 1.8	145 ± 109	2.2 ± 1.3	2 ± 0.3	51 ± 29	1.3 ± 0.8
With/Ball	14 ± 4	12 ± 3.0	154 ± 52	3.2 ± 0.9	3 ± 0.4	38 ± 9	1.0 ± 0.3
Static	72 ± 22	-	-	-	7 ± 2.7	525 ± 343	13.6 ± 8.8

Table 4. Work-rate and time-motion profile for senior club players (n = 14) without a carbohydrate supplement.

Activity	Frequency	Mean Distance (m)	Total Distance (m)	% Total Distance	Mean Time (s)	Total Time (s)	% Total Time
Walking	218 ± 16	12 ± 2.4	2561 ± 458	36.9 ± 5.9	8 ± 1.8	1806 ± 299	48.5 ± 6.1
Jogging	175 ± 31	17 ± 2.8	3010 ± 819	42.8 ± 7.5	5 ± 0.8	922 ± 231	24.6 ± 6.1
Striding	37 ± 10	4 ± 1.3	413 ± 117	5.9 ± 1.2	3 ± 0.4	95 ± 19	2.5 ± 0.5
Sprinting	12 ± 3	11 ± 2.7	34 ± 48	2.0 ± 0.7	2 ± 0.7	27 ± 10	0.7 ± 0.3
Sideways	38 ± 8	4 ± 0.8	145 ± 53	2.1 ± 0.7	2 ± 0.7	69 ± 23	1.8 ± 0.6
Walk Back	97 ± 15	4 ± 0.4	397 ± 60	5.9 ± 1.4	4 ± 0.5	358 ± 56	9.6 ± 1.6
Jog Back	31 ± 15	5 ± 1.8	160 ± 106	2.4 ± 1.6	2 ± 0.5	61 ± 34	1.7 ± 0.9
With/Ball	11 ± 6	11 ± 2.9	123 ± 58	1.7 ± 0.8	3 ± 0.5	28 ± 16	0.8 ± 0.4
Static	72 ± 37	-	-	-	5 ± 1.7	379 ± 230	10.1 ± 6.3

Table 3 records the work-rate profile of senior club players who took the CHO supplement (mean overall distance ± SD = 6178 ± 1238 m), while Table 4 presents the work-rate profile of players (mean distance 6952 ± 1071 m) who did not have a CHO supplement. Striding proved to be the only significantly different activity between the groups (P < 0.05). In this case the players without the supplement recorded significantly higher values for the various indices employed, i.e., they spent more time striding, and covered more distance when striding during the game than players who did ingest the solution. No other distances between the groups were found.

4 Discussion

In this study an attempt was made to examine the effects of carbohydrate supplementation on the performance of Gaelic footballers. The amount ingested was well below the maximum rate at which carbohydrate can be utilised during exercise (Coggan and Coyle, 1981) but was realistic in a competitive context. The utilisation of a cross-over design was not feasible during this research; besides, such a design can still have errors associated with variability between games and with the quality of opposition. The cross-sectional design is based on the efficacy of separating the 'experimental' group from the typical performance profile determined in reference subjects by boosting the capabilities of the former. The outcome was that the performance profiles were no different in the experimental subjects, except in a minor way. The total distance covered by the players using the supplement was 7898 ± 1177 m which contrasts with the 7575 ± 1208 m by the reference group, the difference being non-significant (P > 0.05).

Fatigue, as evidenced by a fall in the work-rate, occurred irrespective of whether a supplement was used or not. In the present circumstances it is unlikely that this decrement was attributable to reduced muscle glycogen stores or altered blood glucose concentrations, since the performance profile was similar in the two groups. The change in work-rate profile may reflect the impending culmination of the match and a switch to a defensive strategy by the team winning. Such a strategy for protecting a lead would reduce the effective area of play and players would cover less distance in this period. Additionally, the losing team might transfer the ball more quickly into attacking areas by long kicks, avoiding the necessity of players to travel with the ball and therefore decreasing the distance covered.

In conclusion, fatigue (as reflected in a decline in work-rate) was evident towards the end of a Gaelic football game. A team that could maintain a high tempo of play for the whole game may therefore gain an advantage. It is likely that improving the standard of physical fitness, particularly aerobic fitness, is the means for offsetting fatigue while competitive matches are restricted to 60 min duration.

5 References

Coggan, A.R. and Coyle, E.F. (1991) Carbohydrate ingestion during prolonged exercise: Effects on metabolism and performance, *Exercise and Sport Science Reviews,* 19, 1–40.

Florida-James, G. and Reilly, T. (1995) The physiological demands of Gaelic football, *British Journal of Sports Medicine*, 29, 41–45.

Keane, S., Reilly, T. and Hughes, M. (1993) A work rate and match analysis of Gaelic football, *Australian Journal of Science and Medicine in Sport*, 25, (4), 100–102.

Kirkendall, D.T. (1993) Effects of nutrition on performance in soccer, *Medicine and Science in Sports and Exercise*, 25, 1370–1374.

Leatt, P. and Jacobs, I. (1989) Effect of glucose polymer ingestion on glycogen depletion during a soccer match, *Canadian Journal of Sports Sciences*, 14, 112–116.

Saltin, B. (1973) Metabolic fundamentals in exercise, *Medicine and Science in Sport*, 5, 137–146.

39 EFFECTS OF β-HYDROXY β-METHYLBUTYRATE ON MUSCLE METABOLISM DURING RESISTANCE EXERCISE TRAINING IN RUGBY UNION PLAYERS

W. SAMBROOK
University College, Chester, UK

1 Introduction

Whilst the anabolic effects of leucine have been well documented, it is not until recently that these effects have come under scrutiny. Research by Frexes-Steed et al. (1992) has shown that the anabolic effects of leucine have only been seen during conditions of high stress in the muscle. It is hypothesised, however, that the anabolic effect upon muscle occurs post-training (Biolo et al., 1995), and thus Nissen et al. (1996) suggested that a metabolite of leucine may be responsible for the anabolic effect upon muscle through the inhibition of protein breakdown. This product, as suggested by Nissen et al. (1996), is β-hydroxy β-methylbutyrate (HMB).

The production of HMB, through leucine transamination via a mid-product of α-ketoisocaproate (KIC), occurs mainly in the liver, but also in the cytosol and mitochondria of muscle cells. The production of HMB is a very inefficient process forming only 5–10% of the metabolism of KIC (Van Koevering and Nissen, 1992), therefore providing only a small endogenous source. It is thus necessary to supplement the diet with exogenous HMB, if the proposed anabolic effects of HMB on muscle mass are to be studied (Nissen et al., 1996).

This study used a group of previously resistance trained subjects to investigate the anabolic effects of lean muscle mass. The lack of a sport specific ergogenic aid in Northern hemisphere rugb, made Rugby Union players an ideal subject group. This was later substantiated in a newspaper article (Cain, 1998) published after the completion of this study, highlighting that the England team is increasingly outgunned by the striking physical development of opponents using dietary supplements. The article named both creatine and HMB as ergogenic aids currently used by Southern hemisphere teams.

This study investigated the effects of the dietary supplement HMB on muscle metabolism. This involved measuring the extent of lean muscle anabolism and investigating the process through which such lean muscle anabolism occurs. The potential increase in lean muscle mass was investigated through the manipulation of the independent variable, the dietary supplementation of HMB, and run in conjunction with a resistance training programme. The increase in lean muscle mass was estimated through two dependent variables measured. The hypothesised process of HMB inhibition of muscle proteolysis, by which the increase in muscle anabolism occurs, was assessed through the activity of two further dependent variables, the muscle isoforms creatine phosphokinase (CPK) and lactate dehydrogenase (LDH).

2 Methods

2.1 Subjects
Twelve male back row Rugby Union forwards with similar resistance training histories, from the North of England First Division participated in the study.

2.2 Design
A placebo controlled repeated measures design with a double-blind supplementation programme was used.

2.3 Measurements
Dependent variables were body composition, muscle girth, CPK and LDH activity. Independent variables were HMB supplementation (to six subjects, with six subjects fed a carbohydrate placebo), standardised resistance training programme.

2.4 Procedure
The subjects either ingested 1.5 g of HMB or dextrose monohydrate placebo, with 150 ml of water before breakfast each day over the four-week supplementation period. A standardised resistance training programme was run simultaneously to the supplementation period. The resistance training programme consisted of 10 exercises performed over 4 sets of 8 repetitions at 85% of the 1RM (determined at the initial testing session). Subjects followed a diet consisting of 60% carbohydrate, 20% fat and 20% protein. A food diary was kept by each subject for 24 h prior to each session of testing, and checked by the investigator. Subjects were tested at 2-week intervals before, during and post-supplementation for the following measures. Body composition was measured using skinfold and bioelectrical impedance techniques. Body fat was calculated using Durnin and Womersley's (1974) four-site method. Both CPK and LDH measurements were taken from blood serum measures via finger pricks. Muscle girth was measured using a non-elastic tape measure on four different sites. Statistical analysis was calculated through a two-way repeated measure on one-factor ANOVA.

3 Results

The lean muscle values and the body fat percentages are an average of the bioelectrical and skin fold methods used. The units of CPK activity are convertible to the change in micrograms of Pi, using the Sigma (1997) tables. One unit of LDH activity is defined as that amount of enzyme that will catalyse the formation of one micro mole of NADH per minute under the conditions of the assay procedure (Sigma, 1997). The significance value (P) is calculated using one-factor repeated measures ANOVA.

Table 1. Summary of mean body weight, lean muscle mass, body fat and muscle enzyme activity changes over the four-week test period.

Variable	Pre	HMB Group (n = 6)			Pre	Placebo Group (n = 6)			Sig (P)
		Post	Diff kg	Diff %		Post	Diff kg	Diff %	
Weight (kg)	83.7	83.8	+0.1	+0.1	89.5	89.7	+0.2	+0.2	ns
Lean Muscle (kg)	70.9	71.7	+1.0	+1.4	71.7	71.8	0	0	ns
Body Fat (%)	14.3	14.0	−0.3	−2	19.5	19.3	−0.2	−1	0.05
CPK Activity, (U·ml⁻¹)	217.2	161.8	−161.5	−74.7	218.6	220.3	+1.7	+0.8	0.001
LDH Activity, (U·l⁻¹)	161.8	72.8	−89	−55	161	154.7	−6.3	−3.9	0.001

(ns = not significant)

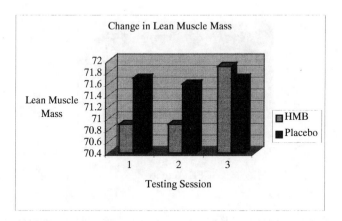

Figure 1. Change in lean muscle mass over the four-week supplementation period.

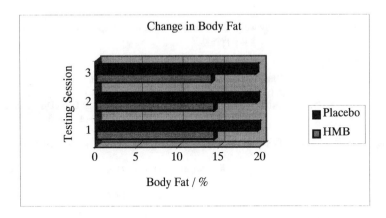

Figure 2. Change in body fat over the four-week supplementation period.

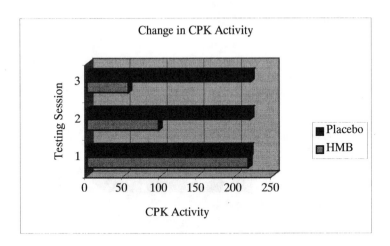

Figure 3. Change in CPK activity over the four-week supplementation period.

4 Discussion

The significance of the decrease in both the CPK and LDH activity levels in the HMB group is indicative of a decrease in levels of muscle damage (McDowall, 1996). It follows that if the level of muscle damage decreased, then by association the level of muscle protein damage decreased. However, the continuous repair process of synthesis and breakdown maintains the integrity of muscle protein A decrease in the concentration of protein breakdown, occurring concurrently with an increase in protein synthesis, caused by resistance training (MacDougall et al., 1980), leads to a net positive balance of protein synthesis. It is through this reckoning that Nissen et al. (1996) suggested that HMB is associated with the inhibition of muscle protein breakdown (proteolysis). Nissen et al. (1996) further stated that it is through the inhibition of muscle protein breakdown that HMB forms a net muscle protein level and thus an increase in lean muscle mass development.

The HMB group experienced a mean increase of 1.4% in lean muscle mass, whilst the placebo group level remained constant. Despite the difference in the results, there was no statistical difference between the groups, as only 75% of the subjects experienced increases in lean muscle mass. This study is unable to support the current theory that HMB increases lean muscle development at a rate above that of normal (Nissen et al., 1997).

As with Nissen and co-workers' (1997) study, all the subjects showed a decrease in body fat. This study, however, found a 1% greater decrease in the HMB than placebo group. This has not previously been shown and clearly requires further research to establish the process by which HMB catalyses this effect. In terms of rugby players, the increase of lean muscle mass at the expense of fat mass is crucial to the players' development of power. However, the advantages of an ergogenic aid, which decreases fat mass, with little or no effect upon lean mass are less specific to Rugby Union.

5 Conclusion

The significant decreases in CPK and LDH activity encountered through HMB supplementation in this study were not paralleled by a significant concomitant increase in lean muscle mass that was experienced in the study of Nissen et al. (1996) study. The lack of statistical significance in the increase in lean muscle mass indicates that although a significant decrease in muscle plasma enzymes was encountered, this study is unable to support the theory of muscle anabolism through HMB suppression of protein breakdown (Nissen et al., 1996).

Further detailed research into the biochemical effect of HMB is recommended. Such research may well be aimed at the effect of the HMB butyrate ion on gene expression. A great deal of research has surrounded butyrate and gene expression, but not the HMB butyrate ion and gene expression. This may provide a possible pathway through which muscle anabolism has previously been shown to be enhanced.

5 References

Biolo, G., Maggi, S.P., Williams, B.D., Tipton, K.D. and Wolfe, R.R. (1995) Increased rates of muscle protein turnover and amino acid transport after resistance exercise in humans, *American Journal of Physiology*, 268, E514–E520.

Cain, N. (1998) Why our boys were lost in land of giants, *Sunday Times*, 12–04–998.

Durnin, J.V.G.A. and Womersley, J. (1974) Body fat assessed from total body density and its estimation from skinfold thickness, *British Journal of Nutrition*, 32, 77–92.

Frexes-Steed, M., Lacy, D.B., Collins, J. and Abumrad, N.N. (1992) Role of leucine and other amino acids in regulating protein metabolism in vivo, *American Journal of Physiology*, 262, E925–E935.

MacDougall, J.D., Elder, G.C.B., Sale, D.G., Moroz, J.S. and Sutton, J.F. (1980) Effects of strength training and immobilisation on human muscle fibres, *European Journal of Applied Physiology and Occupational Physiology*, 43, 25–34.

McDowall, I.L. (1996) Clinical Laboratory Techniques, Personal Communication, unpublished manuscript.

Nissen, S.L., Sharp, R., Ray, M., Rathmacher, J.A., Rice, D., Fuller, Jr., J.C., Connelly, A.S. and Abumrad, N. (1996) Effect of leucine metabolite β-hydroxy β-methylbutyrate on muscle metabolism during resistance-exercise training, *Journal of Applied Physiology*, 81, 2095–2104.

Sigma Diagnostics. (1997) Quantitative determination of serum creatine kinase and lactate dehydrogenase, Sigma Diagnostics, unpublished manuscript.

Van Koevering, M. and Nissen, S. (1992) Oxidation of leucine and alpha-ketoisocaproate to β-hydroxy β-methylbutyrate in vivo, *American Journal of Physiology*, 262, 27–31.

40 DIETARY ANALYSIS OF ENGLISH FEMALE SOCCER PLAYERS

D. SCOTT*, P.J. CHISNALL* and M.K. TODD**
*   Faculty of Health and Exercise Sciences, University College Worcester,
     Worcester, UK
**   Department of Exercise Sciences, University of Southern California,
     Los Angeles, USA

1 Introduction

Girls' and women's soccer is enjoying a period of unprecedented popularity in England, with 34,000 females currently registered with clubs (English F.A., 1998). Soccer participation results in an increased energy demand that must be matched by an increased energy intake in order to maintain performance and desired body weight. Despite the increased participation of females in soccer, there is a clear lack of research concerning the nutritional needs and practices of female soccer players. There has been some research into the dietary intake of male soccer players (Jacobs et al., 1982; Bangsbo, 1994; Maughan, 1997), although this is also limited and it is acknowledged that little is known about the nutritional status of elite performers in soccer (Maughan, 1997). The sparse information regarding the nutritional habits of female soccer players is often adapted from research into male soccer players or from other female sporting populations. In order to further optimise performance in training and competitions, there is clearly a need to investigate the demands and dietary habits of female soccer players.

2 Methods

Seventy-two English female soccer players took part in the current investigation, 26 from National Premier Division clubs (DP) (22.2 ± 6.2 years) and 46 from National Division One clubs (D1) (22.8 ± 6.0 years). The collection and analysis of data occurred during the second half of the 1997/98 competitive season, at a time when all players were engaged in full training and match play. Players were issued with food diaries and a comprehensive set of instructions on the procedures to be followed, and were required to record all food and drink consumed for two consecutive days. The completed dietary records were analysed using a computerised version of food composition tables (CompEat). Upon collection of the food diaries, height (Seca stadiometer), body mass (Seca scales) and skinfold thickness (Harpenden calipers) values were measured on all players in the evening before a training session. Percent body fat was estimated from skinfold thickness measured at four sites (Durnin and Womersley, 1974).

Boxplots of the data revealed the presence of outliers, and so 20% trimmed means ($\bar{x}_t$) are presented. The two groups were compared using trimmed means according to the method of Yuen (1974), which maintains good control over the probability of making a type I error when comparing heavy-tailed distributions (Wilcox, 1996).

3 Results

The physical characteristics of the players are shown in Table 1. There were no significant differences (P < 0.05) between any of these variables for players from the two groups.

Table 1. Physical characteristics of players from the National Premier Division (DP) and National Division One (D1) clubs. Data presented as trimmed means ($\bar{x}_t$).

	Premier Division (DP)		Division One (D1)	
	$\bar{x}_t$	SD	$\bar{x}_t$	SD
Age (years)	22.2	6.2	22.8	6.0
Height (cm)	163.2	5.7	163.8	5.9
Body mass (kg)	63.3	6.2	61.3	6.3
Body fat (%)	24.2	3.8	26.0	3.7

Results for dietary energy and macronutrient intake, relative to body mass, are shown in Table 2. There were no significant differences (P < 0.05) between the two groups for energy intake or any of the macronutrients when expressed relative to body mass. However, relative energy intake and carbohydrate were lower than baseline recommended intakes for females involved in sport, 47–60 kcal·day^{-1} (197–251 kJ·kg^{-1}·day^{-1}) and 6 g·kg^{-1}·day^{-1} respectively (Economos et al., 1993; Burke, 1995)

When the dietary intake of the macronutrients was expressed as a percentage of total energy intake (Table 3), carbohydrate contributed a significantly greater percentage of total energy for Premier Division players (54.6 ± 5.9%) than Division one players (47.8 ± 9.8%, P < 0.001). There were no significant differences in the fractions of total energy intake accounted for by fat, protein or alcohol between the two groups.

Table 2. Daily energy and macronutrient intake, relative to body mass, of players from the National Premier Division (DP) and National Division One (D1) clubs.

	Premier Division (DP)		Division (D1)	
	$\bar{x}_t$	SD	$\bar{x}_t$	SD
Energy (kcal·kg^{-1})	28.1	11.3	29.3	9.9
Protein (g·kg^{-1})	1.0	0.4	1.1	0.4
Fat (g·kg^{-1})	1.0	0.5	1.1	0.6
Carbohydrate (g·kg^{-1})	3.8	1.2	3.6	1.2

Table 3. Percent contribution of the macronutrients to energy intake for the National Premier Division (DP) and National Division One (D1) clubs.

Macronutrient	Premier Division (DP)		Division (D1)	
	$\bar{x}_t$	SD	$\bar{x}_t$	SD
Protein	13.9	4.0	14.6	3.2
Fat	31.0	6.9	33.3	9.5
Carbohydrate	54.6	5.9	47.8	9.8
Alcohol	0.1	1.5	3.8	7.4

Table 4 shows daily micronutrient intake for the two groups. There were no significant differences for the intakes of iron, vitamin C or calcium between the two groups. However, both groups consumed mean iron intakes approximately 40% below the Reference Nutrient Intake (RNI) of 14.8 mg·day^{-1} (COMA, 1991).

Table 4. Daily micronutrient intake of players from the National Premier Division (DP) and National Division One (D1) clubs.

Micronutrient	Premier Division (DP)		Division (D1)	
	$\bar{x}_t$	SD	$\bar{x}_t$	SD
Vitamin C (mg)	78.7	109.5	61.9	51.6
Iron (mg)	8.81	3.27	8.70	3.16
Calcium (mg)	707	355	669	284

4 Discussion

The results of the current investigation suggest that the nutritional practices of English female soccer players are sub-optimal. Borgen and Corbin (1987) stated that a low energy intake is less common in sports such as soccer where a very low body fat percentage is not usually perceived as being crucial to performance. However, the current results do not support this view, and on average the two groups consumed an energy intake, relative to body mass, less than the recommended range of 47–60 kcal·kg⁻¹·day⁻¹ (Economos et al., 1993). This is similar to the findings of other studies investigating the nutritional practices of female sporting populations. Hickson et al. (1986) found female basketball players consumed energy intakes that were lower than estimated energy expenditure. Similarly, Clark (1993) reported that female athletes from a range of sports appeared to maintain high training loads on a lower energy intake than might be expected. Low energy intakes will cause a negative energy balance, which could lead to undesired weight loss, and adverse effects on performance due to lower energy levels. The intermittent nature and the high physical demands of the game are such as to require a high rate of energy production. If this energy expenditure is not matched by energy intake, then performance levels will have to be lowered and fatigue will occur earlier.

In the current study, Premier Division players achieved a significantly greater percentage of energy from carbohydrates than Division One players ($P < 0.05$), although both groups only consumed a level recommended for the general population (BNF, 1992). Repeated bouts of high intensity exercise during training and competition will deplete the muscle and liver glycogen stores, and carbohydrate depletion may contribute to fatigue and reduced capability for performance during a soccer match (Bangsbo, 1994). Some authors have recommended that athletic populations should consume at least 6 g·kg⁻¹·day⁻¹ of carbohydrate for low intensity exercise (Economos et al., 1993; Burke, 1995), and thus, from Table 2, it can be seen that on average both groups were consuming below this value. The use of central measures of location in the statistical analysis slightly obscures the extremely large inter-individual variability observed within the sample. Some players in the present investigation reported a carbohydrate intake of less than 40% of total energy intake, with the lowest individual report of 14.2%. Accounting for error, it is apparent that some of the players analysed are clearly consuming a diet that does not provide sufficient carbohydrate to maintain optimal performance in training and competition. Burke (1995) recommended 1 g·kg⁻¹·day⁻¹ protein intake for participation in general sports activity. Both groups, it seems, achieved these protein intake criteria. Protein and fat intake were also highly variable between individual players, and it is noted that some individuals were consuming low protein intakes, 0.3 g·kg⁻¹·day⁻¹ being the lowest intake. Since some of the subjects in this study were 16–18 years old girls, this could pose problems for growth and development.

There were no significant differences between the two groups for the intakes of any of the micronutrients analysed. However, iron intake for both groups was approximately 40% below the Reference Nutrient Intake of 14.8 mg (COMA, 1991). Low levels of iron intake among female athletes have been commonly reported in previous research (Banister and Hamilton, 1985; Nutter, 1991). Low iron intakes can lead to lower haemoglobin levels and in the long term, anaemia, subsequently causing fatigue and lethargy. However, iron supplementation should not be recommended solely on dietary analysis, and physiological or haematological assessment of the status of the individual should always be made before an increased dietary intake of any vitamin or mineral is recommended.

5 Conclusion

These results show dietary intakes for female soccer players, which are similar to that of the general population in the United Kingdom. Indeed, Pate et al. (1993) found sedentary females to consume 26 kcal (109 kJ)·kg^{-1}·day^{-1}, which is only slightly less than those reported for the current groups. Furthermore, the contributions from the macronutrients are similar to those for normal, healthy individuals. It is apparent that some soccer players consume a diet that supplies less carbohydrate than is normally recommended to sustain performance in training and competition. Female soccer players should be encouraged to consume energy intakes which match levels of expenditure, in order to achieve a eucaloric profile and thus maintain high levels of performance and desired body weight. Furthermore, carbohydrate intakes should be increased to at least 6 g·kg^{-1}·day^{-1} (Economos et al., 1993; Burke, 1995) in order to provide fuel for repeated bouts of multiple sprint exercise. It is also clear that the players, families, and coaching staff should be educated in the field of nutrition, and how the diets of players can be adapted accordingly. Research on female soccer is in its infancy and further research can only serve to enable the current playing standards to flourish.

6 References

Bangsbo, J. (1994) *Fitness Training in Footbal,* HO+Storm, Bagsvaerd.

Banister, E.W. and Hamilton, C.L. (1985) Variations in the iron status with fatigue modelled from training in female distance runners, *European Journal of Applied Physiology*, 54, 16–23.

Borgen, J.S. and Corbin, C.B. (1987) Eating disorders among female athletes, *Physician and Sportsmedicine*, 15, 89–95.

British Nutrition Foundation (BNF) (1992) *Nutritional Requirements*, BNF.

Burke, L. (1995) *The Complete Guide to Food for Sports Performance*, Allen and Unwin, NSW.

Clark, N. (1993) Nutritional problems and training intensity, activity level and athletic performance, in *The Athletic Female* (ed. A.J. Pearl), Human Kinetics, Champaign, IL, pp. 165–168.

Committee on Medical Aspects of Food Policy (COMA) (1991) *Dietary Reference Values for Food Energy and Nutrients for the United Kingdom*, HMSO, London.

Durnin, J.V.G.A. and Womersley, J. (1974) Body fat assessed from total body density and its estimation from skinfold thickness: measurements on 481 men and women aged from 16 to 72 years, *British Journal of Nutrition*, 32, 77–97.

Economos, C., Bortz, S.S. and Nelson, M.E. (1993) Nutritional practices of elite athletes: Practical recommendations, *Sports Medicine*, 16, 381–399.

English Football Association (1998) *Talent Development Plan*, Football Association, London.

Hickson, J.F., Schrader, J. and Trischler, L.C. (1986) Dietary intakes of female basketball and gymnastic athletes, *Journal of the American Dietetic Association*, 86, 251–253.

Jacobs, I., Westlin, N., Karlsson, J., Rasmusson, M. and Houghton, B. (1982) Muscle glycogen and diet in elite soccer players, *European Journal of Applied Physiology*, 48, 297–302.

Maughan, R.J. (1997) Energy and macronutrient intakes of professional football (soccer) players, *British Journal of Sports Medicine*, 31, 45–47.

Nutter, J. (1991) Seasonal changes in female athletes' diets, *International Journal of Sports Nutrition*, 1, 395–407.

Pate, R.R., Miller, B.J., Davis, J.M., Slentz, C.A. and Klingshirn, L.A. (1993) Iron status of female runners, *International Journal of Sports Nutrition*, 3, 222–231.

Wilcox, R.R. (1996) *Statistics for the Social Sciences*, Academic Press Inc., California, USA.

Yuen, K.K. (1974) The two sample trimmed t for unequal population variances, *Biometrika*, 61, 165–170.

PART SIX

Paediatric Science and Football

41 DETERMINING THE ROLE OF THE PARENTS IN PREPARING THE YOUNG ELITE PLAYER FOR A PROFESSIONAL A. F. L. CAREER

B. GIBSON
Edith Cowan University, Perth, Western Australia

1 Introduction

With the expansion of Australian Rules Football into a national competition and the development of the Australian Football League (AFL), the opportunity for players to undertake a full time professional career as an Australian Rules player is increasingly attractive. Sponsorship, television rights and increased membership have provided the finance that is necessary for players to be contracted on a full time basis. This new structure has seen the introduction of the drafting system and the competitive recruiting of the most promising young players across Australia. The intensity of the competition has led to players as young as 16–17 years of age becoming the targets of the AFL recruiting agents. Being drafted by a club results in a significant change to the lifestyle and routines of these young players. In some instances it may even mean a change of location as they move interstate to take up the appointment with their new club. Because of the demands that this sudden transition makes, the role of the parents is significant in shaping the level of success that the player can experience. The problem was to define that role and establish guidelines for parents to follow.

Therefore the purpose of this study was twofold. The first was to identify the role that the parents and family play in assisting the young player at the junior level. The second was to examine the role of the parents in assisting their son in making the transition and pursuing a professional football career in the AFL.

2 Methods and Procedures

The sample included the parents of 30 players who had been drafted and were playing with AFL clubs. The specific breakdown of the parents interviewed is seen in Table 1. The purpose of the study was explained to the parents when they were first approached and asked to become participants in the study. The parents were interviewed (together where possible) using the same standardised set of questions and all interviews were tape recorded so as to permit an accurate post-interview analysis to be performed. Interview times varied from 60-90 min.

Table 1. Breakdown of parents interviewed

Number of families = 30
Number of mothers = 28
Number of fathers = 30

3 Results and Discussion

3.1 The early years
Parents in this study were asked to reflect on the early years of their son's involvement in sport and to indicate what factors they emphasised during that time. The parents indicated that during the early years at the junior level:
- they all encouraged their son to play a wide range of games and sports;
- they endeavoured to make participating in the game situation a fun experience;
- they placed emphasis on involvement and enjoyment rather than the winning aspect;
- they viewed sport as an environment for developing confidence, and self esteem as well as social and physical skills.

  During the early stages of participating in sport, the parents' role was mainly described as one of supporting the young son through his new experiences. The parents were intent on providing a positive environment, fostering the youngster through the stages of development and enabling their son to follow his interests in a range of sporting activities. Stress was not a major consideration as the emphasis was on enjoyment and reducing social evaluation. The experiences of these parents are supported by research which has demonstrated that there is every likelihood of 'burn out' occurring amongst young children who become over-involved in a single sport (Loehr and Kahn, 1987; Gould, 1993)

3.2 The transitional stage to senior football
The transition to considering football as a career opportunity occurs around 16–18 years. In a number of cases the parents indicated that their son gave up all other commitments at this age to concentrate on his physical preparation and skill development in an attempt to be successful in the annual draft. The decision to choose a professional career in football was viewed as a serious commitment and in the majority of cases the decision was made only after extended consultation between the parents and the son where a number of issues were considered.

3.3 Issues for consideration

The issues considered were as follows:

- the demands and sacrifices that needed to be made to reach the top;
- the strong competition for the limited number of places offering significant financial rewards;
- the pressures that are applied from numerous directions including parents, coaches, spectators, sponsors and fans;
- the continual evaluation to which the young player is subjected.

3.4 Parents' role

While providing support and encouragement for their son to take up the professional playing career, parents were conscious of not placing undue pressure on their son.

In making the decision the parents' role included assessing a number of factors that would have an important bearing on future outcomes.

These factors included:

- the required financial outlay;
- the real and imagined gains;
- the chances of success;
- the implications of failure;
- the impact on the family;
- the player's physical capabilities;
- the level of commitment;
- possible career paths after retirement.

3.5 Required financial outlay

This aspect must be considered as substantial costs may be inflicted on the family who may choose to relocate, change employment or even move interstate to be with their son. While the club drafting the young player will meet the majority of costs there will be other costs to the family that need to be taken into account.

3.6 The real and imagined gains and implications of failure

There is a cost of not 'making it', and the statistics demonstrate that a significant number who accept the initial offer, will fail to fulfil their potential and are eventually delisted.

It is difficult to quantify the loss for those who don't make the grade. Often individuals who do not make it, not only lose out on the recognition and financial rewards associated with success, but they also lose a large part of their youth to rigorous training schedules (Lee, 1997). On the other hand success will bring with it financial security, life-long career prospects and notoriety.

3.7 Impact on the family
The family of successful athletes plays a supporting role in each stage of their development (Bloom, 1982; Hellstedt, 1990). This was also true for the parents involved in this study. In an intense sporting environment there are going to be additional commitments for the family and this could result in disruptions to their lifestyle, neglect of other siblings and even create a situation where it becomes impossible for the family to continue as one unit. These factors need to be considered prior to making the final decision.

3.8 Players' physical capabilities and level of commitment
As the Australian Rules Football code is both physically and psychologically demanding, it is important to assess the player's capabilities and level of commitment. It is essential that the player is able to cope with the setbacks, injuries and disappointments which may occur on a regular basis and still maintain his enthusiasm and dedication to the training and playing programme. Parents raised this as a major area of discussion during the lead up to deciding which direction to take.

3.9 Career paths after retirement
Considering that the average time that a player can maintain a playing career at the top level is 5–7 years (Lee, 1997), a career after football is a real consideration. As there is no guarantee that players will be able to take up a football related career with the media, public relations or marketing, it is important to maintain other forms of education and training in preparation for a second career. Numerous parents expressed this concern and saw it as their role to assist in organising ongoing education and training for their son.

The parents in this study were supportive of the decision to play professional AFL football. The following are reasons they provided for their support:
- an AFL football career will provide their son with every opportunity to reach his full potential;
- this profession would provide financial security;
- it was an opportunity to capitalise on his natural ability and experience success.
- it was a chance to fulfil a lifelong goal.

Once the decision was made to take up the offer through the draft system the role of the parent became particularly important in the first few years of their son's playing career. The major role was to establish a support system to facilitate a smooth transition to the new lifestyle. This involved:
- providing financial guidance, selecting a manager and negotiating contracts;
- assisting with a planned education programme or relevant training courses;
- providing guidance through times of loss of form and injury;
- dealing with the politics of sport;
- assisting with the social adjustments and maintenance of a balanced lifestyle;
- assisting with diet and nutritional needs.

While the role of the parents did vary across the group, the factors that have been mentioned here were common in a significant number of cases. It was evident that the role was largely determined by a number of common factors including the player's age, experience, dependence, family relationships, parent's professional expertise and the particular AFL club that the son played with.

4 Summary

The results of this study demonstrated that parents do have a significant role to play, particularly in the early years of professionalism. The most important role was found to be one of support where parents were there to advise and assist when required. While this role varied from family to family, all parents did have a major supporting role to play. It was apparent that this role was dependent on the family's expertise and experience, and the relationship that existed between the parents and the son. This role will naturally vary over time and needs to be revised as the son matures and establishes himself as a full time professional.

5 References

Bloom, B. (1982) The role of gifts and markers in the development of talent, *Exceptional Children,* 48, 511-18.

Gould, D. (1993) Intensive sports participation and the pubescent athlete: Competitive stress and burnout, in *Intensive Participation in Children's Sports* (eds B.R. Cahill and A.J. Pearl) Human Kinetics, Champaign, IL, pp. 19-38.

Hellstadt, J.C. (1990) Early adolescent perceptions of parental pressure in the sport environment, *Journal of Sport Behaviour,* 13, 135-144.

Lee, M. (ed) (1997) *Coaching Children in Sport: Principles and Practice.* E. and F.N. Spon, New York.

Loehr, J.E. and Kahn, E.J. (1987) *Nett Results,* The Stephen Greene Press, Lexington Massachusetts.

42 ANTHROPOMETRIC CHARACTERISTICS OF 11–12 YEAR OLD FLEMISH SOCCER PLAYERS

M. JANSSENS, B. VAN RENTERGHEM and J. VRIJENS
Department of Movement and Sport Sciences, University of Ghent, Belgium

1 Introduction

Large differences in anthropometric characteristics occur during prepuberty and puberty between children of the same chronological age, due to the difference in biological maturity. In several sports, especially in contact sports as soccer, those differences may be of decisive importance. As such, the question arises if certain characteristics accord with the performance level of young soccer players. Do players with a specific body type have a larger chance to be selected for a higher competition level? Or even more, do the body characteristics of young Flemish soccer players show any sport specific profile?

Therefore, the purposes of this study were (a) to investigate the anthropometric profile of 11–12 year old male Flemish soccer players, (b) to investigate if differences in playing level are linked to anthropometric differences, (c) to compare this anthropometric profile with that of the average Flemish boy of the same age, and (d) to compare the profile with that of other soccer populations of the same age.

2 Methods

2.1 Subjects
The subjects were all participants in the Ghent Youth Soccer Project (GYSP), a longitudinal study on growth and performance of youth soccer players. In this study, longitudinal data are available for 165 subjects (11–12 year old male Flemish soccer players) observed annually. The group was divided into 3 levels: level 1 (1st and 2nd league, n = 63), level 2 (3rd and 4th league, n = 65) and level 3 (regional league, n = 37).

2.2 Anthropometric variables
Stature was measured to the nearest 0.1 cm with the subject being stretched upwards to the fullest extent. Body mass was measured to the nearest 0.1 kg. Bi-acromial breadth, bi-cristal breadth, bi-condylar breadth of humerus and femur were used to evaluate breadth development. Muscle development was evaluated by biceps and calf girth. Five skinfold measurements were

obtained with a Harpenden caliper (biceps, triceps, subscapular, supraspinal and calf). Somatotype of the subjects was calculated using the Heath–Carter method (Heath and Carter, 1967). Body mass index (BMI) was calculated. Skeletal age (SA) was assessed using the TW2 system derived by Tanner et al. (1975).

2.3 Statistical analysis
Descriptive statistics (mean ± SD) were estimated for subject characteristics. One-way ANOVA was applied when comparing players of different playing levels. Probability levels of 0.05 were used for statistical significance.

3 Results

Descriptive statistics for all variables according to playing level are given in Table 1. The 11–12 year old male soccer players exhibited a mean overall height of 150.7 cm and a body mass of 40.1 kg. Their mean skeletal age was 12.3 years, while the mean calendar age was 12.2 years. The mean Body Mass Index was 17.4 and the mean somatotype was 2.4 – 4.0 – 3.8.

When comparing soccer players of different playing levels, significant differences were found for body mass, sum of skinfolds, biacromial breadth, bicondylar breadth humerus, BMI and endomorphy. Level 1 players exhibited significantly lower body mass, sum of skinfolds, biacromial breadth, bicondylar breadth humerus, BMI and endomorphic component, when compared to level 3 players. Additionally, level 1 players showed a significantly lower sum of skinfolds and endomorphic component than level 2 players.

No differences in skeletal age could be found between players of different playing levels.

The anthropometric profile of 11-12 year old Flemish soccer players is compared with age-related norms (Beunen et al., 1988 and 1991) in Table 2.

The only remarkable difference between a soccer player and an average boy of the same age, is the sum of skinfolds (35.6 mm for the soccer player vs 45.4 mm for the average Flemish boy). Values for stature, body mass, bicondylar breadth humerus, bicondylar breadth femur, biceps girth and calf girth are similar for soccer players and average boys of the same age.

Table 1. Mean values (± SD) for anthropometric characteristics.

	All subjects (N = 165)	Level 1 (n = 63)	Level 2 (n = 65)	Level 3 (n = 37)
Age (years)	12.2 (0.7)	12.2 (0.6)	12.2 (0.8)	12.2 (0.5)
Skeletal age (years)	12.3 (1.5)	12.3 (1.3)	12.2 (1.5)	12.6 (1.6)
Stature (cm)	150.7 (7.6)	150.6 (7.0)	150.0 (7.9)	152.4 (8.0)
Body mass (kg)	40.1 (7.0)	38.8 (5.7) °	40.1 (7.6)	42.8 (8.0)
Sum of skinfolds (mm)	35.6 (14.8)	30.4 (9.1) *	36.9 (14.3)	43.3 (20.4)
Biacromial breadth (cm)	32.4 (2.1)	32.0 (2.0) °	32.4 (2.0)	33.2 (2.4)
Bicristal breadth (cm)	24.0 (1.6)	23.9 (1.4)	24.0 (1.7)	24.4 (1.9)
Bicon br humerus (cm)	6.0 (0.5)	6.0 (0.4)	5.9 (0.4)	6.2 (0.5)
Bicon br femur (cm)	9.0 (0.5)	8.9 (0.4) °	9.0 (0.5)	9.2 (0.5)
Biceps girth (cm)	22.5 (2.0)	22.3 (1.8)	22.3 (2.1)	23.3 (2.2)
Calf girth (cm)	29.6 (2.3)	29.2 (1.9)	30.0 (2.4)	30.0 (2.8)
BMI	17.4 (2.8)	17.1 (1.5) °	17.7 (2.1)	18.3 (2.5)
Endomorphy	2.4 (1.1)	2.0 (0.7) *	2.5 (1.1)	3.0 (1.4)
Mesomorphy	4.0 (1.0)	3.9 (1.0)	4.0 (1.0)	4.2 (0.9)
Ectomorphy	3.8 (1.2)	4.1 (1.0)	3.7 (1.1)	3.6 (1.5)

° significantly different between level 1 and level 3 (P < 0.05)
* significantly different between level 1 and level 2 and between level 1 and level 3 (P<0.05)

Table 2. Comparative table with age-related norms [mean (SD)].

	GYSP ('96)	Eurofit ('91)	Beunen ('88)
Age (years)	12.2 (0.7)	12	12.5
Stature (cm)	150.7 (7.6)	153.4 (7.2)	148.8 (6.2)
Body mass (kg)	40.1 (7.0)	43.0 (9.2)	38.6 (6.2)
Sum of skinfolds (mm)	35.6 (14.8)	45.4 (26.7)	
Bicon br humerus (cm)	6.0 (0.5)		6.1 (0.4)
Bicon br femur (cm)	9.0 (0.5)		9.0 (0.5)
Biceps girth (cm)	22.5 (2.0)		21.8 (1.9)
Calf girth (cm)	29.6 (2.3)		29.3 (2.1)

Figure 1 shows the somatotype of our population and other youth soccer populations. Our soccer players exhibited a mean somatotype of 2.4–4.0–3.8, indicating that they can be classified as ecto-mesomorph. When comparing the somatotype of our population to those of other youth soccer populations, they all show the same somatotype.

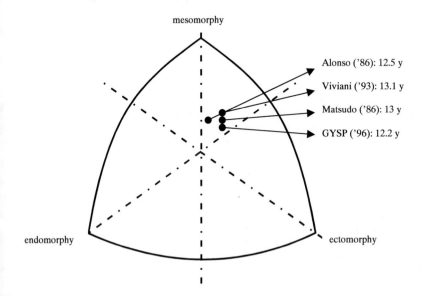

Figure 1. Somatotype: comparison with other youth soccer populations.

4 Discussion

The 11–12 year old soccer players had normal values for stature, weight, breadth measurements, girths of biceps and calf, and skeletal maturation. They had below average values for sum of skinfolds, indicating that the subcutaneaous fat development is the only specific anthropometric characteristic at the age of 12.

When comparing soccer players of different playing levels, no differences in skeletal age could be found. One might expect to find the more mature players in the highest playing levels. This is clearly not the case in 11–12 year players, but might be expected at later stages of puberty, when the maturational differences become more obvious.

As expected, the differences in body fat found between 11–12 year boys playing in the 1st or 2nd national league and their contemporaries playing at lower levels, coincide with the differences found in body mass, BMI, and endomorphy.

The lower (although not significant) mesomorphic component in level 1 players, compared to level 3 players could be expected, based on the differences in breadth and circumference measurements.

In comparison with other young soccer populations, our subjects (GYSP '96) showed the same somatotype. Young soccer players can be classified as ecto-mesomorph.

In conclusion, Flemish 11–12 year old soccer players show average anthropometric values, except for the subcutaneous fat development. According to the Heath-Carter method, they can be classified as ecto-mesomorph.

5 References

Alonso, R.F. (1986) Estudio del somatotipo de los atletas de 12 años de la EIDE occidentales de Cuba, *Boletin de Trabajos de Anthropologial,* 3-18.

Beunen, G.P., Malina, R.M., Van 't Hof, M.A., Simons, J., Ostyn, M., Renson, R. and Van Gerven, D. (1988) *Adolescent Growth and Motor Performance: A Longitudinal Study of Belgian Boys.* Human Kinetics, Champaign, IL.

Beunen, G.P., Borms, J., Vrijens, J., Lefevre, J., Claessens, A.L. (1991) Fysieke fitheid en sportbeoefening van de Vlaamse jeugd van 6 tot 18 jaar, *IOS rapport.*

Heath, B.H. and Carter, J.E.L. (1967) A modified somatotype method, *American Journal of Physical Anthropology,* 27, 57-74.

Matsudo, V.K.R. (1986) Effects of soccer training on adolescents and adults physical fitness characteristics, in *Celafiscs – Des Anos de Contribuição às Ciências do Esporte.* Laboratorio de Aptidao Fisica de Sao Caetano do Sul. Brasil, SP, 298-304.

Tanner, J.M., Whitehouse, R.H., Marshall, W.A., Healy, M.J.R. and Goldstein, H. (1975) *Assessment of Skeletal Maturity and Prediction of Adult Height (TW 2 Method).* Academic Press, New York.

43 SOMATOTYPE CHARACTERISTICS OF YOUNG FOOTBALL PLAYERS

M.TOTEVA
National Sports Academy, Sofia, Bulgaria

1 Introduction

The scientific approach to the selection of adolescent athletes for different sports focuses the attention of physicians and coaches on the morphological features of athletes and on the somatotype characteristics of the respective sports, with respect to eye.

The aims of the present study were to draw up a somatotype characteristics of young football players with respect to age, and to establish the specific somatotype changes occuring during growth with respect to the improvement of methods of selection.

2 Methods

Subjects for the study were 80 boys from sports schools, aged 12–17 years who were engaged in soccer. The somatotype anthropometric method of Heath-Carter (Duquet and Carter, 1990) was used for the purpose of the investigation. The individuals were classified in six age groups, at an interval of one year. The data were submitted to basic statistical analyses. The SDD criterion for assessment of the significance of differences between the somatotypes, was used.

3 Results

The memendomorphy was relatively low (Table 1). The development of this component indicates that no great changes have been found, although a moderate tendency to increase was observed. The significant differences were in the group of 13 year-old boys who have entered maturity and due to this they are less endomorphic.

The mesomorphy is the dominant component which increases slightly with age. The differences in component means are up to one-half unit. Players 15 years old were more mesomorphic than the others (5.05).

The ectomorphy component shows a slowly significant decrease of one unit over the whole age period.

The young athletes at the age of 12 and 13 have a mean mesomorph-ectomorph somatotype. Changes in the somatotype characteristics occurred at the age of 14 when young football players become ecto-mesomorphs. This is the period when the muscle mass increases, especially in the lower extremities. This somatotype is the same for the other age groups as well. Comparing our data with the results of young Brasilians, Cubans and Hungarians, we find that they are more endomorphic than our Bulgarian contingent.

Having knowledge of the age when somatotype features become stable and stay similar to the adult age is of major interest and concern for sports practice. According to our study the Bulgarian football players have an ecto-mesomorph somatotype and this somatotype is already steady in 15-year old adolescents.

Table 1. Age, size and somatotype of young football players.

Age	N	Height		Weight		Endomorphy		Mesomorphy		Ectomorphy	
		X	SD	X	SD	X	SD	X	SD	X	SD
12	15	158.14	10.1	44.18	8.05	1.86	0.55	4.52	0.91	4.27	0.88
13	15	165.94	5.28	52.06	6.68	1.63	0.23	4.66	0.62	4	0.8
14	15	166.57	6.47	57	8.58	2.08	0.67	4.92	0.6	3.17	0.98
15	15	174.07	3.61	66.25	5.12	2.25	0.52	5.05	0.9	3	0.55
16	10	176.14	4.48	69.04	7.13	2.4	0.39	4.85	0.89	3.2	0.61
17	10	174.85	5.12	66.5	6.75	2.3	0.48	4.92	0.65	3.25	0.83
Total	80	-	-	-	-	2.09	0.47	4.82	0.76	3.48	0.78

4 Conclusion

There are specific age dynamics of somatotype development in young football players. These data can be used in the morphological control and have to be taken into consideration in the final selection of young athletes.

5 Reference

Duquet, W. and Carter, J.E.L. (1990) Somatotyping, in *Kinanthropometry and Exercise Physiology Laboratory Manual: Tests, Procedures and Data* (eds. R. Eston and T. Reilly), E. and F.N. Spon, London, pp. 35-50.

44 TALENT IDENTIFICATION IN ELITE YOUTH SOCCER PLAYERS: PHYSICAL AND PHYSIOLOGICAL CHARACTERISTICS

A.M. FRANKS, A.M. WILLIAMS, T. REILLY and A.M. NEVILL
Research Institute for Sport and Exercise Sciences, Liverpool John Moores University, Liverpool, UK

1 Introduction

Identifying, developing and nurturing players have become priorities in soccer. The spiralling costs of purchasing players on the transfer market as well as the 'Bosman Ruling' by the European Court of Human Rights in 1995, which precludes clubs from withholding a player's registration at the completion of his contract, have reinforced the need for clubs to implement appropriate talent identification and development structures. Identifying soccer potential at an early age ensures that players receive specialised coaching to accelerate the talent development process. With the increased need to develop and nurture young talented players, it is essential to determine key elements of the talent identification process in soccer.

Perceptions of talent in soccer are diverse and complex. Talent may be genetically determined, complicated in structure and subject to environmental conditions. There is no consensus of opinion, nationally or internationally, regarding the theory and practice of talent identification, selection and development in sport, although research identifying the required characteristics for elite performance continues.

Talent identification is the recognition of potential by means of certain measures accepted as markers of future high performance (Borms, 1996). A talented individual has features which are distinctive from less talented individuals. These features can be measured to form a base for the prediction of performance. The aim in talent identification is to increase the probability of selecting a future elite player at an early age.

Players' anthropometric and physiological characteristics (e.g., height, weight, body fat, bone diameter, muscle mass, aerobic and anaerobic performance) are recognised as important determinants of performance (Borms, 1996). Consequently, physical and physiological assessment procedures may assist in the identification of young talent (Carter, 1982; Panfil et al., 1997). For example, physiological tests were employed by Jankovic et al. (1997) to identify key predictors of talent in soccer players aged 15 to 17 years. They reported that the more successful players were taller, heavier, had higher maximal oxygen uptake (VO_{2max}) and anaerobic power values than unsuccessful players. Jankovic et al. (1997) concluded that these measurements (i.e., physical assessment, VO_2 max, anaerobic performance) could be useful in predicting later success in football (Carter, 1985; Panfil et al., 1997).

Malina (1983) has suggested that successful young athletes have similar somatotypes/physiques to older successful athletes who display physical characteristics that favour good performances. Similarly, Panfil et al. (1997) found that elite youth soccer players have a higher morphological age (i.e., more physically mature) than their less elite counterparts and that coaches favour players advanced in morphological growth during the selection process. Hawes and Sovak (1994) argued that morphological age may not be an indicator of potential ability and that performers should be analysed over a period of time to determine the key predictors in identifying talent. They suggested that somatotype could be used, but not in isolation. However, although certain anthropometric and physiological measures may be factors limiting performance in soccer (eg., height in goal-keepers and aerobic performance in midfield players), players may be able to compensate by developing strengths in other areas (e.g., agility, power).

Research concerned with the physical and physiological predictors of talent has highlighted a number of potentially important measures such as somatotype, aerobic and anaerobic characteristics. As yet, there is no firm consensus regarding the relative importance of these measures in predicting talent in soccer. Therefore, the aim of this study was to investigate whether future success in soccer could be predicted from physical and physiological measurements.

2 Methodology

2.1 Participants
Historical data from 64 international youth soccer players (14–16 years) were analysed. Players were categorised according to playing position and, on a retrospective basis, their eventual level of success in the game. Data were gathered from the international coaches' subjective assessments with regard to each player's primary playing position. On the basis of this information, players were categorised as goal-keepers (n = 8), defenders (n = 24), midfielders (n= 22) and forwards (n = 10). Success was determined retrospectively by using international coaches' subjective ratings of playing ability and, more importantly, on the basis of whether the player in question received a professional contract. Those who signed a professional contract were categorised as 'successful' (n = 32), while those who did not turn professional were categorised as 'unsuccessful' (n = 32). These categorisations enabled differences to be investigated across playing positions (i.e., goal-keepers, defenders, midfielders, forwards) and players' eventual success in the game (i.e., successful vs unsuccessful).

2.2 Measurements
Measurements included height, body mass, body fat (%), aerobic performance and anaerobic performance. Body fat (%) was calculated by the sum of four skinfold measurements (triceps, biceps, subscapula and suprailiac). Aerobic performance were measured using a progressive 20-m shuttle run test protocol as described by Leger and Lambert (1982) for the prediction of maximum oxygen uptake (VO_{2max}). Anaerobic performance was measured by means of a test which players performed three 15 and 40 m sprints.

2.3 Analysis of data
A two-way ANOVA was used to analyse the data. The two factors were 'success' (successful vs unsuccessful) and 'playing position' (goalkeeper, defender, midfielder and attacker). The dependent variables were body mass, height, % body fat, aerobic performance and anaerobic performance.

3 Results

Analysis of variance indicated no significant differences between successful and unsuccessful players in all dependent variables. However, there were differences across playing positions in height ($F_{3,59}$ = 9.78, P < 0.001), body mass ($F_{3,59}$ = 10.87, P< 0.001) and body fat (%) ($F_{3,42}$ = 6.12, P< 0.001). Goalkeepers were the tallest (183.5 ± 1.8) (see Figure 1), heaviest (79.4 ± 1.8) (see Figure 2) and had largest body fat (%) (14.0 ± 0.72) (see Figure 3), while forwards were the smallest (171.8 ± 1.715) and midfield players had the lowest body mass (67.61 ± 1.11) and body fat (%) (10.52 ± 0.43).

Although there were differences across playing positions in height, body mass, body fat (%), all players on average, had a high aerobic and anaerobic performance regardless of playing position (see Figures 4-6).

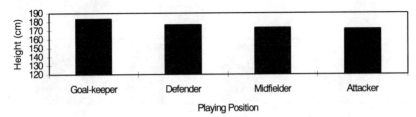

Figure 1. Mean height (cm) for international youth soccer players across playing positions.

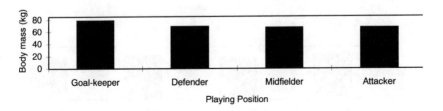

Figure 2. Mean body mass (kg) for international youth soccer players across playing positions.

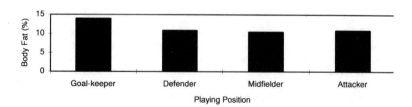

Figure 3. Mean estimated body fat (%) for international youth soccer players across playing positions.

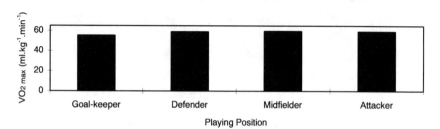

Figure 4. Mean estimated VO$_2$ max (ml·kg^{-1}·min^{-1}) for international youth soccer players across playing positions.

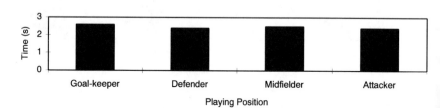

Figure 5. Mean 15-m sprint times for international youth soccer players across playing positions.

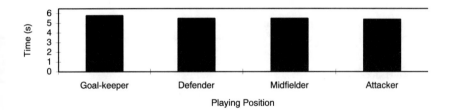

Figure 6. Mean 40-m sprint times for international youth soccer players across playing positions.

4 Discussion

The retrospective analysis of data indicated that there were no statistically significant differences between players who were deemed to have been successful in the profession and those who were considered unsuccessful on any of the physical or physiological measurements. There were differences across playing positions in height, body mass and body fat (%). These results conflict with those of Jankovic et al. (1997) who indicated that successful players were taller, heavier, had higher maximal oxygen uptake and anaerobic power values than unsuccessful players.

The absence of any significant distinguishing characteristics between successful and unsuccessful players was not altogether unexpected, though perhaps disappointing. Although research on talent identification has typically produced equivocal findings, the weight of available evidence suggests that this area of work may have some practical utility in attempting to guide the talent identification process. It may be that the nonsignificant findings may be due to several familiar problems in the analysis of historical data (e.g., reliability and accuracy of measurement, incomplete data sets), the fact that players had already been pre-selected on the basis of certain performance characteristics, or because the battery of tests employed were not extensive and lacked sensitivity.

It may be that the type of data available for this analysis may be more effective in talent detection (i.e., identification of potential performers who are currently non-participants) as opposed to talent identification (i.e., recognising current participants with the potential to become top performers). The majority of measures used, failed to separate individual players who were similar in ability. As a whole the group was relatively homogeneous compared to the variability that is found in elite adult teams. This homogeneity may reflect the talent detection process in this instance, with the players identified for specialist training on the basis of body size, growth indices and fitness.

Although all these measurements may be useful for initial talent identification, present observations suggest that none of the physical and physiological characteristics successfully predicted subsequent success in soccer. To enhance sensitivity, future research should adopt a multidisciplinary approach to identify talent in soccer, by taking into account anthropometric, physiological, psychological, educational and sociological factors (see Williams and Franks, 1998). Such an analysis would enable researchers and coaches to account for a greater proportion of the variance between talented and less talented players, thereby promoting greater accuracy and improving the understanding of the talent identification process in soccer.

5 Acknowledgements

This project was funded by The Football Association.

6 References

Borms, J. (1996) Early identification of athletic talent, *Paper presented at the 1996 International Pre-Olympic Scientific Congress,* Dallas, USA.

Carter, J.E.L. (ed) (1982) *Physical Structure of Olympic Athletes.* Part I. The Montreal Olympic Games Anthropological Project, S. Karger, Basel.

Carter, J.E.L. (1985) Morphological factors limiting human performance, in *Limits of Human Performance,* (eds D. Clarke and H. Eckert), American Academy of Physical Education Papers No. 18, Human Kinetics, Champaign, IL, pp. 106-117.

Leger, L. and Lambert, J. (1982) A maximal multistage 20-m shuttle run test to predict VO$_2$max, *European Journal of Applied Psychology,* 49, 1-12.

Hawes, M.R. and Sovak, D. (1994) Morphological prototypes, assessment and change in elite athletes, *Journal of Sports Sciences,* 12, 235-242.

Jankovic, S., Matkovic, B.R. and Matkovic, B. (1997) Functional abilities and process of selection in soccer, in *The 9th European Congress on Sports Medicine Program and Abstract Book,* Porto.

Malina, R.M. (1983) Menarche in athletes: a synthesis and hypothesis, *Annals of Human Biology,* 10, 1-12.

Panfil, R., Naglak, Z., Bober, T. and Zaton, E.W.M. (1997) Searching and developing talents in soccer; a year of experience, in *Second Annual Congress of the European College of Sports Science: Sport Science in a Changing World of Sports Book of Abstracts II,* pp. 649.

Williams, A.M. and Franks, A. (1998) Talent identification in soccer, *Sports Exercise and Injury,* 4, 159-165.

45 A COMPARISON OF SELECTED PHYSICAL, SKILL AND GAME UNDERSTANDING ABILITIES IN FINNISH YOUTH SOCCER PLAYERS

P. LUHTANEN*, T. VÄNTTINEN*, M. HÄYRINEN* and E. W. BROWN**
*   Research Institute for Olympic Sports, Jyväskylä, Finland
** Institute for the Study of Youth Sports, Michigan State University, East Lansing, Michigan, USA

1 Introduction

Soccer coaches have often considered the importance of the individual soccer skills, psychomotor skills, physical abilities and tactical understanding of the game. At junior level, it has been shown that age, physical abilities, psychomotor skills, and game-like skills influence success in matches. As evidence for this, it was found that the losers were behind the winners by 8% in soccer skills tested, 3-8% in physical strength and speed and 1% in psychomotor choice reaction time (Luhtanen, 1988). In addition, the losers had 13% less time to control the ball in different match situations. However, the losers were better than the winners in their tactical understanding of the game. It can be appreciated that knowledge of how tactical situations should be solved is not enough if the abilities for executing the skills of the game are insufficient.

Game understanding has not been defined clearly. The concept represents the quality of responses of a player to a variety of either actual or simulated game events. Expert judgements, made about the level of appropriateness of these responses, may be used as measures of the level of game understanding that an athlete possesses on a game by game basis. The understanding of constructs such as time, space, and movement (speed, direction, and timing) are pertinent to many games in solving problems in each game-like task.

Tactical awareness plays an essential role in game understanding. Bunker and Thorpe (1986) indicated that the uniqueness of games is in the decision-making process that precedes the execution aspect of performance in the game. They contended also that each game situation poses a unique problem and that this element of games lies within the cognitive area of learning. Thomas, French and Humphries (1986) have explained that sport performance is a complex product of cognitive knowledge about a current situation and past events, combined with a player's ability to produce the sport skills required. In basic football situations, players work in three major roles: (a) on offence with or without the ball in order to keep possession of the ball, (b) on offence to move on to the goal scoring area and to score a goal, and (c) on defence to get possession of the ball and to prevent a goal.

The purpose of this study was to examine and compare physical abilities, technical skills and game understanding in Finnish youth players who compete at the national level.

2 Methods

Soccer players (N = 106) from six First Division clubs, representing two teams in each age (16 and 18 years) and gender (boys and girls) group, served as subjects for the entire project. The number of subjects by group were as follows: Boys – 16 (n = 32); Boys – 18 (n = 28); Girls – 16 (n = 32); and Girls – 18 (n = 24). The number of valid subjects for each test is shown in Tables 1 and 2.

Three physical tests were administered: (1) sprinting time over 30 m, measured with photocells ($\pm$ 0.01 s); (2) shuttle running time for four 10-m lengths, measured using a video timer ($\pm$ 0.04 s); and (3) vertical jumping height in the counter-movement jump (CMJ), measured with the ERGOJUMP contact mat (Bosco et al., 1983).

Three skill tests were also administered: 1) a ball control test in which players used their feet, thighs and head; 2) a dribbling test which used a slalom track; and 3) a passing test against two walls (Luhtanen, 1988). In each skill test, the subjects attempted to perform the specific task as quickly as possible. The output of each of the skill tests was the time of execution in seconds, measured with a video timer ($\pm$ 0.04 s).

Additionally, an advanced test battery was constructed to measure game understanding (offensive and defensive play) in soccer. The test included 76 video segments of selected offensive (with and without ball) and defensive match situations. The subjects had to orient themselves to the situation and then try to understand play, space and time for both speed and direction of movements of the players and ball. The subjects had to respond for the video segments and quickly (< 10 s) decide what they would do in each situation (three options) and then justify (< 15 s) their decision from a list of eight multiple choices (Luhtanen et al., 1997). For validation of the game understanding test, a group of national youth team coaches and two expert researchers graded the priority order of the options on a three-point scale (2, 1, 0 points). The same group evaluated the grading for the arguments of the eight multiple choices n the same three-point scale.

The means and standard deviations of the game understanding scores were calculated. One- and two-way analysis of variance (ANOVA) and the Kruskal–Wallis one-way ANOVA were applied to the data. Least significant difference (LSD) analysis was applied to compare separate groups. Pearson and Spearman's correlation coefficients were calculated between the game understanding variables and performance test variables.

3 Results

The results of the physical and skill tests are shown in Table 1. As expected for the physical tests, the boys performed significantly better in the 30-m sprint (F = 171.39, P < 0.001), vertical jump (F = 67.39, P < 0.001) and shuttle run (F = 66.13, P<0.001). Similarly, the older subjects performed significantly better in 30 metre sprint (F = 3.78, P = 0.056), vertical jump (F = 3.55, P = 0.064) and shuttle run (F = 5.75, P = 0.019). In the skill tests, significant differences, favouring the boys, were also found in ball control (F = 15.82, P < 0.001), passing (F = 55.40, P < 0.001) and dribbling (F = 41.55, P < 0.001). Age effects were significant in passing (F = 3.53, {P = 0.065) and dribbling (F = 6.05, P = 0.017).

Table 1. Mean (± SD) times for different physical and skill tests in each gender and age group.

Groups	Physical Tests			Skills Tests		
	30 m sprint (s)	Shuttle run (s)	CMJ (cm)	Ball control (s)	Dribbling (s)	Passing (s)
G-16, n=18	4.92 ± .17	10.52 ± .49	30.5 ± 4.2	68.6 ± 66.3	30.1 ± 2.5	49.9 ± 5.6
G-18, n=17	4.90 ± .27	10.37 ± .70	32.8 ± 4.3	60.3 ± 51.5	29.7 ± 3.1	47.6 ± 5.3
B-16, n=21	4.41 ± .11	9.65 ± .52	40.4 ± 6.5	27.9 ± 21.1	27.1 ± 2.2	41.6 ± 3.7
B-18, n=17	4.26 ± .15	9.20 ± .30	42.7 ± 3.7	18.5 ± 6.9	24.5 ± 2.6	39.8 ± 2.7

The results of the game understanding tests are shown in Table 2. In game understanding, significant differences, favouring the boys were only found in the understanding of the defensive play ($F=25.22$, $P<0.001$).

Table 2. Mean (± SD) scores for the game understanding test for offensive, defensive and in total play in each gender and age group.

Groups	Offensive with ball	Offensive without ball	Defensive	Total
G-16, n=18	89.1 ± 9.6	45.5 ± 6.4	34.9 ± 5.6	169.5 ± 16.2
G-18, n=17	91.7 ± 9.4	46.2 ± 6.2	37.2 ± 5.1	175.0 ± 15.1
B-16, n=21	90.3 ± 9.4	45.7 ± 5.2	42.2 ± 8.5	178.2 ± 17.2
B-18, n=17	93.6 ± 8.6	43.1 ± 6.7	45.1 ± 4.9	181.8 ± 10.7

The highest correlation coefficients among the physical test items and among the skill test items were found in the girls: between vertical jump and sprinting speed ($r = 0.699$, $P < 0.001$) and between dribbling and passing test ($r = 0.655$, $P < 0.001$). In game understanding, the highest correlation coefficient also occurred in girls, between the understanding tests of offensive play with and without ball ($r = 0.393$, $P = 0.019$).

4 Discussion

Game performance has been analysed in soccer by Luhtanen (1988) using quantitative (number of executions) and qualitative (percentage of successful executions) game performance variables. These variables were the number of passes, balls received, runs with the ball, scoring trials, interceptions and tackles. The total game performance was the numerical involvement in passes, 'receivings', runs with ball, scoring trials, interceptions and tackles. The corresponding analysis was made also on the percentages of the successful executions. Selected skill (ball control, passing, dribbling, scoring), physical (sprinting speed, jumping power), psychomotor (reaction time, choice reaction time) and game understanding (knowledge of rules, tactics of basic situations and movements on the field) variables were also related to the game performance measures. For teaching and coaching purposes, it is important to understand in boys' and girls' teams the relationships between different player abilities and real game performance. If there is not a balance in training programmes, this could result in programmes for the footballer that do not focus on the appropriate domains for the total development of the players.

In this study, the comparison between gender and age groups showed, as expected, that the oldest boys were best in all physical and skill variables. Relatively, the difference between the oldest boys and girls (Girls–16 and Girls–18) was in the vertical counter-movement jump (24–29 %). In the sprinting time of 30 m and four times 10-m shuttle run, the differences were 13–14 % and 11–13 %, respectively. Correspondingly, the comparison between gender groups showed that the oldest boys were better than girls in ball control, dribbling and passing by 69–73 %, 17–19 % and 16–20 %, respectively.

Concerning game understanding tests, the oldest boys were best in game understanding of offensive play with ball and defensive play. The relative difference between gender was largest in the defensive understanding (18–23 %). The difference between gender in the understanding of the offensive play with ball was 4–7 %. The difference between boys and girls was lowest in the game understanding of the offensive play without the ball.

It could be suggested that when coaches are planning future training programmes that a better balance between the skill, tactical and physical training should be kept in mind. A general conclusion of this study is that game understanding was the least discriminating variable in assessing player abilities in soccer between boys and girls at the age of 16–18 years.

5 References

Bosco, C., Komi, P.V. and Luhtanen, P. (1983) A simple method for measurement of mechanical power in jumping, *European Journal of Applied Physiology,* 50, 273-282.

Bunker, D. and Thorpe, R. (1986) Is there a need to reflect on our games teaching? in *Rethinking Games Teaching* (eds R. Thorpe, D. Bunker and L. Almond), Loughborough University of Technology, England, pp. 25-33.

Luhtanen, P. (1988) Relationships of individual skills, tactical understanding and team skills in Finnish junior soccer players, in *New Horizons of Human Movements,* Proceedings of the 1988 Seoul Olympic Scientific Congress, Volume II, Seoul, pp. 1217-1221.

Luhtanen, P., Blomqvist, M., Häyrinen, M. and Vänttinen, T. (1997) Game understanding and game performance in modified soccer and throwing game at primary school level. Abstracts of the AIESEP Singapore, *World Congress on Teaching, Coaching, and Fitness Needs in Physical Education and the Sport Sciences,* p. 56.

Thomas, R.T., French, K.E. and Humphries, C.A. (1986) Knowledge development and sport skill performance: Directions for motor behaviour research, *Journal of Sport Psychology,* 8, 259-272.

46 A GAME PERFORMANCE ANALYSIS BY AGE AND GENDER IN NATIONAL LEVEL FINNISH YOUTH SOCCER PLAYERS

P. LUHTANEN*, T. VÄNTTINEN*, M. HÄYRINEN* and E.W. BROWN**
* Research Institute for Olympic Sports, Jyväskylä, Finland
** Institute for the Study of Youth Sports, Michigan State University, East Lansing, Michigan, USA

1 Introduction

All actions in soccer vary in time, space, speed and direction of the movement. Thus, each individual action by a player can be considered as a random test for individual skills, physical abilities, tactical understanding and team skills. The traditional approach to soccer coaching is technical and focuses on physical development of players. This approach has produced mainly skilful players who may not be able to use their skills in the game because the skills were learned in isolation from the context of the game. It is obvious that to be successful in games demands more than just physical skilfulness. Therefore, a growing interest has been focused on the importance of the players making correct decisions in the light of tactical awareness. Game understanding can be defined as a player's ability to interpret tactical offensive and defensive problems by selecting appropriate solutions in different game situations based on justifiable arguments for these solutions.

Game understanding has been used as a concept in the teaching of team games (Bunker and Thorpe, 1982). The teaching programmes have traditionally started with basic skills to master the ball with the appropriate sports equipment. Sport knowledge has also been used to describe a part of game understanding. Sport knowledge has been divided into declarative, procedural and strategic knowledge (Thomas et al., 1993). Young players may lack sufficient strategic knowledge and understanding in a variety of domains in the game of soccer. However, strategic and tactical aspects can be added progressively through teaching programmes in soccer. For this study, a test battery system, including video items, was created. Subjects had to make decisions about appropriate play in response to the game simulations on video. These simulated situations included: (a) offensive play where an athlete works with or without the ball to maintain possession, move the ball toward the opponent's goal or score; and (b) defensive play where an athlete works to prevent a goal from being scored and helps the team regain possession.

The purpose of this study was to examine game performance and its relationship with game understanding, physical abilities and technical skills in Finnish national level youth players.

2 Methods

Players (N = 106) from six First Division clubs served as subjects. Boys and girls formed two teams in each age (16 and 18 years) group and they played within their own age and gender group. The number of valid subjects for each variable is shown in Tables 1 and 2.

The official Finnish Football Association rules were applied to each match. All matches were recorded with a JVC camcorder. The matches were analysed using SAGE Game Manager™ for Soccer software. The game performance was considered as output of the players in the match conditions for both offensive (receiving the ball, passing the ball, running with the ball, shooting) and defensive (tackling, intercepting) manoeuvres. The total and effective playing times were recorded. The game performance results were standardised for 90-min playing time. Traditional test batteries were used to measure technical skills (ball control, passing and dribbling) (Luhtanen, 1988), physical abilities (time to sprint 30 m, maximum vertical jump) (Bosco et al., 1983) and time to complete a shuttle run.

An advanced test battery was constructed to measure game understanding (offensive and defensive play) in soccer. The test included 76 video segments of selected offensive (with and without the ball) and defensive match situations. The subjects had to orient themselves to the situations and then try to understand play, space and time for both speed and direction of movements of the players and ball. The subjects had to respond for the video segments and to decide quickly (< 10 s) what they would do in each situation (three options) and then justify (< 15 s) their decision by selecting from a list of eight multiple choices (Luhtanen et al., 1997). For validation of the game understanding test, a group of youth national team coaches and two expert researchers graded the priority order of the options on a three-point scale (2, 1, 0 point). The same group evaluated the grading for the arguments of eight multiple choices on the same three-point scale.

A linear stepwise regression analysis (criterion P = 0.05) was applied to the successful action that occurred under match conditions and to the variables tested. In the regression analysis, the dependent variables were the percentages of successful manoeuvres in receiving the ball, passing the ball, running with the ball, shooting, tackling, intercepting and total average of all successful manoeuvres obtained by the SAGE Game Manager™ for Soccer software analysis system. The independent variables were the individual results in the physical tests, skill tests and game understanding tests.

3 Results

The effective playing time was highest (74 min) in girls aged 18 and lowest (57 min) in girls aged 16. The average number of the offensive manoeuvres per player ranged between 32-34 executions in boys and 31-37 in the girls. Corresponding figures for executions of defensive play ranged between 20-21 in the boys and 24-25 in the girls.

The dependent variables for offensive match analysis are shown in Table 1.

Table 1. Mean (± SD) percentage of successful offensive executions of the analysed matches in each gender and age group.

Group	Receiving	Passes	Runs with ball	Shots	Offensive play
Girls -16, n=18	82.4 ± 12.7	56.2 ± 14.5	50.2 ± 33.9	10.0 ± 14.9	65.9 ± 8.8
Girls -18, n=17	88.1 ± 9.6	49.6 ± 17.8	49.3 ± 26.2	17.5 ± 32.3	63.5 ± 9.1
Boys-16, n=21	80.4 ± 13.3	58.2 ± 12.4	46.3 ± 25.3	16.7 ± 35.6	66.0 ± 10.0
Boys-18, n=17	83.8 ± 12.8	55.2 ± 9.3	54.8 ± 26.0	16.7 ± 40.8	66.6 ± 8.2

The dependent variables for defensive match analysis are shown in Table 2.

Table 2. Mean (± SD) percentage of successful defensive executions of the analysed matches in each gender and age group.

Group	Interceptions	Tackles	Defensive play
Girls-16, n=18	86.9 ± 15.9	51.0 ± 20.2	74.5 ± 13.9
Girls-18, n=17	97.2 ± 4.4	61.2 ± 18.9	81.7 ± 9.7
Boys-16, n=21	89.5 ± 14.4	47.6 ± 21.0	69.5 ± 17.3
Boys-18, n=17	94.1 ± 7.3	46.8 ± 21.7	69.5 ± 24.7

The independent variables were outcomes in physical, skill and game understanding tests explained in the Methods section. The results of these tests have been described elsewhere (Luhtanen et al., this volume).

In the boys, game understanding of defensive play significantly explained successful manoeuvres in receiving in the matches ($F = 9.82$, $P = 0.004$, $r^2 = 0.253$), and in successful runs with ball ($F = 6.20$, $P = 0.020$, $r^2 = 0.205$). In the girls, the passing test significantly explained successful runs with the ball during the matches ($F = 10.21$, $P = 0.004$, $r^2 = 0.290$) and sprinting speed successfully explained tackling ($F = 9.25$, $P = 0.005$, $r^2 = 0.242$).

In the girls' age groups, the physical and skill tests mainly explained significantly ($F = 4.88$, $P < 0.046$, $r^2 = 273$) successful manoeuvres in match conditions. In the boys' age groups, the skill, physical and game understanding test results significantly explained ($F = 4.97$, $P < 0.043$, $r^2 = 0.262$) successful actions in match conditions.

4 Discussion

Basic concepts in evaluating game understanding are as follows: task in a specific situation; time for execution of the task; space and time to use for the execution, direction and speed of the movement concerning own players, opponents and ball; and timing of all these factors. Also the anticipation of the coming situation, related to the skills of the athletes, play certain roles in solving problems in offensive and defensive match situations.

A purpose of this research was to try to develop a better understanding of teaching and coaching of games in a developmentally appropriate fashion, employing both a technical and tactical focus. Through this approach teachers and coaches can improve their players' game performance before starting to play competitive 11 vs 11 regulation games. Maintaining possession of the ball occurs by selecting and executing appropriate passing, ball control and support skills. Mitchell et al. (1994) have identified the major tactical problems that must be solved for teams to score, prevent scoring and restart play effectively. To score, a team must progressively solve complex problems of how to maintain possession of the ball, attack goal, create space while attacking and use that space effectively. For example, to maintain possession of the ball, players must support team mates with the ball and be able to pass and control the ball over various distances. It should be recognised that movements, made by players without the ball, have to be considered in the teaching of games. Some tactical problems and movements without the ball are complex for novice players to understand. In teaching, movements without the ball are often ignored.

This study indirectly showed through the video tests that it was easiest to understand what to do as defender in the different situations where the opponent had the ball. In defensive play, it is quite easy first to recognise a proper position, then to move to mark a player or a zone and to pressurise for winning or clearing the ball. This means that the children understand well how to decrease time and space available to the player with the ball. The most difficult problems existed in the situations where their own team had the ball and they had to decide what to do without the ball in order to support the player in possession. This may be due to the lack of time for teaching and coaching soccer in the youth teams. At these age levels in girls, the technical skills are not, in general, sufficient in ball control, passing, dribbling and especially in one-touch passes. Problems also exist for players in creating space and time for team-mates or for themselves. This simply means that it is difficult to teach players how to create and utilise free space. More difficulties will arise if patterns of play such as wall passes, overlapping, or scissors are included in the teaching programme.

It can be concluded that all player abilities (physical, skill and game understanding) in the boys, in these age categories, were in a better balance than in girls.

5 References

Bosco, C., Komi, P.V. and Luhtanen, P. (1983) A simple method for measurement of mechanical power in jumping, *European Journal of Applied Physiology,* 50, 273-282.

Bunker, D. and Thorpe, R. (1982) A model for the teaching games in secondary schools, *Bulletin of Physical Education,* 18, 5-8.

Luhtanen, P. (1988) Relationships of individual skills, tactical understanding and team skills in Finnish junior soccer players, in *New Horizons of Human Movements,* Proceedings of the 1988 Seoul Olympic Scientific Congress, Volume II, Seoul, pp. 1217-1221.

Luhtanen, P., Blomqvist, M., Häyrinen, M. and Vänttinen, T. (1997) Game understanding and game performance in modified soccer and throwing game at primary school level, in Abstracts of the AIESEP Singapore, *World Congress on Teaching, Coaching, and Fitness Needs in Physical Education and the Sport Sciences,* p. 56.

Mitchell, S.A., Griffin, L.L. and Oslin, J.L. (1994) Tactical awareness as a developmentally appropriate focus for the teaching of games in elementary and secondary physical education, *The Physical Educator,* 51, 21-28.

Thomas, J.R., Thomas, K.T. and Gallagher, J.D. (1993) Developmental considerations in skill acquisition, in *Handbook of Research in Sport Psychology* (ed R. Singer), Macmillan, New York, pp. 73-105.

PART SEVEN

Psychology and Motor Behaviour in Football

47 GROUP COHESION IN ENGLISH PROFESSIONAL FOOTBALL: A STUDY OF YOUTH TRAINEES

B. HEMMINGS* and A. PARKER**
* University College Northampton, Northampton, UK
** University of Warwick, Coventry, UK

1 Introduction

Professional football is an established and celebrated part of English popular culture, yet insightful and substantive revelations regarding its inner-workings are few and far between. This report presents the main findings of a critical and detailed ethnographic study of youth training within English professional football. It presents the sub-cultural experiences and verbalised accounts of a group of first and second year football youth trainees at one English Premier League club. Key questions and issues emerged, but in particular notions of group cohesion, collectivity and team solidarity feature as the central element of this narrative.

2 Researching life at Colby Town: Ethnography and Qualitative Method

In this research Colby Town was the name given to the club for confidentiality reasons. During the season under study the club supported a youth team squad of 19 players, of which eight were first year youth trainees and 11 were second year youth trainees. All first year trainees were between the ages of 16 and 17 and had arrived at Colby straight from school. Accordingly second years were embarking on their second full year of paid work after leaving compulsory education, and were all between the ages of seventeen and nineteen.

In the original study (see Parker, 1998) a qualitative/ethnographic research approach was adopted whereby data were gathered in a variety of ways. Observation and participant observation were central. Semi-structured interviews were based on broad interview agenda to focus on key aspects of institutional/occupational life (see May, 1993). Rather than being subjected to direct, structured interview questions, respondents were encouraged to talk around a number of general issues, all of which related in some way to their life histories at Colby and/or to their experiences of football in previous years. In terms of interview technique, such issues constituted loosely arranged themes and topics ('points of departure') which encouraged informal discussion and from which related issues ('emergent themes') surfaced and were subsequently followed up. The design of interview agenda was decided upon after entry into the research setting so as to allow some measure of institutional context to be gained via the establishment of individual and group relationships. In turn, key conceptual themes were identified from early observations in the field from which subsequent theoretical inferences were generated. In this sense, the analysis of data occurred throughout the research period, running parallel to fieldwork and thus informing data collection (Bryman and Burgess, 1994). Additionally, a sociometric questionnaire (Sanderson, 1996) was distributed amongst trainees in order that some form of data 'triangulation' might take place with regard to information concerning peer group relations.

Trainees were interviewed at least twice over the course of the research period which lasted the full duration of a football season – from early July until May. For the most part the researcher attended the club for three days each week as participant observer, spending two days training, working and socialising with trainees, and one day at local colleges of Further Education as a fellow student. After the initial three months of participant observation, interviews were conducted with trainees. The vast majority of these interviews were carried out on a personal one-to-one basis. On only one occasion, towards the end of the fieldwork period, did a group interview take place in accordance with the collective requests of seven second-year boys. Interviews were also conducted with the youth team coach. All interviews were tape-recorded. To supplement these data, a detailed fieldwork diary was kept throughout the research period, which, in an attempt to limit institutional suspicion, was written-up each evening on return from club, college and/or social settings.

3 Key Findings

Cohesion, or a lack of it, emerged as one key finding. A culture of authoritarianism was found, where the need for togetherness in trainees was stressed on a daily basis. Little cohesion (social or task) was evident, with the existence of a clear first and second year split, clique formation, social isolates, and favouritism shown toward some individuals. The interview quotes which follow have been selected to provide examples of these findings.

The first quote is from the youth team coach about aspects of *togetherness*.

> 'Well, I think it's important that the lads stick together both on the pitch and off the pitch to a certain degree, although they need their own privacy. And I think if, – like youth team players do at any football club, they do things together, whether it's the cleaning up, the boots, whether it's the jobs, the youth team spend more time at the football club than probably any other person – because their job demands that, and you've got to be a team to do all the different jobs and all the different things required of you. And whether its training or coming up from the training ground, I like the lads to be together, y'know, I like us to work as a team, play as a team, try and stick together, and try and promote a little bit of comradeship really."

This trainee quote refers to the existence of a *first and second year split* between trainees.

> 'A lot of things that we do, we tend to do it in first and second years. I think that's carried round from't club – I think Terry (name of youth team coach) splits us up too much. You can feel, y'know, that they think that you shouldn't be there, and that one or two of their mates should be there. But it's just like all't second years stickin' together. It's like sayin' y'know, "only us second years should be playing". It's like every position, people are fighting for it, and first years will say, "Well I think he should be in", and second years will say, "well second years are more experienced and better players."

The following trainee quote refers to the notion of *favouritism* shown toward some trainees.

> 'I didn't think that at a club like this, or at any other club, that they'd have people who'd be, y'know, their favourites, or have favouritism. I don't think it's right. And it all comes down to like whether you're a favourite or not, because you don't know how to become a favourite, you're either liked or you're not...I can't believe that there's so much bentness in it. Y'know, you wouldn't imagine it. You think, well, professional people, in professional jobs, and they've still got favourites. But when it comes down to that it should be out ot' window, it should be straight, who's playin' well and who's not. But it don't go down like that...I know you're gonna get favourites in any job, but not in football cos it's not a thing you can do, cos if you're not a good player, you shouldn't be making it.'

This quote is from the youth team captain who seemed to be *isolated* from both first and second year trainees.

> 'Oh yeah its hard cos you get a lot of stick an' that, but you just get through it don't you, that's the sort of person I am...it ain't gonna bother me...I just get on with my job and that's it, that's probably why, y'know, I've done so well, cos I just get on with my job, y'know what I mean, I'm not bothered. But if someone says something to me I just let it pass, I wouldn't like, start back at him.'

Where the following trainee quote refers to 'England' this was a small *clique* of players who had come to the club after spending time at the 'English National School of Excellence' at Lilleshall.

> 'Well, it, like, gets in the papers an' that, like, "Three England lads at Colby", an' y'know, people come to watch and they say, "Who's the England lads", an' stuff like this...It's wrong, cos it made me feel bad last year. I wasn't jealous of them at all, it just made me think, well, why should they get it. I mean its only because they're England, and a lot of the other lads just felt terrible. I mean we thought we were worthless compared to them. I mean, they just all stuck together.'

Participant observation over the season also highlighted many inconsistencies about togetherness and equality. Discussions between players frequently centred on preferential treatment for some including signing on fees, free boots and favourites being allowed home more readily. First years completed the more degrading jobs (e.g., cleaning toilets) and were responsible for carrying equipment to and from practice, as well as having to prepare kit for more of the professional players than second years on a daily basis. There was also differing squad responsibility whereby second years would take warm-ups and cool-downs. There was also evidence of output restriction imposed by second years on first years such as lessening training intensity on runs and abusing first years who seemed to be showing too much effort in practice. Second years were frequently seen walking to the training pitch with the coach and the discussion of team affairs with second years was overheard on many occasions. In training itself, cliques tended to pass to one another and first and second years often complained of the predictability of team selection. The first and second years were also regularly split for the purpose of practice matches.

Triangulation of data collection was achieved through sociometric analysis. The following sociogram affirms the existence of a first and second year trainee split, clique formation and evidence of social isolates.

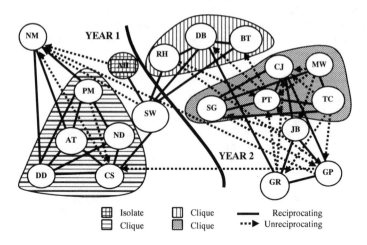

Figure 1. Sociogram showing trainee social relations.

4 Discussion

In this study, the coaching rhetoric about togetherness and cohesion was not matched by daily practices. Quantitative approaches to cohesion have found similarity of individual experience as being a salient antecedent of cohesion (Carron, 1982), and it seems clear that in this study this similarity did not exist for each trainee. Cohesion in the life of the trainee was non-negotiable and imposed in almost militaristic fashion, with youth team players paying lip service to the idea of togetherness when in reality self-preservation and individualism seemed to be more beneficial. Suitably, in this instance, education of the coach about the development of cohesion was necessary. Leading researchers have suggested more qualitative and longitudinal research is necessary to progress the field and enable a greater understanding of the development or erosion of cohesion (Widmeyer, Carron and Brawley, 1993). This study provides such an account. The authors recognise that the generalisation of findings is problematic with the qualitative approach used, but biographical testimonies from within football seem to suggest similarities (e.g., Dougan, 1980). Future research should perhaps attempt to integrate methodological approaches in order to further the understanding of team cohesion in professional football.

5 References

Bryman, A. and Burgess, R. (1994) *Analysing Qualitative Data,* Routledge, London.

Carron, A. (1982) Cohesiveness in sports groups: Interpretations and considerations, *Journal of Sport Psychology,* 4, 123-138.

Dougan, D. (1980) *Doog,* Readers Union, Newton Abbot.

May, T. (1993) *Social Research Methods,* Open University Press, Buckingham.

Parker, A. (1998) Staying on-side on the inside: Problems and dilemmas in ethnography, *Sociology Review,* 7, 10-13.

Sanderson, F. (1996) Psychology, in *Science and Soccer,* (ed T. Reilly), E. and F.N. Spon, London, pp. 273-312.

Widmeyer, W.N., Carron, A. and Brawley, L. (1993) Group cohesion in sport and exercise, in *Handbook of Research in Sport Psychology,* (eds R. Singer, M. Murphey and L. Tennant), McMillan, New York, pp. 672-691.

48 BILATERAL MOTOR PERFORMANCE EFFECTS FROM TRAINING THE NON-DOMINANT FOOT IN COMPETITIVE SOCCER PLAYERS

J. HOFF* and E. HAALAND**
* Norwegian University of Science and Technology, Department of Sport Sciences, Trondheim, Norway
** Stadion Medical Practice, Bergen, Norway

1 Introduction

Soccer players are encouraged to learn to dribble, shoot or kick the ball with either foot. Players seldom use both limbs with equal emphasis, and develop a preferred side representing the subjects 'handedness' or rather 'footedness' in terms of soccer performance. Provins and Glencross (1968) have shown that professional typists have slightly better performances using their left hand, whereas non-typists have significantly better performances using their right hand. When the same groups were given a simple handwriting task, both groups recorded a highly significant difference in performance between the two sides in favour of the preferred hand. These findings are strong indications that differences between hands might be a function of practice rather than genetic predispositions. Whether the development of a dominant side of the body is a function of nature or nurture is still under debate (Provins, 1997).

The ability to learn a particular skill more easily with one hand or foot after the skill has been learned with the opposite hand or foot is known as bilateral transfer (Magill, 1993) or cross-education (Cook, 1933, 1936). Two explanations are used for bilateral transfer, a cognitive explanation and a motor control explanation. The cognitive explanation is related to the knowledge of 'what to do' to achieve the goal of a skill. Support for the cognitive element was given by Kohl and Roenker (1980) where bilateral transfer was shown both for a unilateral training group and a mental training group, but not for a control group familiarised with the task.

There are two ways of considering the motor control explanation. The first is the traditional motor programme perspective, where the memory representation is responsible for the class of movements or actions and would in principle be available for both limbs. The second is founded in a dynamical systems perspective where time and space features of the movement act as control mechanisms in an integrated perception-action coupling. Training of the non-dominant hand has shown that this hand could improve to a higher level of tapping skill than the dominant hand, and this training also resulted in improved tapping using the dominant hand (Peters, 1981).

Peters (1976) suggested that the level of skill in the dominant hand might be restricted by the level of the same skill on the non-dominant side. This is a departure point in this experiment.

The first hypothesis was that increased use of non-dominant foot in soccer training improves soccer specific skills when using the non-dominant foot. The second hypothesis was that training the non-dominant foot enhances soccer specific skills using the dominant foot.

2 Methods

2.1 Subjects
Forty seven male competitive soccer players, aged 15–21 years, served as subjects, matched by age groups, and randomly assigned to a training group (n = 24) and a control group (n = 23). They were all right-handed and right-footed, based on a self-report.

2.2 Tests
Three soccer performance tests were carried out, using preferred and non-preferred foot consecutively.

Test A: A dribble test between five markers in line with a distance of 1 m. Inside and outside of one foot was used. A 4-min period was allowed between each trial with two trials for each foot. Time was measured using a hand-held stopwatch. Collapsed time for both trials on each foot was used.

Test B: A receiving and direct volley shot test. The subject received the ball at chest height in front of the goal, with the back towards the goal, at a distance of 10 m. The ball was received and a volley shot was the second touch at the ball. Points were given for where the shot was placed in the goal. The goal itself and a zone 30 cm outside the goal were divided into zones giving from 6 points in the top corners to 1 point for a shot where the goal-keeper normally stands, and 1 point also for a hit on the ground, within 30 cm outside the goal. Points were added for 15 + 15 shots with each foot. A 4-min rest was allowed between trials.

Test C: A one-touch passing test. Distance was 10 m from a mini-goal 1 m broad and 40 cm high. A pass was received from a player behind, from both the right and left sides respectively. The ball was passed on one touch towards the goal. Altogether, 15 + 15 trials were conducted with each foot. One point was scored per hit into the goal. Both trials were counted.

2.3 Procedure
A pre-test–post-test control group design was carried out. The intervention was increased volume of soccer training with left, non-preferred foot for a period of eight weeks. The training was not specific training on the tests used in the experiment, but general use of left foot in all individual technical training. Dependent variables were the performance scores of the tests described. A standardised 10-min warm-up period using different forms of running was carried out prior to the tests. Subjects were kept ignorant of the hypotheses and purpose of the training intervention. The experimenter did not intervene during training to avoid possible 'Hawthorne' effects.

3 Results

The results from pre- and post-tests for the soccer specific tests using non-dominant leg are presented in Figure 1. A significant group x time interaction was found for test A, the dribbling performance using the trained left leg. For test B, the volley shot at a goal, a significant group x time interaction was found using left leg, and also for test C, the mini goal point shot, a significant group x time interaction for the left leg was found.

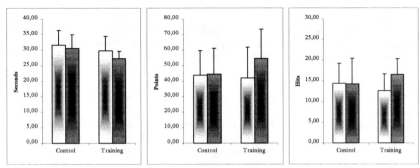

Figure 1. Pre- and post-test for control and training group in the soccer specific tests using non-dominant foot.

A T-test for paired samples showed no initial differences between the groups. For the dominant foot (right), which was not emphasised more than normal during soccer training, the results from pre- and post-test for the soccer specific tests are shown in Figure 2. A significant group x time interaction was found for test A, the dribbling test, when using the dominant (right) foot. For test B, the volley shot against goal, a significant group x time interaction was found using the dominant (right) foot, and for test C, the one-touch pass against a mini-goal using the right foot, a significant group x time interaction was found.

When the three football specific tests A, B, and C are collapsed and transformed to a standard score, the training group showed an improvement over the control group of 6.1% when using the non-dominant foot.

Using the dominant foot, the training group improved 6.2% compared to the control group. The control group did not improve significantly in any of the soccer specific tests used as dependent variables.

Total training time in the eight-week intervention period was 2768 min for the training group, of which the left foot was used for 705 min. The control group trained during the intervention period for 2803 min and reported having used the non-dominant foot for 73 min.

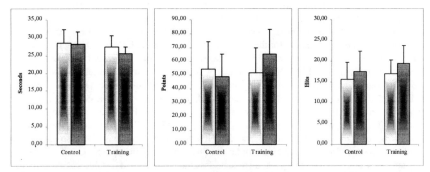

Figure 2. Pre- and post-test for control and training group in the soccer specific tests using non-dominant foot.

4 Discussion

Enhanced training using the non-dominant foot improves soccer specific skills in tests using this foot. This is in line with the first hypothesis. The improvements are not a test–retest effect, as the improvement was compared to a control group also performing test and retest.

That the training group also showed statistically significant improvements over the control group when using the dominant (right) foot was in line with the second hypothesis. The results showed an improvement in performance using the dominant foot after a period of emphasising training on the non-dominant foot. This finding is not entirely in line with general scientific findings which often show a high degree of specificity in motor learning (Magill, 1993). The results are in line with Peters' (1976, 1981) findings, and could be viewed as supporting the suggestion that the level of skill using the dominant side might be restricted by the level of skill on the non-dominant side.

Overall, the tests for the training group showed that the skills using the non-dominant (left) foot develop to about the level of skill that was shown using the dominant foot in the pre-test. It should be pointed out that the difference between right and left leg was of a similar magnitude after the training period. It might be surprising that the control group did not show improvements in any of the soccer specific tests following eight weeks of soccer training. One might question the quality of soccer training, the extent to which the chosen tests represented soccer play, or the skill elements that were emphasised during that particular period of soccer training.

A simple interpretation of the results might lend support to a motor programme perspective (Schmidt, 1976) where one imagines that training improves a general motor program that is available to both the right and left sides. Still, the control group had spent an equal amount of time, and used the same exercises as the training group, and an eventual generalised motor programme should have more opportunities to be improved by use of the dominant foot, since the skills using this foot were of a higher level from the beginning.

On the other hand, one can imagine that training the non-dominant foot creates higher attention to the training situation for the subjects. A training effect from a higher degree of attention might be easier to understand from a theoretical point of view within a dynamical systems approach (Kelso, 1995). This approach, which is based on a self-organisation of brain and behaviour, is currently the dominating model for explaining development of motor learning and control. What seems to be learned in motor skills is dependent upon the coupling together of perception and action, which from a dynamical systems perspective involves picking up relevant information within the optical flow field (Gibson, 1979). This relevant information contains not only physical variables such as velocity, displacement or mass, but also higher order information. Variables of a higher order can be invariants in the optical flow field, that is, they appear with shifts both in time and context. Invariant variables of a stable character in the environment might be texture gradients (Gibson 1979), the direct enlargement of an object on retina (tau information) (Lee, 1976; Lee et al., 1983, 1991, 1993; Bootsma et al., 1994), relative mass (Jacobs, Michaels and Runeson, 1998), or other variables. What seems to be important for the learner is to perceive relevant information in the situation, which contains both environment and subject. This sum of information forms the situation from which the subject chooses actions to improve a motor performance. What the soccer player does should be viewed in the context that is related to invariants in perceptual information and where the player has learned to pick up relevant information for the specific situation. Even if the action itself is crucial to improve performance, a dynamical systems approach would view the picking up of relevant information as the basis for muscular self-organisation, which is available for both the dominant and non-dominant sides. Improved performance is, in this context, a function of the player's ability to learn to relate to relevant information on which the task is dependent, and that the action or behaviour itself is primarily self-organised. Increased attention when training the non-dominant foot, might imply that the subjects are trained to pick up relevant information in the environment during the training session, with this information being available for a high degree of self organising, for use both with the dominant and non-dominant sides. From a dynamical systems perspective, the conclusion and explanation of cross education is that the soccer player learns to pick up relevant information in the interaction between self and the environment, rather than an improvement in a generalised cognitive structure, such as a motor program.

Whatever theoretical model is used in attempting to understand cross-education and the findings in this experiment, there is reason to be aware of the strong effects from using the non-dominant foot in training. The practical applications from the findings in this experiment are that soccer players and coaches should put more emphasis on training the 'wrong' foot to improve soccer skills with both right and left foot.

5 References

Bootsma, R.J., Mestre, D.R. and Bakker, F.C. (1994) Catching balls: how to get the hand to the right place at the right time, *Journal of Experimental Psychology: Human Perception and Performance,* 3, 591-612.

Cook, T.W. (1933) Studies in cross education II. Further experimentation in mirror tracing the star shaped maze, *Journal of Experimental Psychology,* 16, 670-700.

Cook, T.W. (1936) Studies in cross education V. Theoretical, *Psychology Review,* 43, 149-178.

Gibson, J.J. (1979) *The Ecological Approach to Visual Perception,* Houghton Mifflin, Boston.

Jacobs, D.M., Michaels, C.F. and Runeson, S. (1998) Perceptual learning and its importance for the visual perception of relative mass, in *Advances in Perception-Action Coupling* (ed B. Bril), EDK, Paris.

Kelso, J.A.S. (1995) *Dynamic Patterns: The Self-Organisation of Brain and Behavior,* MIT Press, Cambridge..

Kohl, R.M. and Roenker, D.M. (1980) Bilateral transfer as a function of mental imagery, *Journal of Motor Behavior,* 12, 197-206.

Lee, D.N. (1976) A theory of visual control of braking based on information about time-to-collision, *Perception,* 5, 437-459.

Lee, D.N., Davis, M.O., Green, P.R. and van der Weel, R. (1993) Visual control of approach by pigeons when landing, *Journal of Experimental Biology,* 180, 85-104.

Lee, D.N., Young, D.S., Reddish, P.E., Lough, S. and Clayton, T.M.H. (1983) Visual timing in hitting an accelerating ball, *Quarterly Journal of Experimental Psychology,* 35A, 333-346.

Lee, D.N., Reddish, P.E. and Rand, D.T. (1991) Aerial docking by hummingbirds, *Naturwissenshaften,* 78, 526-527.

Magill, R.A. (1993) *Motor Learning: Concepts and Applications.* Brown and Benchmark, Indiana.

Peters, M. (1976) Prolonged practice of a simple motor task by preferred and non-preferred hands, *Perceptual and Motor Skills,* 43, 447-450.

Peters, M. (1981) Handedness: Effect of prolonged practice between hand performance Differences, *Neuropsychology,* 19, 587-590.

Provins, K.A. (1997) The specificity of motor skill and manual asymmetry: A review of the evidence and its implications, *Journal of Motor Behavior,* 29, 183-192.

Provins, K.A. and Glencross, D.J. (1968) Handwriting, typewriting and handedness, *Quarterly Journal of Experimental Psychology,* 20, 282-289.

Schmidt, R.A. (1976) The schema as a solution to some persistent problems in motor learning theory, in *Motor Control: Issues and Trends* (ed G.E. Stelmach), Academic Press, New York, pp. 41-65.

49 REFLECTIONS AND CONSIDERATIONS OF PROVIDING SPORT PSYCHOLOGY SERVICES WITH PROFESSIONAL FOOTBALL PLAYERS

P.J. KREMER* and D.B. MARCHANT**
*   University of Melbourne and Western Bulldogs Football Club
**   Victoria University of Technology

1 Introduction

The Australian Football League (AFL) has grown rapidly in recent years with increased media coverage, game attendances, number of teams participating, and revenue generated. This growth has heralded an era of increased professionalism as reflected in increased player payments and the time that players need to dedicate to playing at the elite level. In addition, this increased professionalism is evident through the employment of various personnel having specialist skills often in a full time or part time capacity. Professionals that have long been regarded as essential employees at the elite level of Australian Rules include medical practitioners, physiotherapists, fitness and conditioning experts, and dieticians. Another profession that has recently been absorbed into the AFL system is sport psychology.

Although sport psychology has had some presence in the AFL since the 1960s, this has been ad hoc. Through the period of the late 1980s and early 1990s sport psychology gradually became more accepted as a valuable arm of a club's personnel. From these early beginnings, sport psychology has expanded, for instance, Kremer, Shen and Tonn (1999) reported that during the 1997 season 75% of the 16 AFL clubs employed a sport psychologist. Unfortunately, there is still considerable confusion about what sport psychology is, and what services a sport psychologist can provide in professional football. Moreover, there exists a poor understanding of the issues that potentially compromise the productivity of a sport psychologist working with professional football clubs.

The present purpose is to provide an elaboration of the employment of a sport psychologist with a professional football club by drawing on the experiences of the authors, both of whom have been employed with AFL clubs over the past three seasons. The focus is on three key areas: (1) the services that a sport psychologist can offer; (2) suggestions on how an organisation can gain maximum benefit from a sport psychologist; (3) the difficulties and barriers that typically confront a sport psychologist working for a football club.

2 Services that a Sport Psychologist Offers

The sport psychologist can potentially provide a broad range of services summarised under the following headings: (1) mental skills training; (2) clinical issues; (3) lifestyle/personal issues; (4) group/team dynamics; (5) coaching skill development; (6) organisational and cultural change issues.

Sport psychologists are often employed by football clubs to improve the performance of players by teaching them mental skills and techniques. Much of the face-to-face counselling that a sport psychologist provides includes such techniques as goal setting, imagery training, anxiety/arousal regulation, stress management, concentration and attention, self-confidence, cognitive style retraining, pre-match preparation and performance routines. Aside from developing the more sport specific mental skills, sport psychologists may provide skill development in broader life skills, for instance, time management, and communication skills. A sport psychologist may also assist in areas of general player welfare. Although many professional football clubs employ a welfare officer, the sport psychologist can contribute in welfare issues by assisting the welfare officer with general player counselling, crisis counselling, formal assessment of support systems, vocational and career counselling.

There are numerous other areas in which a sport psychologist might contribute. Individual counselling may assist a player during the rehabilitation period of a career threatening injury. Sport psychologists can perform 'at risk' assessment of players who may be susceptible to over-training and burnout. Assisting a player through the period of a performance slump is another area where the sport psychologist's role may be overlooked. An area that has until recently been overlooked is helping athletes deal with major transitions within their sport, such as beginning and ending professional careers (Fortunato and Marchant, 1999). Sport psychologists may also contribute to reducing the incidence and impact of substance abuse through the initiation of education programmes. One recent example of this is the 'buddy' programme introduced by Australian Rugby League clubs (Clews, 1999). Clinical issues where a sport psychologist may assist, at least in the assessment and referral stage, are eating disorders, personality disorders, and pathological addictions. Player profiling and mental skill testing are activities that are performed to varying degrees by sport psychologists. Similarly a sport psychologist with a strong background in interviewing and psychometrics may have much to offer in the area of player recruitment.

Group and team activities also make up an important aspect of activity of sport psychologists. The objective of the sport psychologist in this area is to minimise potential productivity losses by the group of players. One of the principal team activities is conducting team 'goal setting'. Other activities include facilitating activities where codes of conduct, player rules, and player expectations from each other are clearly defined. Another related area focuses on developing criteria for within group peer accountability – both when playing and training. Development of appropriate leadership skills is often incorporated into team activities. This aspect is often neglected in team sports – with the assumption being that leadership skills will accrue with experience.

The sport psychologist may also contribute to the development of more effective coaching skills. Providing training to enhance listening and communication skills may be invaluable in some situations such as training coaches to listen to messages from players and to utilise more effective counselling techniques when interacting with players. Sport psychologists can also assist coaches with an understanding of the best methods with which to provide information to players and when the optimal time to present this information would be. The provision of confidential feedback (reports, audio tapes, video tapes) to coaches is a useful method for the sport psychologist to effect skill development in this area. The sport psychologist may also assist coaching staff to handle the many pressures that accompany their positions.

Finally, another area where sport psychologists may contribute within the football club is involvement at a broader organisational level. Activities involved might include the development of a club mission statement where the club's objectives are encapsulated in a clear and concise statement. Another important, yet often overlooked area is the setting up of an appropriate communication framework that will facilitate communication across diverse levels within the organisational structure and provide opportunity for conflict to be resolved more efficiently. Two further areas for potential involvement include bench-marking and cultural change. The sport psychologist, in conjunction with other club personnel may help to establish a set of criteria for which club personnel (e.g. recruitment officer, fitness coordinator, and so on) may be evaluated.

Generally, the range of services provided by sport psychologists is dependent on their unique set of skills and competencies. Consequently, although we have outlined a diverse range of possible activities, the sport psychologist's actual activities will be governed by a principle of competence (Australian Psychological Society Code of Professional Conduct, 1995). That is, sport psychologists should understand their limitations (and strengths) and ensure they practice only in areas in which they have appropriate skills.

3 Difficulties and Barriers

Numerous practitioners have acknowledged the difficulties arising from myths and misconceptions associated with the field of sport psychology (e.g., Bond, 1990; Kirkby, 1995). Our experiences suggest that at the elite level of Australian Rules football there are many myths and misconceptions that persist and consequently present a challenge for the practitioner to overcome.

One of the most prevalent misconceptions is that sport psychologists work mainly with 'problem' athletes – (i.e., 'shrink mentality'). It is important to eliminate this limited and largely inaccurate perception, to ensure that athletes feel comfortable about seeing a sport psychologist for the full range of consulting possibilities. The reality is that although sport psychologists may assist athletes experiencing specific behavioural or emotional 'problems', the majority of time is spent working with 'normal' athletes. Despite the so called professionalism of AFL football, some coaches, players and administrators still make ignorant, inappropriate or derisive comments about sport psychology work, thus reinforcing the 'shrink mentality'. We recommend that sport psychologists undertaking work with football clubs make a concerted effort to educate those around them about what sport psychology is (and is not!). Similarly, key personnel responsible for the employment of sport psychologists should ensure they are well briefed on what sport psychologists do.

Another misconception about applied sport psychologists working with football clubs is the belief that sport psychologists can provide a magical 'quick fix' for individuals and teams. Sport psychologists do have specialised skills that can sometimes help athletes overcome problems rapidly. Generally, however, mental skills like physical skills need considerable practice before performance improvements are likely to result. Employing a sport psychologist on the basis of a single session or a few sessions in the belief that this can bring about some meaningful change in performance is likely to result in disappointment and doubts about the merit of sport psychology. A related misconception is the belief that the sport psychologist can resolve an issue through a second hand description (e.g., coach or other third party). Sport psychologists are trained to gather first hand information from a primary source (e.g., observation, interview, or statistics), rather than speculate on the basis of secondary information.

A further misconception is the commonly held belief that sport psychology work is highly visible, lavishly resourced, universally respected and highly paid. Our experiences suggest that this is rarely the case. Very few sport psychologists make a full time living from consultancy work and often derive an equivalent full time wage by combining sport psychology with more traditional areas of psychology (e.g., educational psychology, counselling psychology). To our knowledge, no sport psychologists are currently employed full time by AFL clubs. It is our view that most clubs would be prepared to utilise the services of a sport psychologist on a full-time basis, but the budgets at most clubs do not permit this 'luxury'. Consequently, clubs often engage a sport psychologist on a fractional or sessional basis. Invariably, these circumstances will lead to frustration for both parties. The coach (or club) may wish for the sport psychologist to provide a greater commitment to achieving results despite being aware that the club is unable or unwilling to pay for the increased workload. Conversely the sport psychologist may identify a range of areas requiring attention but with other commitments (e.g., employment elsewhere) may need to reassess the activities they undertake.

There are numerous logistical issues that may prevent the sport psychologist from working to full potential. One of the foremost of these is the difficulty in scheduling time for contact with coaches or players. Timetabling contact around (physical) training sessions, personal commitments (including work, study and family), media and promotional activities, meetings and other activities is an extremely challenging exercise. Similarly, finding an appropriate location to perform the work is difficult. Few clubs provide sport psychologists with a dedicated space or office. Usually the sport psychologist will have access to a room that doubles for other purposes. Invariably situations arise when there is a clash of needs for the room, and the sport psychologists find themselves performing their activities in a range of other locations including another room or location (e.g., coaches room, or other office), local cafe, weights room, playing field, in a corridor). Sometimes these environments present a nuisance, most often they will severely compromise the effectiveness of the work. For example, carrying out arousal control and imagery interventions is virtually impossible in a make-shift setting. Having a room that is private and equipped with basic furniture (table and chairs) will enhance the likelihood of more productive work. Other equipment needs include access to a telephone, computer, printer, photocopier and stationery. In our experience, these basic needs are only partially provided.

The potential case load at most professional football clubs is large. In the AFL for instance, sport psychologists may be required to work with a list of 42 senior players, as well as rookies, supplementary list players, coaching staff, and other personnel. The resultant workload coupled with financial and scheduling constraints can produce situations where it is difficult to accommodate all parties. Under these circumstances it may be necessary to limit the range of contact to a defined subset of these personnel.

4 Getting the Most from a Sport Psychologist

There are various factors that football clubs should consider to ensure that the club gets maximum benefit from employing a sport psychologist. We have collapsed these issues into six specific recommendations:

1. A time specific contract between the organisation and sport psychologist should include a detailed description of duties, expected time commitment, and financial arrangements.
2. A demonstrated support for the position should be reflected in the following ways; (1) infrastructure needs are considered, (2) the sport psychologist receives reasonable access to players, staff and meetings, (3) regular meetings between senior coach, team manager and sport psychologist, (4) opportunities for the sport psychologist to negotiate how he/she can best use his or her skills, and (5) remuneration is commensurate with qualifications and experience.
3. Consensus between the organisation and sport psychologist on appropriate performance expectations and realistic assessment procedures.
4. A clear communication process is developed between sport psychologist, management, coaches and players.
5. Organisations should be aware of the professional obligations for sport psychologists in regard to their professional code of conduct (e.g., confidentiality, ethical behaviour, referral records, access to assessments and case notes, maintaining professional relationships, self promotion).
6. Whether the sport psychologist's role is to include involvement in other activities for which he or she may possess specialised skills.

5 Conclusions

Sport psychology is now reasonably well entrenched in Australian Rules football at the elite level. The effectiveness of sport psychologists is often compromised by long standing misconceptions about the field. To gain maximum benefit from the unique skills that sport psychologists provide, we recommend that football organisations intending to employ a sport psychologist consider the issues we have discussed at the outset of the working alliance. Following this approach should ensure a more productive arrangement for both parties.

6 References

Australian Psychological Society (1995) *Code of Professional Conduct,* Melbourne, Australia.

Bond, J. (1990) Sport psychology in Australia: An overview, in *Australian Sport Psychology: The Eighties* (eds J. Bond and J. Gross), Australian Institute of Sport, Australian Sports Commission, Canberra, pp. 3-5.

Clews, G. (1999) Help a mate: Developing an educational awareness program for professional football players around anger, gambling, alcohol and drug use, Communication to *Fourth World Congress of Science and Football,* February, Sydney, Australia.

Fortunato, V. and Marchant, D. (1999) Forced retirement from elite football in Australia, *Journal of Personal and Interpersonal Loss,* 3.

Kirkby, R.J. (1995) Sport psychology in Australia: Past myths and future directions, *Australian Psychologist,* 30,75-77.

Kremer, P.J., Shen, C.J. and Tonn, N. (1999) A profile and typical practices of mental skills practitioners working with elite Australian Football League teams during the 1997 season. Communication to *Monash University Sport Psychology Conference,* January, Melbourne, Australia.

50 THE RELIABILITY AND VALIDITY OF TWO TESTS OF SOCCER SKILL

S.J. MCGREGOR, M. HULSE and A. STRUDWICK
Loughborough University, Loughborough, UK.

1 Introduction

The amount of work performed by soccer players is lower in the second half of a game (Karlsson, 1969) and muscle glycogen has been observed to decline over the course of a soccer match (Saltin, 1973). Therefore, there is evidence of an association between low work rate and low muscle glycogen stores. Within the game of soccer, work-rate is insufficient for success unless allied to and co-ordinated with the fundamental skills of the game (Reilly and Holmes 1983). Although there is an increasing amount of literature concerning work-rate, there is little research on the influence of match play and fatigue on soccer skill. One of the reasons for this is probably the difficulty in assessing skill performance in a reliable way. Soccer demands a range of skills, including, passing, controlling, trapping and dribbling with the ball. Therefore, it is reasonable to include a dribbling and passing test as a reflection of one of the many skills required by soccer players. The purpose of the present study was to devise two valid and reliable tests that could be used in the investigation of fatigue and soccer skill.

2 Methods

Sixty-seven male university level soccer players, aged 18–24 years, volunteered to participate in the study. Playing experience and ability ranged from county representative to semi-professional. Thirty players were randomly assigned to the Loughborough Soccer Passing Test (LSPT) and 37 to the Loughborough Soccer Dribbling Test (LSDT).

  To complete the LSDT, the players dribbled a ball between a line of six cones, 3 m apart, as fast as possible. An additional cone was used to indicate the starting line of the test. The players started with feet behind the starting line as illustrated in Figure 1. The examiner timed the player from the moment the ball was touched in a forward direction after the signal 'go' was given. The time for the test finished when the player stopped the ball on the finishing line. This test was also timed manually using a stopwatch (Accusplit 725 XP). The player had to finish with the ball in his control. Ten such attempts were made with a 1-min break between each and the total of all 10 times was added to give a final accumulated score.

The LSPT involved the performance of 16 passes made from a central zone to four 0.5 m targets that were marked onto benches. The benches were used to allow the ball to rebound and the targets were colour-coded for identification. Small disc cones were placed around the central zone. The passes were randomly selected and called out to the player by the test examiner. The time to complete the test was recorded using a stop watch (Accusplit 725 XP). If the marked areas were missed, or one of the cones was hit during the test, additional 'penalty seconds' were added. If the player ventured out of the outer zone, missed the 0.5-m target or the player hit a cone, 2 s were added. If the bench was missed then a 5-s penalty was added. The LSPT is shown below (Figure 2).

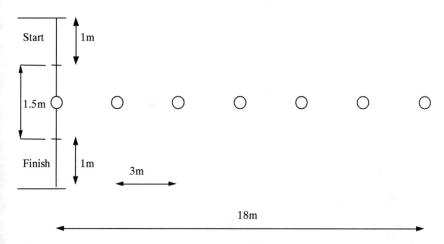

Figure 1. Diagrammatic representation of the Loughborough Soccer Dribbling Test (LSPT).

Following familiarisation, subjects performed the skill tests twice so that the level of agreement between the two trials could be calculated (Bland and Altman, 1986) and the reliability of each test determined. The score generated from each test was correlated (Spearman's rank order) against the group's skill ability ranking, as assessed by the players' coach to determine the validity of the tests.

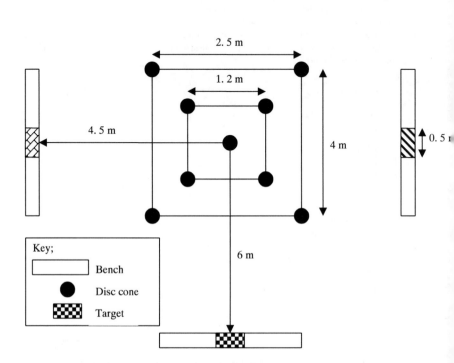

Figure 2. Diagrammatic representation of the Loughborough Soccer Passing Test (LSPT).

3 Results and Discussion

The mean (± SD) skill test scores for the LSDT and the LSPT were 149.0 (18.4) s and 55.0 (4.5) s, respectively. The mean (± SD) difference between trial 1 and trial 2 for both tests are presented in Table 1, together with some more commonly used statistical tests. The validity coefficient was significant for both tests, LSDT r = 0.78 (P < 0.01); LSPT r = 0.64 (P < 0.05).

Table 1. Test – retest reliability using different statistical techniques.

Method	LSDT	LSPT
Mean (± SD) difference in score between test 1 & test 2 (s)	0.08 (± 3.28)	- 0.03 (± 3.13)
95% agreement limits (s)	-6.35 to 6.51	-6.17 to 6.11
t-test	t = 0.89	t = 0.95
Pearson's correlation	r = 0.99	r = 0.83

The results of this study suggest that the LSDT and LSPT are reliable indicators of dribbling and passing skill for the group of players chosen. Additional research is required to identify if these tests can be applied to other groups of the population. The LSDT and LSPT have been useful in the investigation of fluid balance and soccer skill (McGregor et al., 1997).

4 References

Bland, J.M. and Altman, D.G. (1986) Statistical methods for assessing agreement between two methods of clinical measurement, *Lancet,* i, 307-310.

Karlsson, H.G. (1969) Kolhydratomasttning under en fotbollsmatch. Report Department of Physiology III, reference 6, Karolinska Institute, Stockholm (cited in Ekblom, B., 1986, Applied physiology of soccer, *Sports Medicine,* 3, 50-60).

McGregor, S.J., Nicholas, C.W. and Williams, C. (1997) The influence of prolonged, intermittent shuttle running and fluid ingestion on soccer skill, *Proceedings of the Second Annual Congress of the European College of Sports Science,* 20-23 August 1997, Copenhagen, Denmark.

Reilly, T. and Holmes, M. (1983) A preliminary analysis of selected soccer skills, *Physical Education Review,* 6, 64-71.

Saltin, B. (1973) Metabolic fundamentals in exercise, *Medicine and Science in Sports,* 5, 137-146.

51 FATIGUE DURING GAME PLAY: A REVIEW OF CENTRAL NERVOUS SYSTEM ASPECTS DURING EXERCISE

R. MEEUSEN
Faculty of Physical Education and Physiotherapy, Free University Brussel, Brussels, Belgium

1 Introduction

During a soccer match, players exercise at a relatively high intensity. It is therefore not surprising that fatigue occurs, causing performance to fall off in the later stages of the game. Since decision making, as well as technical and tactical aspects are important elements in soccer, fatigue towards the end of the game could cause deteriorations in mental performance. This 'mental' aspect of fatigue is poorly understood, but it is clear that central nervous system functioning plays an important role.

Decrements in prolonged exercise performance are traditionally attributed to peripheral (i.e. muscular) factors. This peripheral fatigue has been well studied and can involve impairments at the neuromuscular transmission, at the cellular level, involve substrate depletion, or accumulation of metabolites (Davies and Bailey, 1997; Davis, 1995; Green, 1997; Meeusen and De Meirler, 1995). Recently, there has been some interest in the possible role of the central nervous system in fatigue. A few mechanisms have been suggested to cause central fatigue, such as accumulation of ammonia in the brain (Bannister and Cameron, 1990), and changes in neurotransmitter concentrations especially changes in brain 5-hydroxytryptamine (5-HT serotonin) concentration and metabolism (Newsholme et al., 1987).

2 Central Fatigue

Most of the research data deal with the possible role of 5-HT as an attractive candidate for the 'central fatigue hypothesis'. This is probably due to the fact that 5-HT has been shown to induce sleep, affect tiredness, depress motor neuron excitability, influence autonomic and endocrine function, suppress appetite, and is involved in mood and depression (Meeusen, 1999). It is supposed that increased concentrations of brain 5-HT, will be the key factor in producing central fatigue (Blomstrand et al., 1989; Newsholme et al., 1987). Increases in brain 5-HT synthesis occurs in response to an increase in the delivery to the brain of tryptophan (TRP).

Since the introduction of the microdialysis technique to collect extracellular fluid, the precursor-induced release of 5-HT has been studied. Some studies (Carboni et al., 1989; Westerink and De Vries, 1991) found increases, while others (Sharp et al., 1992) did not find any increase in 5–HT release following L-TRP administration. However it is noteworthy that in these studies TRP was administered to control undisturbed animals, while there are several lines of evidence to suggest that the effect of L-TRP on 5-HT function depends on the level of serotonergic neuronal activity (Meeusen, 1999). For example, after TRP load extracellular 5-HT release will be more effective in food deprived animals (Meeusen et al., 1996, Schwartz et al., 1990), and L-TRP load in combination with electrical stimulation of the raphe nucleus, will increase 5-HT release in the hippocampus dose-dependently on the stimulus frequency (Sharp et al., 1992). It was also shown that L-TRP administration, acute exercise or the combination of both treatments, as well as restraint stress were able to increase hippocampal 5-HT release (Meeusen et al., 1996; Thorre et al., 1997).

In our microdialysis study (Meeusen et al., 1996), we examined whether exercise-elicited increases in brain tryptophan availability (and in turn 5-HT synthesis) alters 5-HT release in the hippocampus of food-deprived rats. To this end, we compared the respective effects of acute exercise, administration of tryptophan, and the combination of both treatments, upon extracellular 5-HT and 5-hydroxyindole-acetic acid (5-HIAA) levels. All rats were trained to run on a treadmill before implantation of the microdialysis probe and 24 h of food deprivation. Acute exercise (12 m·min^{-1} for 1 h) increased in a time-dependent manner extracellular 5-HT levels, these levels returning to their baseline levels within the first hour of the recovery period. Acute administration of a tryptophan dose (50 mg·kg^{-1} i.p.) that increased extracellular 5-HIAA (but not 5-HT) levels in fed rats, increased within 60 min extracellular 5-HT levels in food-deprived rats. Whereas 5-HT levels returned toward their baseline levels within the 160 min that followed tryptophan administration, extracellular 5-HIAA levels rose throughout the experiment. Lastly, treatment with tryptophan (60 min beforehand) before acute exercise led to marked increases in extracellular 5-HT (and 5-HIAA levels) throughout the 240 min that followed tryptophan administration. This study indicates that exercise stimulates 5-HT release in the hippocampus of fasted rats, and that a pre-treatment with tryptophan (at a dose increasing extracellular 5-HT levels) amplifies exercise-induced 5-HT release. It should be noted that in this study none of the animals showed any sign of fatigue during the exercise session, although extracellular 5-HT levels increased markedly, especially in the L-TRP and exercise trial. Furthermore, since it was shown that during exercise 5-HT, DA, NA and GLU release increases in striatum (Meeusen et al., 1994, 1995, 1997), as well as 5-HT release in hippocampus (Meeusen et al., 1996) without affecting running capacity of the animals, the direct relationship between increased 5-HT release and fatigue could not be established. This indicates that much more research is necessary to discover the possible relationship between 5-HT, exercise and fatigue. Changes in neurotransmission caused by eating, precursor loading, and by exercise can thus affect all of the behavioural and physiological functions that precursor dependent neurons happen to subserve (Meeusen and De Meirleir, 1995).

3 Fatigue in Soccer

Recently, we started a study with Third Division soccer players who were followed prospectively in order to establish the training load during a season. The players had to fill out a training log to evaluate not only their physical activity, but also other parameters, such as muscle soreness, fatigue, well-being and so on. During the period the players were followed, a number of them showed signs of fatigue, but no one became really 'overtrained' (Piacentini et al., in press). Until present, the training log that we administered gave us good indications of their training status and physical and mental well-being. The comparison of their subjective feelings with objective data showed a good correlation.

4 Conclusions

The results of these studies indicate that although it is believed that the so-called 'central fatigue hypothesis' evolves around increased 5-HT concentration in the brain, it seems that more transmitters are involved, and that the interaction between several neurotransmitters could play a key role in physical and mental fatigue during prolonged exercise. In order to monitor the physical and mental well-being of soccer players, a specific training log can be used; however, these logs need to be collected very frequently in order to avoid loss of data.

5 References

Bannister, E. and Cameron, B. (1990) Exercise-induced hyperammonemia: peripheral and central effects, *International Journal of Sports Medicine,* 11(Suppl 2): S129-S142.

Blomstrand, E., Perret, D., Parry-Billings, M. and Newsholme, E. (1989) Effect of sustained exercise on plasma amino acid concentrations and on 5-hydroxytryptamine metabolism in six different brain regions in the rat, *Acta Physiologica Scandinavica,* 136, 473-481.

Carboni, E., Cadoni, C., Tanda, G. and Di Chiara, G. (1989) Calcium dependent, Tetrodotoxin-sensitive stimulation of cortical serotonin release after tryptophan load, *Journal of Neurochemistry,* 53, 976-978.

Davis, M. and Bailey, S. (1997) Possible mechanisms of central nervous system fatigue during exercise, *Medicine and Science in Sports and Exercise,* 29, 45-57.

Davis M. (1995) Central and peripheral factors in fatigue, *Journal of Sports Sciences,* 13, S49-S53.

Green, H. (1997) Mechanisms of muscle fatigue in intense exercise, *Journal of Sports Sciences,* 15, 247-256.

Meeusen, R. (1999) Overtraining and the central nervous system. The missing link?, in *Overload, Performance Incompetence, and Regeneration in Sport* (eds M. Lehmann, C. Foster, U. Gastmann, H. Keizer and J.M. Steinacker), Kluwer Academic/Plenum Publishers New York, pp. 187-202.

Meeusen, R., Sarre, S., Michotte, Y., Ebinger, G. and De Meirleir, K. (1994) The effects of exercise on neurotransmission in rat striatum, a microdialysis study, in *Monitoring Molecules in Neuroscience* (eds A. Louilot, T. Durkin, U. Spampinato and M. Cador), pp. 181-182.

Meeusen, R., Smolders, I., Sarre, S., De Meirleir, K., Ebinger, G. and Michotte, Y. (1995) The effects of exercise on extracellular glutamate (GLU) and GABA in rat striatum, a microdialysis study, *Medicine and Science in Sports and Exercise,* 27(5), S215.

Meeusen, R., Smolders, I., Sarre, S., De Meirleir, K., Keizer, H., Serneels, M., Ebinger, G. and Michotte, Y. (1997) Endurance training effects on striatal neurotransmitter release, an *'in vivo'* microdialysis study, *Acta Physiologica Scandinavica,* 159, 335-341.

Meeusen, R., Chaouloff, F., Thorré, K., Sarre, S., De Meirleir, K., Ebinger, G. and Michotte, Y. (1996) Effects of tryptophan and/or acute running on extracellular 5-HT and 5-HIAA levels in the hippocampus of food-deprived rats, *Brain Research,* 740, 245-252.

Meeusen, R. and De Meirleir, K. (1995) Exercise and brain neurotransmission, *Sports Medicine,* 20, 160-188.

Newsholme, E., Acworth, I. and Blomstrand, E. (1987) Amino acids, brain neurotransmitters and a functional link between muscle and brain that is important in sustained exercise, in *Advances in Myochemistry* (ed G. Benzi G). John Libby Eurotext, London, pp. 127-138.

Piacentini, M., Magnus, L., Shonnon, M. and Meeusen, R. (in press) Monitoring training and overtraining in different sports, *Vlaam Tijdschrift voor Sportgeneeskunde and Sportwetenschappen.*

Schwartz, D., Hernandez, L. and Hoebel, B. (1990) Tryptophan increases extracellular serotonin in the lateral hypothalamus of food-deprived rats, *Brain Research Bulletin,* 25, 803-807.

Sharp, T., Bramwell, S. and Grahame-Smith, D. (1992) Effect of acute administration of L-tryptophan in the release of 5-HT in rat hippocampus in relation to serotonergic neuronal activity: an in *vivo* microdialysis study, *Life Sciences,* 50, 1215-1223.

Thorré, K., Chaouloff, F., Sarre, S., Meeusen, R., Ebinger, G. and Michotte, Y. (1997) Differential effects of restraint stress on hippocampal 5-HT metabolism and extracellular levels of 5-HT in streptozotocin-diabetic rats, *Brain Research,* 772, 209-216.

Westerink, B. and De Vries, J. (1991) Effect of precursor loading on the synthesis rate and release of dopamine and serotonin in the striatum: a microdialysis study in conscious rats, *Journal of Neurochemistry,* 56, 228-233.

52 CAN CROWD REACTIONS INFLUENCE DECISIONS IN FAVOUR OF THE HOME SIDE?

A.M. NEVILL, N.J. BALMER and A.M. WILLIAMS
Research Institute for Sport and Exercise Sciences, Liverpool John Moores University, Liverpool, UK

1 Introduction

The documentary evidence is now overwhelming, home advantage exists in major team sports (e.g., Schwartz and Barsky, 1977; Varca, 1980; Snyder and Purdy, 1985; Pollard, 1986; Courneya and Carron, 1992; Agnew and Carron, 1994; Nevill, Newell and Gale, 1996). Based on their comprehensive review, Courneya and Carron (1992) argued that further verification of the existence of home advantage (i.e., the 'what' of home advantage) is no longer a sufficient rationale to justify game location research. They recommended that future research needs to explore the reasons 'when' and 'why' home advantage occurs.

Several authors have observed that officials consistently make more subjective decisions in favour of the home team (Lefebrve and Passer, 1974; Varca, 1980; Sumner and Mobley, 1981; Greer, 1983; Glamser, 1990). In an attempt to answer the questions 'when' and 'why' does this occur, Nevill et al. (1996) were able to confirm that not only do officials in English and Scottish soccer make more subjective decisions (penalties and sendings-off) in favour of the home side, but also the observed imbalance appears to increase in league divisions with larger crowd sizes.

Thirer and Rampey (1979) and Greer (1983) also recognised the influence that crowds might have on home advantage. When studying home advantage in college basketball, Thirer and Rampey (1979) found that during typical crowd behaviour the visiting teams committed more infractions, (i.e., committed more fouls and lost possession more frequently). During antisocial crowd behaviour (swearing, chanting obscenities), however, home teams committed more infractions. The authors concluded that 'anti-social behaviour from the crowd had a detrimental effect on the home team' (p. 1051).

Greer (1983) also assessed the effect of crowd behaviour (spectator booing) on home and away teams' performance (points scored, turnovers, violations, and a composite score comprising of points scored minus turnovers and violations). Greer observed that during typical crowd behaviour, home teams were better on all four performance measures. During those instances when the crowd was booing (for longer than 15 seconds), the home teams' superiority increased on all four performance measures, two being significant. Greer speculated that the observed improvement in home teams' performance was due either to a decrement in the visiting teams' performance or to referee bias resulting from intimidation by the home crowd (since most of the booing was directed at the officials).

Both studies used quasi-experimental designs to identify the effect of various aspects of crowd behaviour (cheering, booing), and the degree (intensity), on performance outcomes (e.g., fouls). However, by adopting such quasi-experimental designs, researchers recognise that it is almost impossible to untangle other associations that might confound the observed performance outcomes. For example, differences in the number of observed fouls in favour of the home side could be due to a number of home advantage factors, such as frustration or aggression on the part of the away side or the use of more defensive tactics by the away team's coach.

In order to help explain why officials consistently make more subjective decisions in favour of the home team, study 1 investigated whether knowledgeable observers' opinions of 52 tackles/challenges in football matches, recorded on videotape, could be influenced by a partisan crowd's reactions. By isolating the tackles/challenges from their 'real life' setting, the authors recognise that the present study may lose some external validity but, on the other hand, gains the ability to control all the confounding effects associated with the various alternative quasi-experimental designs discussed above.

The purpose of the present study was to determine whether experienced subjects' decisions could be influenced by the noise and reactions of a partisan crowd. It is hypothesised that the presence of such noise leads to some tendency to penalise the home team less, and the away team more.

2 Methods

Eleven participants, made up of experienced footballers, qualified coaches and referees, were asked to assess the legality of 52 incidents from the 1998 Champions' League match between Lens (home) and Panathinaikos (away). This single match was used in order to control the effect of variable crowd sizes between matches, whilst taking advantage of a partisan home crowd of approximately 40,000.

The 11 participants were randomly assigned to either a noise group, receiving background noise (n = 5), or a 'no noise' group, who received only visual stimuli (n = 6). Each participant then judged the legality of the 52 edited incidents, which were presented chronologically, on a 21 in (53 cm) monitor screen, the experimenter recording the adjudications. Coincidentally, of the 52 incidents, 26 were initiated by a home player and the remaining 26 initiated by an away player.

The observers' responses (foul vs no foul), were collapsed into two mean proportions of fouls awarded to the 26 home and 26 away players' challenges. These binomial proportions were analysed using a two-way analysis of variance (ANOVA) with repeated measures, one factor between-subjects (the noise condition; noise vs no noise) and one factor within-subjects (team representation; home vs away player). In recognising the limitations of using traditional ANOVA to compare binomial proportions, the analysis was repeated using an arcsine transformation of the data to stabilise the variances, as recommended by Winer (1971).

3 Results

Table 1 shows the percentages of fouls (± SD) awarded by the two groups of participants, and the referee in response to challenges initiated by either home or away players. This relationship is illustrated graphically in Figure 1. Applying ANOVA to the mean proportions for the two experimental groups, excluding the match referee, revealed a significant two-way interaction between 'noise group' and 'team representation' ($F_{1,9}$ = 8.20, P = 0.019). Analysis incorporating the arcsine transformation yielded similar values ($F_{1,9}$ = 8.03, P =0.020), though reassuringly, the residuals were more acceptably normal than with the initial analysis.

Table 1. Percentage (± SD) of fouls awarded by the observers and the referee, for challenges by the home and away players.

	Home	Away
No Noise (N = 6)	57.6 (± 9.8)	48.3 (± 1.9)
Crowd Noise (N = 5)	50.0 (± 4.7)	56.9 (± 5.0)
Referee (N = 1)	50.0	65.4

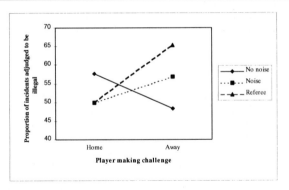

Figure 1. Proportion of fouls awarded by the observers and the referee, for challenges by the home and away players.

The participants, therefore, demonstrated a tendency to penalise the away team significantly more, and the home team significantly less often when exposed to crowd noise alone. Given that the direction of the referee's decisions follow closely that of the 'noise group' (Figure 1), it would be fair to assume that the official could be expected to behave in a manner comparable to the 'noise group'. Moreover, from the referee's perspective, crowd noise would be both louder and more invasive (i.e., directed specifically toward the official), and therefore, if anything their adjudications should demonstrate an increased imbalance. This increased imbalance is suggested in Figure 1, as the referee penalised the away team players to a slightly greater extent than the 'noise group'. This finding should be treated with some caution however, given the distinctly different vantage points of the participants and the referee.

4 Discussion

In the absence of crowd noise, the observers adjudged 57.6% of the home players' challenges to be illegal, compared with 48.3% for the away players. The imbalance in these figures should be attributed to the specific incidents used, as opposed to a trend to favour the home side. In contrast, when exposed to crowd noise, observers awarded 50% of the incidents as fouls by home players and 56.9% by away players. This marked tendency to penalise the away and favour the home players, therefore confirms the earlier hypothesis. These data suggest that decision making is significantly affected by partisan crowd noise, and the direction of the resultant imbalance corresponds with the idea of home advantage.

A number of reports have provided anecdotal evidence to suggest, that referees' decisions can be influenced by the crowds' reactions to favour the home side. For example, Askins (1978) stated:

> 'During the course of any contest there are many incidents which appear ambiguous, even to the most veteran officials. When this occurs, officials do basically what all humans do in such a situation, they seek clarification through any means available at the time. Crowd reaction may sometimes provide the cue which prompts the decision' (p. 18).

The results from the present study have demonstrated how this may occur. Clearly, crowds' reactions to challenges/tackles are capable of influencing observers to be more aggressive or severe when judging challenges by the away players, and more lenient when considering challenges by the home players. Assuming that officials' decisions may be affected in a similar way to the knowledgeable observers, these results provide the first experimental/empirical evidence of how officials' decisions may be influenced by the crowds' reactions in favour of the home side, thus providing a possible explanation for the home advantage phenomenon.

5 References

Agnew, G.A. and Carron, A.V. (1994) Crowd effects and the home advantage, *International Journal of Sport Psychology*, 25, 53-62.

Askins, R.L. (1978). Observations The official reacting to pressure, *Referee*, 4, 17-20.

Courneya, K.S. and Carron, A.V. (1992) The home advantage in sport competitions: A literature review, *Journal of Sport and Exercise Psychology*, 14, 13-27.

Glamser, F.D. (1990) Contest location, player misconduct, and race: a case from English soccer, *Journal of Sport Behavior*, 13, 41-49.

Greer, D.L. (1983) Spectator booing and the home advantage: A study of social influence in the basketball arena, *Social Psychology Quarterly*, 46, 252-261.

Lefebvre, L.M. and Passer, M.W. (1974). The effects of game location and importance on aggression in team sport, *International Journal of Sport Psychology*, 5, 102-110.

Nevill, A.M., Newell, S.M. and Gale, S. (1996) Factors associated with home advantage in English and Scottish soccer, *Journal of Sports Sciences,* 14, 181-186.

Pollard, R. (1986) Home advantage in soccer: A retrospective analysis, *Journal of Sports Sciences,* 4, 237-248.

Schwartz, B. and Barsky, S.F. (1977) The home advantage, *Social Forces,* 55, 641-661.

Snyder, E.E. and Purdy, D.A. (1985) The home advantage in collegiate basketball, *Sociology of Sport Journal,* 2, 352-356.

Sumner, J. and Mobley, M. (1981) Are cricket umpires biased? *New Scientist,* 91, 29-31.

Thirer, J. and Rampey, M. (1979) Effects of abusive spectator behaviour on the performance of home and visiting intercollegiate basketball teams, *Perceptual and Motor Skills,* 48, 1047-1053.

Varca, P. (1980) An analysis of home and away game performance of male college basketball teams, *Journal of Sport Psychology,* 2, 245-257.

Winer, B.J. (1971) *Statistical Principles in Experimental Design.* McGraw-Hill, Tokyo.

53 VISUAL REACTION TIME AND PERIPHERAL VISION IN PROFESSIONAL RUGBY LEAGUE PLAYERS

D. O'CONNOR and M. CROWE
Institute of Sport and Exercise Science, James Cook University, Townsville, Queensland, Australia

1 Introduction

Good visual skills, particularly quick reaction time and being aware of peripheral movement, are important skills in Rugby League. Previous studies in other areas have shown conflicting results with some authors reporting an improvement in visual reaction time (RT) with increased levels of aerobic fitness but others suggesting that RT cannot be enhanced by training. The latter group suggest that training causes changes in cognitive processing rather than any change in RT or other basic visual characteristics (for example, see Abernethy and Wood, 1992). Era et al. (1986) argue for an improvement in RT with exercise and attribute this to an increase in cerebral blood flow and oxygenation. Another study found that university professors who were classified as having high levels of physical activity performed better on simple and choice RT tasks than other professors who had low levels of physical activity (Lupinacci et al., 1993). Sherwood and Selder (1979) also showed that people who ran greater than 42 miles per week did not show the same decrement in RT with age as those who remained sedentary. However, implementation of a six-week walking programme in elderly subjects failed to enhance RT over a non-exercising control group of similar age (Roberts, 1990). This lack of effect of walking on RT was attributed to a failure to investigate previous exercise history but may be more likely due to an inability to significantly change RT. Abernethy and Wood (1992) showed that a four-week visual training programme did not improve general visual performance or sport specific motor performance. Improvements in choice and peripheral RT and accommodation were attributed to familiarity with the test as results were not significantly better than those of a control group (Abernethy and Wood, 1992).

It is commonly accepted that visual skills, particularly RT, decline with increasing age. Era et al. (1986) found that subjects aged between 31–35 years performed better on simple and choice reaction time tasks than subjects aged 51–55 and 71–75 years. Lupinacci et al. (1993) found that professors under 50 years performed better on choice RT tasks but not simple RT tasks compared to professors over 50 yr suggesting that the effect of age is more pronounced with increasingly complex tasks.

During exercise, RT follows a U-shaped curve (Duffy, 1972; Chmura et al., 1994). As exercise starts there is an initial decrease in RT with the activation of the central nervous system but once a critical exercise intensity is reached RT begins to slow (Duffy, 1972). Chmura et al. (1994) associated this critical exercise intensity with the buildup of lactic acid and onset of fatigue.

Little research exists on visual skills in Rugby League players. Quick reaction time is an advantage in both attack and defence in rugby league. Peripheral vision (PV) is important during decisions to jam or slide in defence; offloading in a tackle or going to ground; and during cover defence options. It was therefore the aim of this study to investigate visual performance in Rugby League players using an Acuvision 1000 device in relation to grade, playing position, age, fatigue and training.

2 Methods

2.1 Sample
The subjects used in this study were 63 male Rugby League players recruited from an Australian National Rugby League (NRL) club. The mean (± SD) age, height and weight of the subjects are shown in Table 1 grouped according to grade.

Table 1. Mean age, height and body mass of the three grades of players.

	1st Grade (n = 21)	State League (n = 21)	Under 19s (n = 21)
Age (years)	25.6 ± 1.0	20.9 ± 2.7	18.1 ± 4.4
Height (cm)	180.4 ± 5.9	179.1 ±6.1	179.4 ± 5.8
Body Mass (kg)	94.6 ±7.9	88.7 ± 7:5	88.1 ± 8.3

2.2 Procedure
Ethical approval was granted by the James Cook University Ethics Committee and all procedures explained to the players prior to participation in the study. Visual RT and accuracy and PV time and accuracy were tested using an ACUVISION 1000. The centre of the Acuvision board was mounted at eye level with subjects standing at arm's length from the board. The board has a grid of 1014 squares behind 120 of which lie hidden red lights which can be illuminated in different random sequences. The test protocols included a RT test and a PV test. The RT test consisted of a series of 60 lights to which the subject was to react as quickly as possible by placing a finger on the light (Acuvision code FF60, speed 7). The frequency of lights was 1.072 lights·s^{-1} with each light illuminated for 70% of the time cycle. The subjects were scored on time to complete the test and

the number of correct responses (touched within 70% of the light cycle). The PV test involved the same procedure and scoring as the RT test but included visual fixation on a green light illuminated in the centre of the board. Any visual deviation from the green light was deducted from the number of correct hits.

The subjects initially performed an RT and PV test on day 1 for familiarisation. The same tests were repeated on day 4. On day 8 the subjects underwent a 60 s 'all-out' anaerobic cycle test on a Repco ergometer with blood lactate measured 5 min post test (ACCUSPORT) immediately followed by the RT and PV tests.

The subjects then underwent a six-week training period on the Acuvision which emphasised individual players' weak sectors with a random speed protocol (speed 5–9, mode FF120). The subjects spent an average of 5 min on the board twice a week plus field-based training once a week, which emphasised visual skills such as mirroring, catching of golf balls and use of reaction belts. At the conclusion of the training period the subjects were again tested using the RT and PV tests.

All values are presented as mean ± standard deviation.

3 Results and Discussion

3.1 Effect of learning on RT and PV time and number of correct hits

To establish if there was any learning effect, a comparison was made between the scores on the first trial (Day 1) and the scores from the second trial (Day 4) on both the RT and PV tests. Paired, one-tailed t-tests revealed that performances in the second trial were significantly better than those in the first (see Table 2). This indicates a learning or familiarity component to both tests. Consequently, the best score from these two trials was used as the baseline score for each subject. These findings are in agreement with previous work by Abernethy and Wood (1992) who state that performance on visual tests often improves due to familiarity with the test.

Table 2. Mean (± SD) test time and number of correct hits on the RT and PV tests for trial 1 (Day 1) and trial 2 (Day 4) for all subjects.

	Trial 1	Trial 2
RT (time)	38.05 ± 2.72	36.63 ± 2.44**
RT (hits)	41.98 ± 6.93	45.05 ± 5.38**
PV (time)	43.48 ± 3.34	41.82 ± 3.14**
PV (hits)	25.93 ± 8.0	30.58 ± 7.34**

**$P < 0.001$

3.2 Effect of grade on RT and PV time and number of correct hits

To investigate whether a difference existed in visual skills between players of higher and lower grades, the subjects were grouped according to grade including 1st Grade, State League or Under-19s. Players were required to have played a minimum of 80% of games to qualify for classification to a grade. One-way ANOVA revealed that grade had no significant influence on either the correct number of hits or the time to complete the RT test (see Table 3). However, 1st Graders were significantly quicker and more accurate on the PV test than the Under-19s players. These findings may reflect the longer training history of the 1st Grade players compared to those of the Under 19s or may reflect the fact that subjects with good visual skills are able to compete at a higher standard.

Table 3. Mean (± SD) test time and number of correct hits on the RT and PV tests for 1st Grade, State League and Under-19s groups.

	1st Grade	State League	U/19s
RT (time)	35.91 ± 2.46	36.85 ± 2.25	37.35 ± 2.52
RT (hits)	45.86 ± 5.54	45.71 ± 4.93	43.86 ± 6.39
PV (time)	40.35 ± 3.30	41.59 ± 3.07	43.90 ± 3.01*
PV (hits)	35.48 ± 7.02	33.52 ± 6.98	27.57 ± 6.65*

* $P < 0.01$

3.3 Effect of playing position on RT and PV time and number of correct hits

Players were grouped as either backs, halves and hookers, front row or back row. RT and PV did not differ between these playing positions, possibly indicating the importance of these two parameters for all players regardless of playing position.

3.4 Effect of age on RT and PV time and number of correct hits

The subjects were grouped into the following age groups: 15–19, 20–24, 25–29 and 30–35 year. A one-way ANOVA revealed that test time and the number of correct hits for the RT test did not vary as a result of age (see Table 4). This is consistent with the report of Lupinacci et al. (1993) who only found differences with age on complex RT tasks but not on simple RT tasks. However, the same analysis performed on the PV test data revealed that the 15–19 year group had a significantly slower test time and significantly lower number of correct hits than the 25–29 year group (see Table 4).

Table 4. Mean (± SD) test time and number of correct hits on the RT and PV tests for each age group.

	30–35 years	25–29 years	20–24 years	15–19 years
RT (time)	37.95 ± 2.47	34.77 ± 2.52	36.20 ± 2.05	37.31 ± 2.38
RT (hits)	41.67 ± 6.71	47.57 ± 4.47	46.82 ± 4.72	44.50 ± 5.27
PV (time)	40.90 ± 4.86	38.90 ± 2.42	41.48 ± 2.72	43.03 ± 3.03*
PV (hits)	34.17 ± 10.67	38.71 ± 3.25	34.33 ± 3.01	32.33 ± 6.02*

* P < 0.01

Although the literature indicated that RT deteriorates with age, the maximum age in this study was 33 year, which would classify all subjects as young according to the literature. However, this finding follows the trend above where the first graders, who are predominantly in the older age groups, showed faster times and greater number of hits on the PV test than the Under-19s.

3.5 Effect of fatigue on RT and PV time and correct number of hits

To establish if lactate-induced fatigue had any effect on visual skills, baseline scores were compared to scores gained on the visual tests 5 min after a 60 s 'maximal' cycle ergometer test. Post-exercise blood lactate levels averaged 14.1 ± 1.9 with a range from 10.5 to 18.3 mmol $\cdot$l^{-1} which is comparable to levels reached during games (>10 mmol $\cdot$l^{-1}). Table 5 illustrates a significantly slower time to perform the tests when fatigued (P < 0.05) and significantly lower number of hits (P < 0.001). Consequently, RT and PV are significantly impaired following a 60-s maximal effort on a cycle ergometer. This supports the findings of Chmura et al. (1994) who found an association between an increase in RT and the build-up of lactate.

Table 5. Mean (± SD) test time and number of correct hits on the RT and PV tests at rest and after a 60-s cycle test.

	Baseline	Post-exercise
RT (time)	36.5 ± 2.6	37.5 ± 2.3*
RT (hits)	45.6 ± 5.4	43.8 ± 5.6**
PV (time)	41.1 ± 2.8	42.6 ± 2.8*
PV (hits)	34.4 ± 6.2	28.1 ± 6.8**

* P < 0.05 ** P < 0.001

3.6 Effect of a training programme on RT and PV time and number of correct hits

Test time significantly improved by approximately 2 s on both the RT and PV tests along with an approximate increase of 5 hits on number of correct responses following a six week training programme. Caution should be taken in interpreting these findings as they may just reflect further familiarisation with the Acuvision 1000 rather than a true increase in visual skills.

Table 6. Mean (± SD) test time and number of correct hits on the RT and PV tests at baseline and following a six-week training programme.

	Pre	Post
RT (time)	36.6 ± 2.4	34.2 ± 2.3**
RT (hits)	45.1 ± 5.2	50.3 ± 5.2**
PV (time)	42.0 ± 3.2	39.7 ± 2.7**
PV (hits)	30.3 ± 7.5	36.6 ± 6.1**

** $P < 0.001$

4 Conclusions

This preliminary study into the visual skills of Rugby League players using an Acuvision 1000 has revealed a learning effect with repeated testing; an age and grade effect where PV was generally better in the older more experienced players; no difference in visual skills between different playing positions; deterioration in performance with fatigue; and an improvement in performance on all measured parameters after a six-week Acuvision 1000 and field-based training programme. However, as field-based visual skills are difficult to assess, these preliminary findings may not necessarily translate to improved performance on the field.

5 References

Abernethy, B. and Wood, J. (1992) An Assessment of the Effectiveness of Selected Visual Training Programs in Enhancing Sports Performance, *Australian Sports Commission,* Canberra.

Chmura, J., Nazar, K. and Kaciuba-Uscilko, H. (1994) Choice reaction time during graded exercise in relation to blood lactate and plasma catecholamine thresholds, *International Journal of Sports Medicine,* 15, 172-176.

Duffy, E. (1972) Activation, in *Handbook of Psychophysiology* (eds N.S. Greenfield and R.A. Sternbach), Holt, Rinehart and Winston Inc, New York, pp. 577-595.

Era, P., Jokela, J. and Heikkinen, E. (1986) Reaction and movement in men of different ages: A population study, *Perceptual and Motor Skills,* 63, 111-130.

Lupinacci, N.S., Rikli, R.E., Jones, C.J. and Ross, D. (1993) Age and physical activity effects on reaction time and digit symbol substitution performance in cognitively active adults, *Research Quarterly in Exercise and Sport,* 64, 144-150.

Roberts, B.L. (1990) Effects of walking on reaction and movement times among elders, *Perceptual and Motor Skills,* 71, 131-140.

Sherwood, D.E. and Selder, D.J. (1979) Cardiorespiratory health, reaction time and aging, *Medicine and Science in Sports and Exercise,* 11, 186-189.

54 EFFECTS OF PRACTICE AND KNOWLEDGE OF PERFORMANCE ON THE KINEMATICS OF BALL KICKING

A.M. WILLIAMS, P. ALTY and A. LEES
Research Institute for Sport and Exercise Sciences, Liverpool John Moores University, Liverpool, UK

1 Introduction

A fundamental issue in motor control and learning is how movement patterns change during the acquisition of skill. Newell (e.g., 1985, 1991) proposed that motor skill learning may be divided into two distinct stages, namely co-ordination and control. In the early stage of skill acquisition, the learner must acquire the appropriate topological characteristics of the body and limbs (co-ordination) and then progress, in the second stage, to refine the scaling of the acquired pattern of relative motions (control).

Topological shifts in movement characteristics are assumed to reflect the learner's attempt to master the redundant degrees of freedom involved in a task (Bernstein, 1967). Initially, the learner may 'freeze' some degrees of freedom by keeping certain joint angles, limbs, or the whole body, rigidly fixed. Constraining movement in this way may reduce the complexity of the task by limiting the parameters that need to be incorporated into the control system. As learning progresses, the initially frozen joints develop a greater range and independence of motion. This process is thought to reflect the incorporation of various degrees of freedom into larger functional units or 'coordinative structures' that are able to exploit passive reactive forces in the perceptuo-motor workspace (see Williams, Davids and Williams, 1999).

Another important issue in the acquisition of motor skills is the nature of the information that learners use to discover or facilitate appropriate solutions to the movement problem. Although verbal prescriptive feedback (KP) can provide the learner with information for producing and correcting attempts to perform the skill, few studies have examined how such feedback may change the topological characteristics of movement (Magill and Schoenfelder-Zohdi, 1996). Previous research has tended to rely on outcome rather than process measures of skill learning.

This study attempted to: (i) examine those features of a kicking task which may change as a result of practice; (ii) determine if such changes can be facilitated through feedback based intervention; (iii) ascertain whether, as practice continues, there is a shift in learners' movement patterns towards those exhibited by experts. It was predicted that novice participants would initially constrain or freeze out various degrees of freedom but that these constraints would be lifted as practice continued. Significant increases in lower limb segment velocities and joint ranges of motion were expected to parallel the release of the constraints on the degrees of freedom. Knowledge of performance was expected to facilitate this process by helping to establish appropriate limb movement constraints unique to the kicking skill.

2 Methods

2.1 Participants
Eight males and 16 females volunteered to take part in the study. The female participants (aged 22.3 ± 3.5 years) comprised the novice group since they had no previous experience of any code of football. The males (23.4 ± 2.6 years) were classed as expert soccer players since they had over ten years of playing experience at a semi-professional or varsity level.

2.2 Task
The task to be learnt was the instep drive in soccer. Participants were asked to kick a stationary ball with their right foot towards a 7-m x 2.5-m target at a distance of 11 m. Emphasis was placed on kicking the ball with power rather than accuracy.

2.3 Procedure
Expert and novice participants initially completed a pre-test involving three instep kicks. These trials were filmed using a PAL VHS video camera (Panasonic F-15) positioned to the right of the participant, perpendicular to the plane of motion.

Novice participants were randomly assigned to either a KP or No-KP group. Both groups received 15–20 minutes of practice on the instep kick for four consecutive weeks. Sixty instep kicks were completed each practice period. The KP group received verbal prescriptive feedback every third practice trial. Knowledge of performance was given in the form of a verbal statement derived from Table 1. The feedback statements, which provided a list of corrections related to the key components of the skill, were developed by reference to various coaching texts as well as the opinions of three expert soccer coaches. Statements were organised in a sequential order (1 to 12) beginning with the error that was perceived to be most important to correct for the instep kick to be performed correctly. Each statement prescribed what should be done to perform the skill more effectively rather than just describing the error. The No-KP group merely practised the skill with no corrective feedback.

Novice participants were filmed undertaking a 'post-test' involving a further three instep kicks at the completion of the practice period. Pre-test and post-test performances were assessed via two-dimensional film analysis at a sampling rate of 50 Hz. The points digitised included the hip joint (greater trochanter), the knee joint (lateral epicondyle of the femur), the ankle (lateral mallelous), the tip of the toes (fifth metatarsal) and the ball. Data were smoothed using a 4th order Butterworth filter with a cut off frequency of 7Hz.

2.4 Data analysis
The following measurements were recorded for both groups of novice participants pre- and post-test and comparisons were made to the prototypical pattern displayed by the expert control group:

- Maximum resultant linear foot velocity (MLFV).
- Maximum ball velocity (MBV).
- Maximum ankle (MAAV) and knee (MKAV) angular velocities.
- Foot, upper and lower leg ranges of motion (ROM).
- Timing of upper and lower leg peak angular velocities in relation to ball contact.

These variables were analysed using separate two-way factorial ANOVAs to determine pre- to post-test differences across the two novice groups. In these analyses, Group (KP vs No-KP) was the between-participants variable and Test (pre vs post) the within-participants variable. Separate one-way independent samples ANOVAs were employed to identify any differences between the two experimental groups' post-test performance and that of the expert group.

Angle to angle diagrams of the upper and lower leg and the lower leg and foot were produced to assess qualitatively any changes in movement topology.

Table 1. Priority list of knowledge of performance (KP) statements

Priority	KP Statement
1	Place the support foot to the side of the ball
2	Strike the ball with the instep of the foot
3	Strike the ball in the centre
4	Keep your head down and look at the ball
5	Place the support foot approx. 30cm away from the ball
6	Bring your knee over the ball
7	Follow through with the foot after contact
8	Take a longer last stride to the ball
9	Approach the ball with speed
10	Flex your knee as you bring your leg back
11	Use your arms for balance
12	That was correct

3 Results

3.1 Peak velocities

Maximum resultant linear foot velocity (MLFV). The No-KP (11.46 vs. 11.88 m s^{-1}) and KP (10.98 vs. 12.16 m s^{-1}) participants increased their MLFV from pre- to post-test ($F_{1,14} = 4.50$, $P < 0.05$). However, the mean MLFV values reported for the expert group (16.96 m s^{-1}) were still significantly higher than those reported for the two experimental groups ($F_{2,21} = 43.61$, $P < 0.05$).

Maximum ball velocity (MBV). Increases in ball velocity were observed from pre- to post-test in both the No-KP (13.47 vs 14.58 m s^{-1}) and KP (12.83 vs. 15.23 m s^{-1}) groups ($F_{1,14}$ = 24.32, P < 0.05). However, a larger increase was recorded for the KP compared with the No-KP group, P < 0.05. Expert ball velocity (21.1 m s^{-1}) was significantly higher than that recorded for the novice groups (F $_{2,21}$ = 61.0, P < 0.05).

Maximum knee angular velocity (MKAV). Both the No-KP (11.45 vs 11.88 rad s^{-1}) and KP (10.98 vs 12.16 rad s^{-1}) groups increased MKAV from pre- to post-test ($F_{1,14}$ = 15.92, P < 0.05). The values recorded were lower than those of the expert (19.96 rad s^{-1}) group ($F_{2, 21}$ = 6.53, P < 0.05).

Maximum ankle angular velocity (MAAV). There was a significant interaction between Group x Test ($F_{1,14}$ = 5.48, P < 0.05). Post hoc Scheffé analysis indicated that there was a significant difference between the No-KP and KP groups on the post-test, P < 0.05. The No-KP participants increased their MAAV from 4.43 to 6.35 rad s^{-1}, while the KP participants decreased their MAAV from 4.30 to 4.24 rad s^{-1}. The values reported for the KP group were similar to those of the expert participants (3.50 rad s^{-1}), P > 0.05.

3.2 Range of motion (ROM)
Foot. Increases in foot ROM occurred in both No-KP (1.60 vs 1.83 rad) and KP (1.59 vs 1.80 rad) groups with practice ($F_{1,14}$ = 30.04, P < 0.05). No significant differences were observed between the expert (1.79 rad) and novice groups ($F_{2,21}$ = 0.05, P > 0.05).

Lower leg. There was a significant increase in lower leg ROM from 1.43 to 1.49 rad for the No-KP group and from 1.35 to 1.55 rad for the KP group ($F_{1,14}$ = 4.46, P < 0.05). These values were lower than those reported for the expert (1.88 rads) group ($F_{2,21}$ = 3.70, P < 0.05).

Upper leg. Increases from pre- to post-test were observed for the No-KP group (1.11 to 1.19 rad) and the KP (1.10 to 1.24 rad) group ($F_{1,14}$ = 18.30, P < 0.05). No differences were apparent between the experts' ROM (1.16 rad) and that of the two novice groups, $F_{2,21}$ = 1.47, P > 0.05.

3.3 Timing of actions
Upper and lower leg peak angular velocities in relation to ball contact. All participants reached a maximum lower leg angular velocity at or near ball contact. The No-KP and KP groups reached maximum upper leg angular velocity at 70.9% and 69.8% of maximum lower leg angular velocity on the pre-test, while on the post-test these values had decreased to 64.7% and 64.9% ($F_{1,14}$ = 16.53, P < 0.05). The post-test values were closer to those of the expert (62.8%) group ($F_{2,21}$ =0 .26, P > 0.05).

3.4 Qualitative assessment of performance
Qualitative analysis of angle to angle diagrams of the upper and lower leg and the lower leg and foot indicated that practice and feedback resulted in topological shifts in the characteristics of novices' movements to mirror more closely those of the experts (e.g., see Figures 1 and 2).

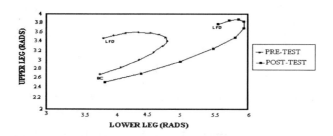

Figure 1. Pre- and post-test angle-angle plots of the lower and upper leg for one participant from the KP group.

4 Discussion

Findings provide support for Newell's (1985, 1991) two-stage model and Bernstein's (1967) notion of freezing and unfreezing degrees of freedom during motor skill learning. Initially, novice participants constrained movement at the hip, knee, and ankle and then following practice these constraints were lifted to produce a more effective kicking pattern. Participants were able to increase the velocities of the knee, foot, and ball significantly and these improvements were associated with increases in joint motion at the hip, knee and ankle. These observations support the claim that degrees of freedom are constrained early in practice and then released as practice continues.

Changes in co-ordination were observed in both novice groups as a result of practice. However, participants who received KP produced a larger increase in MBV and decrease in MAAV compared with those participants who practised the skill without KP. Although not significant, a larger increase in lower leg ROM was also observed for the KP group. The provision of verbal prescriptive statements facilitates the development of a more effective kicking pattern

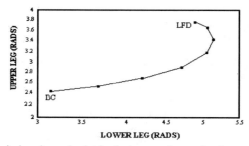

Figure 2. A typical angle–angle plot for the lower and upper leg for one expert participant.

Angle to angle diagrams highlighted topological shifts in the characteristics of the novices' kicking pattern to mirror those of the expert players. However, more direct comparison of the efficacy of KP and practice conditions may require the use of expert soccer players of the same gender as a control group, longer practice periods, and a delayed retention or transfer test.

5 References

Anderson, D.I. and Sidaway, B. (1994) Coordination changes associated with practice of a soccer kick, *Research Quarterly for Exercise and Sport,* 65, 93-99.

Bernstein, N. (1967) *The Coordination and Regulation of Movement.* Pergamon, New York.

Magill, R.A. and Schoenfelder-Zohdi, B. (1996) A visual model and knowledge of performance as sources of information for learning a rhythmic gymnastic skill, *International Journal of Sport Psychology,* 27, 7-22.

Newell, K.M. (1985) Coordination, control and skill, In *Differing Perspectives in Motor Learning, Memory, and Control* (edited by D. Goodman, R.B. Wilberg and I.M. Franks), Elsevier Science Publishing, Amsterdam, pp. 295-317.

Newell, K.M. (1991) Motor skill acquisition, *Annual Review of Psychology,* 42, 213-237.

Williams, A.M., Davids, K. and Williams, J.G. (1999) *Visual Perception and Action in Sport,* E. and F.N. Spon, London.

PART EIGHT

Football Training

55 AGILITY AND SPEED IN SOCCER PLAYERS ARE TWO DIFFERENT PERFORMANCE PARAMETERS

D. BUTTIFANT, K. GRAHAM and K. CROSS
New South Wales Institute of Sport, Sydney, Australia.

1 Introduction

There is a consensus amongst the soccer fraternity that the game is becoming more dynamic. The increased velocity of the game may be attributed to an interplay of influences. It has been proposed that speed and agility are two performance characteristics that positively correlate with the intensity of the game (Buttifant, 1998). However, these two performance characteristics have been purported to be synonymous within the sporting world.

There have been a multiple of tests that have investigated the agility characteristics of athletes; 505, t-test and the Illinois test (Getchell, 1979; Draper and Lancaster, 1985).

It has been previously reported that the Illinois agility test correlates strongly with acceleration and velocity, whereas the 505 test was shown to have no significant correlation with velocity but rather with acceleration (Draper and Lancaster, 1985). Previous agility tests have displayed variances in results which may be attributed to extrinsic influences as well as making it difficult to identify what elements have contributed to a changed result. Agility and speed tests need to be conducted to ascertain whether these performance characteristics are related. The purpose of this study was to determine whether an agility test can discriminate the speed and agility components of a group of National and State representative male soccer players.

2 Methods

A group of 21 male junior national and state representative soccer players (age: 16.1 ± 1.23 years; body mass: 69.2 ± 7.09 kg; height: 175.1 ± 5.83 cm) participated in this study. All subjects completed a 15-min low intensity aerobic warm-up with stretching and striding. All subjects wore football boots throughout the testing. Electronic timing gates were used and all tests were performed on grass. The subjects were tested twice over three days with a one-day rest in between.

The subjects performed 2 x 20-m sprint test followed by a nominal 20-m agility test. The agility test involved an initial movement to the left followed by a movement to the right. This pattern was then repeated.

After finishing this pattern subjects then performed the opposite pattern involving an initial movement to the right. Stands (1.3 m pole) were used to avoid cutting over the witches hats (refer to Figure 1).

The subjects were provided with two familiarisation runs on both sides prior to the agility test. The agility component was derived from the mean differential, which is calculated by deducting the 20-m sprint time from the mean agility time. Technical error of measurement (TEM) values between day 1 and day 2 were calculated for the 20-m sprint and two agility runs (agility right and left) to assess the reliability of the test (Table 1).

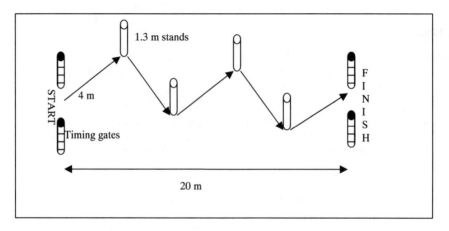

Figure 1. The 20-metre agility test.

Table 1. Technical error of measurement data .

	Agility right	Agility left	Sprint
TEM (s)	0.17	0.12	0.04
%TEM	2.79	1.94	1.36

4 Results

Table 2. Summary data for measures (time in s).

	Sprint 20 m	Agility Left	Agility Right	Mean Agility	Agility Left-Sprint 20 m	Agility Right-Sprint 20 m	Mean Differential
Mean	3.01	6.23	6.12	6.13	3.23	3.12	3.18
Min	2.84	5.96	5.82	5.89	2.88	2.88	3.03
Max	3.13	6.57	6.56	6.57	3.32	3.39	3.50
SD	0.017	0.16	0.16	0.16	0.15	0.15	0.14

The mean ($\pm$ SD) 20-m sprint was 3.01 ± 0.017 s. The mean times for the left and right agility were 6.23 and 6.12 s, respectively. There was no significant difference ($P > 0.05$) between left and right agility.

Table 3. Summary and correlation (r^2) data for measures.

	Sprint 20	Agility left	Agility right	Mean agility	Agility left-sprint 20	Agility right-sprint 20
Sprint 20 m	1					
Agility left	0.092	1				
Agility right	0.093	0.481	1			
Mean agility	0.109	0.849	0.844	1		
Agility left – sprint 20	0.049	0.742	0.299	0.587	1	
Agility right – sprint 20	0.051	0.297	0.735	0.578	0.461	1
Mean differential	0.043	0.59	0.585	0.694	0.842	0.83

Table 3 presents the r^2 correlations for the agility and sprint tests. The table indicates that the 20-m sprint test did not correlate with any of the variables associated with the agility test. The r^2 correlation for the mean agility and the 20-m sprint tests is 0.109, showing that only 10% of the sprint test contributes to the 20-m agility score. In contrast, the mean differential and 20-m sprint data displayed an even lower r^2 correlation of 0.043.

5 Discussion

The agility test that was administered demonstrated a low correlation with the 20-m sprint test. The data indicate that approximately 10% of the mean time in the agility test can be accounted for by the sprint speed. The mean differential value represents the agility component, and could only be accounted by 4% to the sprint speed (Table 3).

It is still to be resolved what contributes to the 'slowing' in the run time in the agility test. There are many contributing factors that may influence the other 90% unaccounted variables. These could be the ability to react quickly, decelerate quickly, anticipate or the eccentric leg strength. However, given the nature of the added components, it may be reasonable to attribute it to the 'agility' of the player.

From the findings of this study it would be acceptable to state that this agility test, which was conducted on a select group of elite junior soccer players, can discriminate the two different performance parameters.

Previous agility tests that have investigated agility have lacked validity and have not accounted for the speed component (Gates and Sheffield, 1940; Hunsicker and Reiff, 1975; Getchell, 1979). The agility test that was conducted in this study was reliable which was demonstrated in the %TEM data (Table 1). Agility was shown to be a different performance parameter; however, there still needs to be further investigation conducted in this area. It would be interesting to determine whether the scores recorded in these tests can be altered by specifically designing a training programme to improve speed or agility.

Another pertinent area that warrants further study is exploring the measurement of reactivity over shorter distances, which may eliminate the speed component and place more emphasis on the agility parameter.

6 References

Buttifant, D. (1998) 5th Applied Physiology Conference, Tasmanian Institute of Sport, Launceston, Tasmania.

Draper, J.A. and Lancaster, G. (1985) The 505 Test: a test for agility in the horizontal plane, The Australian Journal of Science and Medicine in Sport, 17, (1), 15-18.

Gates, D.S. and Sheffield, R.P. (1940) Test of change in direction as measurement of different kinds of motor ability in boys of seventh, eighth and ninth grades, Research Quarterly, 11, 136-147.

Getchell, B. (1979) Physical Fitness a Way of Life, John Wiley and Sons, New York.

Hunsicker, P.R.G. and Reiff, G. (1975) AAHPER Youth Fitness Manual, AAHPER, Washington.

56 EFFECT OF SIX WEEKS OF ISOKINETIC STRENGTH TRAINING COMBINED WITH SKILL TRAINING ON FOOTBALL KICKING PERFORMANCE

P. DUTTA and S. SUBRAMANIUM
Sports Authority of India, Lakshmibai National College of Physical Education,
Trivandrum, Kerala, India.

1 Introduction

It is widely accepted that speed and accuracy in kicking depend upon explosive strength of lower extremities and it is also reported that kicking performance is not affected by different kinds of strength training. Trolle et al. (1993) stated that high resistance strength training did not improve speed in kicking performance. In soccer, strength training can improve kicking performance due to increase in strength (DeProft et al., 1988). Cabri et al. (1988) reported that isokinetic strength training could be effective in increasing the kicking performance of soccer players. Taina et al. (1993) have reported that with maximal strength training of lower limbs of soccer players, speed in kicking performance increased. There are a number of research findings on strength training of soccer players employing isotonic methods and its influence on kicking performance but very few research reports about the effect of isokinetic training on maximum and explosive strength of soccer players. Hence the purpose of the study was to ascertain the effect of six weeks of isokinetic strength training combined with skill training on football kicking performance.

2 Methods

2.1 Subjects
The study was conducted at Netaji Subhash National Institute of Sports, Patiala, India, on 22 national-level players, assigning them randomly to two groups of 11 each. One group was trained by following a specific strength training programme using the Orthotron II isokinetic exercise system (Cybex, Ronkonkoma) and on football skills for six weeks; the other group was trained on football skills only. The pre-training and post-training assessments of peak torque and torque acceleration energy of flexor and extensor muscles of hip, knee and ankle joints were performed using the Cybex 340 isokinetic dynamometer and observing standard protocols. Performance in kicking ability was assessed through selected field tests.

2.2 Variables assessed and tests conducted

Peak torque (PT) is the highest torque value observed from all repetitions and in all points in the range of motion and is indicative of maximum muscular tension capability. Torque acceleration energy (TAE) is the amount of energy expended (work performed) in the first 0.125 s of torque production. The TAE, an indicator of muscular 'explosiveness', was calculated from the isokinetic data.

Kicking for distance and accuracy with the instep and the inner instep of the right and left legs was assessed using the skill tests of Warner (1950), Crew (1968), Cabri et al. (1988), and the national sports talent schemes of the Sports Authority of India (1992). For testing the reliability of all the tests, a test-retest method was used and for evaluation of the validity of the test, the correlation between test score and expert rating was performed. Since there were significant correlations between the performance scores of the subjects in each skill test and expert rating, and also in test–retest score, all the skill tests were considered to be valid tests for further use in this research.

2.3 Statistical procedure

The statistical analysis was performed using the SPSS software package. The effect of training on the two groups was determined using t-tests. Alpha significance level was set at $P < 0.05$.

Table 1. Isokinetic training programme for one week.

Isokinetic exercise	Angular velocity $(rad.s^{-1)}$	Number of repetitions			Number of sets			Recovery time between sets
		Tues	Thur	Sat	Tues	Thur	Sat	
Hip	0.79	6	8	6	3	3	3	2 min
flexion/	1.05	5	10	6	3	3	3	2 min
Extension	3.14	5	10	10	3	3	3	2 min
Knee	0.79	6	8	6	3	3	3	2 min
flexion/	1.05	5	10	6	3	3	3	2 min
Extension	3.14	5	10	10	3	3	3	2 min
Ankle	0.79	6	8	6	3	3	3	2 min
plantar/	1.05	5	10	6	3	3	3	2 min
dorsi-flexion	3.14	5	10	10	3	3	3	2 min
Load		Low	High	Med	Low	High	Med	

3 Results

The mean difference in scores of the isokinetic training group between pre-training and post training at hip, knee, and ankle joints was analysed by applying 't' test for testing of significance.

The results show that isokinetic strength training combined with technical training improved peak torque of right and left hip, knee and ankle joints except in the case of extension of right and left knee joints at 0.79 rad·s^{-1}. Improvements in extension of right and left hip joints (t = 5.86 and 4.40) and in plantar and dorsi-flexion of right ankle (t = 2.28, 3.11) at 0.79 rad·s^{-1} were statistically significant.

The results also revealed that isokinetic strength training with technical training improved torque acceleration energy of hip, knee and ankle joints at 3.14 rad·s^{-1}. Except in knee flexion of the right leg and in dorsi-flexion of the right ankle joint (t = 3.73 and 3.09), the improvement in all other tests was found to be statistically non-significant.

In soccer training test, isokinetic strength training with technical training improved the soccer skills of the subjects in all tests. Results depict that performance in kicking with the instep of the foot for accuracy and distance of right and left leg and in kicking with inner instep of the foot for accuracy with left leg was statistically significant with 't' values 5.25, 2.68, 3.07, 2.87 and 2.27, respectively.

The control group, with only technical training showed improvement in peak torque of plantar and dorsi-flexion of right and left ankle joint, hip joint extension of right and left leg and in knee joint flexion of left leg but in other variables a decrease in performance was seen. A statistically significant improvement was seen only in dorsi-flexion and plantar-flexion of right leg and in dorsi-flexion of left leg.

In torque acceleration energy, significant improvement was evident only in flexion of the hip joint of the right leg, dorsi-flexion and plantar-flexion of right and left ankle joints. In extension of right and left hip joints, knee joint extension of right and left legs and in flexion of the left knee joint, a decrease in performance was found. However, the improvement in kicking performance for distance and accuracy was statistically non-significant.

Table 2. Mean difference between pre-training and post-training performance in peak torque at hip, knee and ankle joints for the isokinetic group.

Joint action	Peak torque (ft lbs)		't' value	P value
	Pre-training (N=11) Mean ± SD	Post training (N =11) Mean ± SD		
Flexion Hip RL	103. 54 ± 22.37	118.63 ± 14.96	1.86	NS
Extension Hip RL	176.63 ± 29.92	265.72 ± 40.60	5.86	S
Flexion Hip LL	103.09 ± 28.23	118.45 ± 19.53	1.48	NS
Extension Hip LL	169.18 ± 32.84	235.81 ± 38.04	4.40	S
Flexion Knee RL	120.27 ± 18.75	124.9 ± 15.76	0.63	NS
Extension Knee RL	180.18 ± 49.01	161.45 ± 24.05	1.14	NS
Flexion Knee LL	109.27 ± 17.97	116.27 ± 12.99	1.05	NS
Extension Knee LL	161.81 ± 24.37	156.54 ± 24.99	0.50	NS
Dorsi-flexion Ankle RL	19.90 ± 5.40	24.72 ± 4.45	2.28	S
Plantar flexion Ankle RL	65.18 ± 17.98	89.81 ± 19.11	3.11	S
Dorsi-flexion Ankle LL	21.36 ± 6.38	24.45 ± 4.14	1.35	NS
Plantar flexion Ankle LL	58.09 ± 22.68	69.72 ± 27.82	1.07	NS

RL – Right leg, LL – Left leg, NS – Not significant, S – Significant.

Table 3. Mean difference between pre-training and post-training performance in torque acceleration energy at hip, knee and ankle joints for the isokinetic group.

Joint action	Torque acceleration energy (ft Ibs)		't' value	Sig
	Pre-training N = 11 Mean ± SD	Post training N = 11 Mean ± SD		
Flexion Hip RL	6.72 ± 2.13	6.90 ± 1.83	0.21	NS
Extension Hip RL	22.9 ± 5.99	28.09 ± 6.05	2.02	NS
Flexion Hip LL	6.45 ± 1.56	7.18 ± 2.2	0.90	NS
Extension Hip LL	23.81 ± 6.39	30 ± 8.73	1.90	NS
Flexion Knee RL	7.9 ± 2.23	13.19 ± 4.14	3.73	S
Extension Knee RL	8.63 ± 2.99	8.18 ± 3.77	0.31	NS
Flexion Knee LL	9.54 ± 2.96	13.54 ± 15.15	.73	NS
Extension Knee LL	8.09 ± 2.42	8.36 ± 3.57	0.21	NS
Dorsi-flexion Ankle RL	0.72 ± .61	1.45 ± .49	3.09	S
Plantar flexion Ankle RL	3.09±1.83	4.27 ± 1.13	1.82	NS
Dorsi-flexion Ankle LL	0.90 ± 1.16	1.54 ± .65	1.62	NS
Plantar flexion Ankle LL	3.18 ± 1.64	4.18 ± 1.99	1.29	NS

RL – Right leg, LL – Left leg, NS – Not significant, S – Significant.

Table 4. Mean difference between pre-training and post-training performance in soccer technique tests for the isokinetic group.

	Soccer technique test			
Name of the Test	Pre-training N = 11 Mean ± SD	Post-training N = 11 Mean ± SD	't' Value	Sig
Kicking with Instep of the foot				
Accuracy:				
Right Leg	3.09 ± 1.23	6.09 ± 1.44	5.25	S
Left Leg	4.09 ± 1.50	5.72 ± 1.35	2.68	S
Distance:				
Right Leg	73.40 ± 7.19	82.16 ± 6.15	3.07	S
Left Leg	64.39 ± 7.68	74.64 ± 9.01	2.87	S
Kicking with Inner Instep of the foot				
Accuracy:				
Right Leg	56.63 ± 5.61	57.63 ± 4.23	0.47	NS
Left Leg	48.54 ± 7.34	55.81 ± 7.67	2.27	S
Distance:				
Right Leg	40.65 ± 5.20	42.92 ± 4.34	1.11	NS
Left Leg	33.89 ± 6.27	37.43 ± 6.17	1.27	NS

S – Significant, NS – Not significant.

4 Discussion

Six weeks of isokinetic strength training on the Orthotron II exercise system through selected exercises caused improvement in all the actions of hip, knee and ankle joints in peak torque. Improvement was statistically significant in extension of right and left hip joints and in dorsi-flexion and plantar-flexion of the right ankle.

Contrary to expectations, the six weeks of isokinetic training caused negative effects in the maximum strength (peak torque) of extensor muscles of both right and left knee joints, although the negative effect was not statistically significant. The decrease in performance in maximum strength for extension of the knee joint may be due to neural activation and muscle pennation which appears to provide some enhancement of force capability for muscle contracting at high speed, particularly in two extreme ranges of muscle motion, but pennation can be somewhat disadvantageous for 'slow speed' concentric force (Scott and Winter, 1991). As the peak torque was measured at a slow angular velocity of 0.79 rad·s^{-1}, it was probably the cause of decreased performance of the knee joint extensor muscles due possibly to muscle pennation altered through training.

As regards muscle 'explosiveness' (torque acceleration energy), significant improvement was seen only with the muscles involved in flexion of the right knee joint and in dorsi-flexion of the right ankle joint. This training also caused a negative effect on the average power of extensor muscles of the right knee joint, which was not significant. Since the training caused a negative effect on the peak torque (maximum strength) of the extensor muscles of the knee joint, a similar negative effect on TAE was also evident, as the maximum strength (peak torque) is considered an essential base for development of explosiveness in the muscle.

In the previous findings kicking performance was unaffected by different kinds of strength training (Trolle et al., 1993). In the current study, six weeks of isokinetic training with technical training caused significant positive effects on kicking for distance as well as accuracy with the instep of right and left foot, whereas accuracy in kicking with inner instep of the left foot only was improved significantly after training. The reason for improvement in performance in kicking for distance and accuracy with the instep can be attributed to the fact that the velocity in kicking was improved due to improvement of peak torque (maximum strength) and torque acceleration energy (explosiveness) with most of the lower limb muscles and due to improvement in timing, motor control and learning of technique caused by technical training.

In this study, there was decrease in peak torque right knee extension but still there was improvement in the distance and accuracy of instep and inner instep kicking. The improvement is probably due to the hip flexor muscles being more important in kicking performance than quadriceps muscles, as reported by Robertson and Mosher (1983) and Narici et al. (1988). It can be further stated that normal soccer training will increase strength but a planned strength training programme along with the normal soccer training emphasizing technique will result in higher strength levels and better kicking performance. The improvement was demonstrated by increased performance in peak torque, torque acceleration energy and soccer technique of the isokinetic training group.

5 Conclusions

The results obtained in this study enable the following observations to be drawn:
1. Isokinetic strength training for six weeks improved the maximum strength (peak torque) and explosive strength (torque acceleration energy) of flexors and extensors of hip joint, flexors of knee joint and flexors (dorsi-flexors) and extensors (plantar-flexors) of the ankle joint.
2. Isokinetic strength training on lower limb muscles for six weeks improved performance of kicking in soccer with respect to distance and accuracy.

6 References

Cabri, J., DeProft, E., Dufour, W. and Clarys, J.P. (1988) The relation between muscular strength and kick performance, in *Science and Football* (eds T. Reilly, A. Lees, K. Davids and W.J. Murphy). E. and F.N. Spon, London, pp. 186-193.

Crew, V.N. (1968) A skill battery for use in service programme soccer classes at the University level, *Thesis,* Eugene: University of Oregon.

De Proft, E., Cabri, J., Dufour, W. and Clarys, J.P. (1988) Strength training and kicking performance in soccer players, in *Science and Football* (eds T. Reilly, A. Lees, K. Davids and W.J. Murphy). E. and F.N. Spon, London, pp. 108-14.

Narici, N.V., Sirtori, M.D. and Mongoni, P. (1988) Maximal ball velocity and peak torque of hip flexors and knee extensors muscles, in *Science and Football* (eds T. Reilly, A. Lees, K. Davids and W.J. Murphy). E. and F.N. Spon, London, pp. 429-433.

Robertson, D.G. and Mosher, R.E. (1983) Work and power of leg muscles in soccer kicking, *Biomechanics IXb,* University Park Press, Baltimore, pp. 533-542.

Scott, S.H. and Winter D.A. (1991) A comparison of three muscle pennation assumptions and their effect on isometric and isotonic force, *Journal of Biomechanics* 24, 163-167.

Sports Authority of India (1992) *National Sports Talent Schemes,* New Delhi, India.

Taina, F., Grehaigne, J.F. and Cometti, G. (1993) The influence of maximal strength training of lower limbs of soccer players on their physical and kicking performances, in *Science and Football* (eds T. Reilly, J. Clarys, and A. Stibbe). E. and F.N. Spon, London, pp. 98-103.

Trolle, M., Aagaard, P., Simonsen, E.B., Bangsbo, J. and Klausen, K. (1993) Effect of strength training on kicking performance in soccer, in *Science and Football III* (eds. T. Reilly, J. Clarys and A. Stibbe). E. and F.N. Spon, London, pp. 95-97.

Warner, G.F.H. (1950) Warner soccer test, *Newsletter of National Soccer Coaches Association of America,* VI, 13, York, A.S. Barnes and Co.

57 QUANTIFYING THE WORK-LOAD OF SOCCER PLAYERS

T. FLANAGAN* and E. MERRICK**
* Sport Science and Fitness Research Unit, Department of Human Biology and Movement Science, Royal Melbourne Institute of Technology, Melbourne, Australia.
** Victorian Institute of Sport, Melbourne, Australia.

1 Introduction

The training programme of soccer players is usually structured to incorporate three main areas of focus. These are physical training, technical skills training and tactical skill training. It is possible to integrate physical, technical and tactical skill training and this is of greater benefit to the holistic development of the players (Von Hollmann et al., 1978; Bangsbo, 1994; Wollstein, 1995; Burwitz, 1997). The major limitation of this holistic approach is the coach's ability to quantify training work-loads and intensities accurately, since training sessions predominately involve on-field, fully integrated ball-centred activities, drills and modified games.

This article steps through the processes undertaken to implement a system to monitor the work-load of soccer players in training and match situations. The system was developed to prevent coaches from altering work-load throughout the training season purely on the basis of subjective observations. This system was designed to allow coaches to utilise fully integrated ball-centred activities, drills and modified games in training, while simultaneously developing superior levels of soccer specific physical fitness.

2 The Problem

Monitoring the work-load of soccer players by coaching and training staff has largely been done on a subjective basis in the past. The problem with quantifying the total work-load stress on players is that training situations involve on-field activities of which the intensities are difficult to quantify. In addition, much of the work-load placed on the athlete during a normal week or microcycle is attributed to the match. When observing the training loads placed on players, it is important not only to consider the overall volume of work done for the week, but also to incorporate the work done in matches.

3 Calculating Work-load

3.1 Principles

Work-load has been universally defined in the past as the product of the duration and the intensity of exercise. When examining the soccer player in training and matches, the duration can be expressed in terms of time. The intensity of any given exercise is more difficult to determine, particularly in modified game situations, which are commonly employed by soccer coaches in the training of players.

A specific objective was to develop a method to estimate intensity accurately and assign a valid intensity value to any training item. Heart rate data collection has been used in the past to observe players' intensities during training and matches (MacLaren et al., 1988; Rohde and Espersen, 1988; Van Gool et al., 1988; Ali and Farrally, 1991; Bangsbo, 1994a,b; Reilly, 1994a, b). Since heart rate data collection is simple and non-invasive, it was determined that heart rate would be the key variable to estimate playing intensities. This method of estimating intensity enables work-load to be easily calculated using the formula: work-load = work duration x work intensity.

The calculation procedure for estimating intensity and assigning an intensity value is very important since the integrity of this system relies on the accurate estimation of intensity. Controlled laboratory testing on a motorised treadmill was used to determine relationships between heart rate and exercise intensity. Players ran at match specific speeds (Bangsbo et al., 1991) of 4, 8, 12, 16 and 21 $km \cdot h^{-1}$. Linear regression equations were used to relate an intensity value (ranging from 0.0 to 1.0) to heart rates ranging from rest to maximal levels of exertion. For any given heart rate measured during a training activity, a corresponding value for intensity can therefore be determined. Work-load is then calculated by multiplying the intensity value for the activity by the total time of the activity in minutes. Table 1 outlines heart rates and their corresponding intensity values.

3.2 Example calculation

Using the table, the work-load of any training item can be calculated. The duration of a modified game, for example, which includes 9 versus 9 players on a 'half pitch' may be 20 minutes. The mean heart rates across players for this drill may be 170 $b \cdot min^{-1}$. Using table 1, the intensity value for this drill is 0.82. Work-load is calculated using the following formula:

$$
\begin{aligned}
\text{Work-load} \quad &= \quad \text{duration x intensity} \\
&= \quad \text{duration of training item x intensity value} \\
&= \quad 20 \text{ minutes x } 0.82 \\
&= \quad 16.4 \text{ units of exercise}
\end{aligned}
$$

A useful application of using heart rate analysis to predict workloads can be seen in Figure 1 where work-load for an elite junior player during a match is expressed in units of exercise.

Table 1. Assigned intensity value to heart rates.

Heart Rate (b·min⁻¹)	Intensity value	Heart rate (b·min⁻¹)	Intensity value
80	0.1	155	6.8
85	0.6	160	7.3
90	1.0	165	7.7
95	1.5	170	8.2
100	1.9	175	8.6
105	2.4	180	9.1
110	2.8	185	9.5
115	3.2	190	10.0
120	3.7	195	10.4
125	4.1	200	10.9
130	4.6	205	11.3
135	5.0	210	11.7
140	5.5	215	12.2
145	5.9	220	12.6
150	6.4		

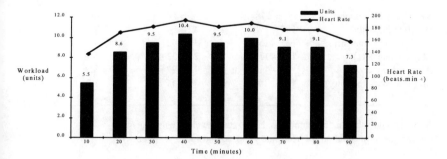

Figure 1. Work-load of a match.

4 Training Categories

Once work-load is estimated during training and competition, it is necessary to organise and periodise training. For this system to be successfully implemented into a programme, it must be further simplified and structured to require minimal administration. Classifying the work-load and intensity ranges using training categories is commonly done in some sports to achieve this goal (Telford, 1991; Craig, 1996). The fundamental aim of training categories is to simplify further the intensity of activities into ranges from rest to maximal levels of exertion. Training categories are particularly useful for coaches to prescribe and modify the amount of time spent in each intensity zone to produce optimal training benefits (Craig, 1996).

A training category system has been derived and can been seen in Table 2. Six training categories were devised. The mean heart rates and ranges for each category corresponded with the mean heart rates achieved during the treadmill trials (running players at match specific speeds of 4, 8, 12, 16, and 21 $km \cdot h^{-1}$). The categories were named with no reference to physiological events or energy systems in order to simplify the system. The category names include SR (stationary recovery), AR (active recovery), LI (low intensity), MI (moderate intensity), HI (high intensity) and VHI (very high intensity). Intensity values for the training categories have been calculated using linear regression equations from the mean heart rates of each category.

Table 2. Training category heart rate ranges, VO_2 ranges and intensity values.

Category	Heart rate range ($b \cdot min^{-1}$)	VO_2 range ($ml \cdot kg^{-1} \cdot min^{-1}$)	Intensity values*
VHI (Very high intensity)	> 178	> 54	10
HI (High intensity)	155 – 178	42 – 54	7.8
MI (Moderate intensity)	135 – 155	32 – 42	5.9
LI (Low intensity)	114 – 135	22 – 32	4.1
AR (Active recovery	93 – 114	11 – 22	3.2
SR (Stationary recovery)	< 93	< 11	1.3

Intensity values calculated from mean heart rate of each category, using linear regression equations of heart rate vs speed and speed vs. intensity

All soccer specific activities can be assigned to a training category using pre-determined mean heart rates as shown in Table 3 on the next page.

Table 3. Examples of heart rates (HR) during training activities (mean ± SD) and matches with assigned training categories (n = 23*).

Activity	Player numbers	Pitch area	Limitations	Mean HR (± SD)	Intensity category (& value)
warm up jog/flexibility	–	–	–	125 ± 5	LI (0.41)
modified game	10 v 10	full pitch	3 touch/pressure game	181 ± 2	VHI (1.0)
modified game	8 v 8	1/2 pitch	3 touch	170 ± 5	HI (0.78)
modified game	8 v 8	1/2 pitch	all in	170 ± 2	HI (0.78)
modified game	7 v 7	3/4 pitch	all in	173 ± 4	HI (0.78)
modified game	7 v 7	3/4 pitch	3 touch:2 floating wing	152 ± 1	MI (0.59)
modified game	6 v 7	3/4 pitch	floating player	161 ± 6	HI (0.78)
modified game	6 v 7	1/3 pitch	all in (coached)	180 ± 6	VHI (1.0)
modified game	6 v 7	2/3 pitch	all in	170 ± 8	HI (0.78)
modified game	6 v 6	2/3 pitch	all in (coached)	173 ± 8	HI (0.78)
modified game	6 v 6	3/4 pitch	3 touch	147 ± 3	MI (0.59)
modified game	6 v 5	2/3 pitch	all in	165 ± 4	HI (0.78)
modified game	5 v 5	1/2 pitch	line ball game	138 ± 10	MI (0.59)
wide crossing drill	–	round pen. box	–	61 ± 9	HI (0.78)
direct shooting drill	–	round pen. box	–	144 ± 12	MI (0.59)
fast break drill	3 v 2	–	4 groups of 3	143 ± 9	MI (0.59)
match	11 v 11	first half	position: strikers	153 ± 5	MI (0.59)
match	11 v 11	first half	position: midfielders	170 ± 12	HI (0.78)
match	11 v 11	first half	position: defenders	171 ± 8	HI (0.78)
match	11 v 11	second half	position: strikers	160 ± 11	HI (0.78)
match	11 v 11	second half	position: midfielders	169 ± 16	HI (0.78)
				169 ± 8	HI (0.78)

*Approximately 6–8 players were monitored during each activity. A total number of 23 different players were used in total. Players' HR responses to each activity were measured approximately four times, then averaged with other players mean HR responses to get an overall squad mean HR.

5 Practical Implementation

As previously stated, the modern coach attempts to integrate physical, technical and tactical training. By creating measurable relationships between low intensity and very high intensity training drills, the coach can calculate overall workloads. Pre-season, the coach would prescribe more low intensity high volume (aerobic) work and would progress to high intensity low volume exercise during the pre-competitive phase. Table 4 is a practical example of a pre-competition training session using the category system. Table 5 and Figure 2 also indicate the monitoring of work-load during a typical microcycle (week) leading up to a weekend match.

Table 4. Example training session and work-load* calculation.

Activity (intensity category)	Duration	Intensity	Units
Warm up jog/flexibility (LI)	15	0.41	6.2
Direct shooting drill (MI)	15	0.59	8.9
Fast break drill (MI)	20	0.59	11.8
Sprint technique (HI)	10	0.78	7.8
Modified game: 6 vs 7 on 2/3 pitch (HI)	20	0.78	15.6
Cool down jog/flexibility (AR)	10	0.32	3.2
		Total units	53.5

*Work-load (units) = (duration x intensity)

Table 5. Work-load done during a typical microcycle leading up to a weekend match.

Day	Session type (am)	Units	Session type (pm)	Units	Total
Mon	Low vol. recovery	8	Skills	35	43
Tues	Aerobic run	13	Strength	38	51
Wed	Pliometrics	22	Tactical/skills	48	70
Thurs	Aerobic run	13	Strength	46	59
Fri	Low vol. recovery	8	Set plays/skills	26	34
Sun	Match	76			76
				Units (total)	333

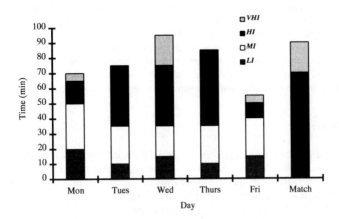

Figure 2. Time spent in each training category during a typical microcycle leading up to a weekend match.

6 Some Considerations

There will be some physiological differences between players in heart rate response. The training category system does allow for this by having quite large heart rate ranges. For example, players can have an average heart rate response to a drill of up to 20 b·min⁻¹ different and can still lie within the same training category. It is not common, however, for heart rate responses to vary greatly during well structured ball centred activities, drills and modified games. Another consideration is that there will be differences in work-rates across different positions. This can be overcome by structuring the drills which make players work at a similar level. For example, players may be asked to sprint 5 m after passing the ball, extra defenders or attackers may be incorporated into activities and even extra balls or goals can be employed to alter the work-rates of selected players.

7 Conclusion

While understanding the limitation associated with estimating work-load from laboratory and field measurements, this is an effective, soccer specific, scientifically based system for quantifying work-load. This system is to be viewed as a model for predicting and monitoring work-load and not as an exact recipe for prescription. The system is designed to be practical and for easy implementation. Special care has been taken to reduce the ambiguity of the system and the training categories. Care was also taken to avoid the need for advanced levels of physiological knowledge in order to implement this work-load monitoring system effectively.

The success of the system depends on the quality of heart rate data measurement of players. It is important that the data collection is comprehensive across a wide selection of players.

8 Future Directions

Future research will focus on refining this system and the possibility of incorporating other physiological variables to increase the validity of intensity prediction. The system will be particularly useful in the future for allowing coaches to trial different models of overload to produce optimal training responses and improvements in performance. The system is also a very effective way to periodise work-load when coaching with the holistic philosophy of simultaneously developing technical, tactical and physical capacities of soccer players.

9 References

Ali, A. and Farrally, M. (1991) Recording soccer players' heart rates during matches, *Journal of Sports Sciences,* 9, 183-189.

Bangsbo, J. (1994a) *Fitness Training in Football – A Scientific Approach.* Ho + Storm, Bagsvaerd.

Bangsbo, J. (1994b) Physical conditioning, in *Football (Soccer)* (ed B. Ekblom), Blackwell Scientific Publications, Oxford, pp. 124-138.

Bangsbo, J., Norregaard, I. and Thorso, F. (1991) Activity profile of professional soccer, *Canadian Journal of Sports Science,* 16, 110-116.

Burwitz, L. (1997) Developing and acquiring football skills, in *Science and Football III* (eds T. Reilly, J. Bangsbo and M. Hughes), E. and F.N. Spon, London, pp. 201-208.

Craig, N. (1996) *Scientific Heart Rate Training,* Pursuit Performance and High Performance Kinetics.

MacLaren, D., Davids, K., Isokawa, M., Mellor, S. and Reilly, T. (1988) Physiological strain of 4-a-side soccer, in *Science and Football III* (eds T. Reilly, A. Lees, K. Davids and W.J. Murphy), E. and F.N. Spon, London, pp. 76-80.

Reilly, T. (1994) Physiological aspects of soccer, *Biology of Sport,* 11, 3-20.

Rohde, H.C. and Espersen, T. (1988) Work intensity during soccer training and match-play, in *Science and Football III* (eds T. Reilly, A. Lees, K. Davids and W.J. Murphy), E. and F.N. Spon, London, pp. 68-75.

Telford, R. (1991) Endurance Training, in *Better Coaching* (ed F. Pyke), Australian Coaching Council, pp. 125-134.

Van Gool, D., Van Gerven, D. and Boutmans, J. (1988) The physiological load imposed on soccer players during real match play, in *Science and Football III* (eds T. Reilly, A. Lees, K. Davids and W.J. Murphy), E. and F.N. Spon, London, pp. 51-59.

Von Hollman, W., Rost, R., Leisen, H. and Mader, A. (1978) Physiologische internistische aspekte zum Fussballsport, *Deutsche Zeitscrift fur Sportmedizin,* 1, 2, 60.

Wollstein, J. (1995) A new model of athletic development: Perceptual motor skill development programs for squash, *Australian Squash Coach,* Spring, 5-8.

58 THE PHYSIOLOGICAL EFFECT OF PLAYING THREE SIMULATED MATCHES IN A WEEK: IMPLICATIONS FOR OVERTRAINING / OVERPLAYING

P. RAINER
University College, Worcester, UK.

1 Introduction

Over a 10-month period incorporating 50–60 football games, players are at risk of becoming overtrained due to 'an imbalance between training and recovery'. Practitioners that use overtraining for adaption periodise so as to ensure a compensatory physiological supercompensation (Bompa, 1983), where ideally subsequent training should not take place until supercompensation has occurred (Harre, 1982).

Football players are now professional athletes and changes within the game itself are creating a drive for higher levels of fitness, which require greater versatility from players, longer playing seasons and greater frequency of games. Despite these expectations players are required to achieve peak fitness at least twice a week unlike in other sports where tapering and periodisation are used widely.

To date although no study has monitored the physiological stress players experience over a season, studies have indicated that fitness levels of players are maintained (Reilly and Thomas, 1977), or slightly reduced. While authors have reported improvements in fitness in the early part of the season (Bangsbo, 1994; Brady et al., 1997; Rebelo and Soares, 1998), they suggested that after the early part of the season players' aerobic capacity and anaerobic capacity may deteriorate. Although these studies were not looking to identify symptoms of training fatigue, it was unclear whether the reasons were due to a reduction in fitness per se, or as a result of accumulated fatigue and inadequate recovery over the season.

The study therefore proposed to monitor the physiological stress placed upon players during a typical week of a season and establish whether a simple battery of tests could be employed to identify players showing signs of accumulated fatigue.

2 Methods

Semi-professional football players (n = 8) performed three simulated football matches using an adapted version of Nevill and co-workers' (1993) intermittent sprint test. Subjects performed three games in the space of eight days (this being reflective of the demands of a professional football player). Prior to training, players' baseline data were established on day 1 and day 9 after completion of training and three days after the final session.

Subjects on arrival at the laboratory were measured for resting heart rate, blood lactate, resting metabolic rate, blood pressure and body weight. Following this, subjects completed a 10-min submaximal steady-state run, with measures of oxygen uptake, heart rate and ventilation taken during the 5th and 10th min and blood pressure 2 and 7 min post-exercise.

Following a 10-min recovery, subjects performed a maximal sprint test; 10 x 6 s sprints followed by 24 s recovery (Lakomy, 1987). Blood samples were taken 5 min post-exercise for determination of maximal lactate values. Heart rate was recorded using heart rate telemetry at the end of each sprint.

3 Results

Resting and submaximal measures yielded no significant differences (P > 0.05) for any of the tests. Maximal lactate and heart rate during the intermittent sprint test decreased between T1 and T2 (P < 0.05), returning to baseline at T3. Peak power output between all three tests was decreased (P < 0.05).

Table 1. Resting measures.

Test		T1	T2	T3
Rest HR	Mean	59.25	60.13	60.00
($b \cdot min^{-1}$)	± SD	10.78	8.87	9.02
Rest Lactate	M	2.01	2.06	1.96
($mmol \cdot l^{-1}$)	± SD	0.33	0.27	0.17
RMR	M	1.82	2.36	1.97
($kcal \cdot min^{-1}$)	± SD	0.34	0.30	0.42
V_E Rest	M	11.6	11.5	11.9
($l \cdot min^{-1}$)	± SD	1.4	1.9	0.8
VO_2 Rest	M	5.18	5.33	4.99
($ml \cdot kg \cdot min^{-1}$)	± SD	0.97	1.14	0.84

No significant differences (P > 0.05) reported.

Table 2. Submaximal measures in response to exercise.

Test		T1	T2	T3
HR 5th	Mean	169	165	164
(b·min⁻¹)	± SD	13	13	11
HR 10th	M	175	173	164
(b·min⁻¹)	± SD	7.5	8.3	5.6
VO₂ 5th	M	53.08	51.01	50.12
(ml.kg.min⁻¹)	± SD	4.70	6.43	7.42
VO₂ 10th	M	56.94	53.83	55.98
(ml.kg.min⁻¹₎	± SD	6.56	7.15	6.50
V_E 5th m	M	87.6	91.4	86.3
(l·min⁻¹)	± SD	12.2	5.7	9.6
V_E 10th	M	96.9	103.1	99.3
(l·min⁻¹₎	± SD	11.9	8.0	10.7

No significant differences (P > 0.05) reported.

Table 3. Performance and physiological responses to the intermittent sprint test.

Test		T1	T2	T3
Max Lactate	Mean	14.5	13.5	14.6*
(mmol·l⁻¹)	± SD	2.35	1.59	1.63
Max HR	M	186	183	186*
(b·min⁻¹)	± SD	9.2	8.5	6.4
PPO	M	870.1	819.1	762.5*
(Watts)	± SD	66.0	127.3	111.1
MPO	M	765.9	706.9	684.7
(Watts)	± SD	73.7	111.4	115.9
Fat Index	M	15.9	18.5	15.5
(%)	± SD	6.5	1.9	5.7

*Significant difference (P < 0.05)

4 Discussion

The results would suggest that following three games in a typical week of a season, players exhibit high levels of training fatigue and stress. It may be that following a three-day recovery players are still exhibiting symptoms of fatigue and bodily functioning has not returned to

4 Discussion

The results would suggest that following three games in a typical week of a season, players exhibit high levels of training fatigue and stress. It may be that following a three-day recovery players are still exhibiting symptoms of fatigue and bodily functioning has not returned to homeostasis. Of the tests employed, resting and submaximal function may not be sensitive enough to highlight symptoms of match fatigue and tests that target neuromuscular and physiological patterns used within the game should be reflected in the test protocol (Fry et al., 1991). Test protocols used to monitor players should reflect the anaerobic intermittent nature of the game (Van Borselen et al., 1992) and as studies have reported different sports may potentially cause different symptoms of training/match fatigue, this may explain why tests of anaerobic function highlighted signs of fatigue (Stone and Keith, 1991).

The study also raises the question of fitness testing during a competitive season, at a time when players may be in a state of accumulated fatigue. Testing during this time could lead to poor scores and wrongful conclusions made about the results. The question is whether it is practical for players to be rested during a season to allow adequate recovery.

5 Conclusion

Managers and coaches may be naive to the fact that players may be compromising fitness over a season by overplaying. Although symptoms developed are not as serious as full-blown overtraining syndrome, accumulated fatigue caused by constant playing and inadequate recovery may place players on the overtraining continuum, such that players enter games not fully recovered from previous engagements. Ultimately this could lead to a decrease in performance, susceptibility to injury and immunosuppression. It would seem currently that methods of fitness assessment need to be re-addressed, so players can be monitored continually to identify the symptoms of fatigue before they develop further.

6 References

Bangsbo, J. (1994) *Fitness Training for Football: A Scientific Approach,* HO + Storm, Copenhagen, Denmark.
Bompa. T.O. (1983) *Theory and Methodology of Training,* Kendall Hunt Publishing Company.
Brady, K., Maile, A. and Ewing, B. (1994) An investigation into the fitness of professional of soccer players over two seasons, in *Science and Football III* (eds T.Reilly, J.Bangsbo and M.Hughes), E and F.N. Spon, London, pp. 118-122.
Fry R.W., Morton A.R. and Keast D.K. (1991) Overtraining in athletes: An update, *Sports Medicine,* 12, 32-65.
Harre. D. (1982) *Principles of Sport Training: Introduction to the Theory and Methods of Training,* Berlin.

Lakomy, H.K.A. (1987) Measurement of human power output, in *High Intensity Exercise* (eds B. Van Gheluwe and J. Atha), Switzerland, pp. 46-57.

Nevill, M.E., Williams, C. and Roper, C. (1993) Effects of diet on performance during recovery from intermittent sprint exercise, *Journal of Sports Sciences,* 11, 119-26.

Rebelo, A.N. and Soares, S.M.C. (1998) Impact of soccer training on the immune system, *Journal of Sport Medicine and Physical Fitness,* 38, 258-61.

Reilly, T. and Thomas, V. (1977) Application of multivariate analysis to the fitness assessment of soccer players, *British Journal of Sports Medicine,* 11, 183-184.

Stone, M.H. and Keith, R.E. (1991) Overtraining: A review of the symptoms, signs and possible causes, *Journal of Applied Science Research,* 25, 35-50.

Van Borselen, F., Fry, A.C. and Kraemer W.J. (1992) The role of anaerobic exercise in overtraining, *National Strength Conditioning Association Journal,* 14, 74-79.

59 EFFECTS OF EXERCISE MODE ON FOOT SKIN TEMPERATURE

A.J. PURVIS and N.T. CABLE
Research Institute for Sport and Exercise Sciences, Liverpool John Moores
University, UK.

1 Introduction

In contrast to the large range of temperatures found in the global environment (-90ºC to +60ºC), the human body is regulated within a very narrow range normally between 35ºC and 40ºC. The accurate regulation of internal temperature is necessary to maintain normal body function. During exercise metabolic rate is elevated above resting levels by as much as 15 times to provide the necessary energy for muscle contractions (Reilly, 1996; Sawka and Wenger, 1988). This increase in metabolism produces heat within the active muscles, which must be dispersed to the environment in order to maintain core temperature within healthy limits.

Sweating with evaporation of sweat from the skin surface is the major mechanism for the dissipation of heat. Blood is distributed to the skin by dilation of peripheral blood vessels; sweat is then secreted onto the surface of the skin where heat is removed by vascular transfer in turn vaporising sweat. There are many regional differences in skin blood flow and sweating characteristics of the skin (Day, 1967) with the foot reported to have a very high density of sweat glands (Sato et al., 1989) and hence a high capacity for evaporative heat loss. However, little is known about changes in skin temperature of the foot particularly during intermittent exercise when there is a greater increase in core and skin temperature compared with continuous exercise (Cable and Bullock, 1996).

Clothing, including footwear, acts as an interactive barrier and affects heat transfer. Thermal discomfort of the feet, due to build up of heat and accumulation of sweat, is a difficulty for athletes. Information about changes in skin temperature of the foot is of importance to footwear manufacturers and athletes, as problems with comfort could affect performance.

Therefore, the aim of this study was to measure changes in foot skin temperature during continuous and soccer-specific intermittent exercise whilst wearing footwear.

2 Methods

Two trials were carried out with 12 subjects aged from 19 to 33 years. The subjects were in good health, regularly participated in exercise and freely volunteered for the study. All subjects were asked to refrain from consuming caffeine for 12 h, alcohol for 24 h and food for 3 h prior to the test.

The experimental design consisted of two exercise protocols. One group (n = 6) followed a soccer-specific intermittent protocol and the other (n=6) performed a continuous protocol, both on a motorised treadmill. Both protocols were performed at the same average intensity (12 km·h^{-1}). The intermittent protocol was previously devised to simulate soccer match-play (Drust, 1997). The protocol, illustrated in Figure 1, represents the actions of a soccer player during match play and uses four different speeds; walk (6 km·h^{-1}), jog (12 km·h^{-1}), cruise (15 km·h^{-1}) and sprint (21 km·h^{-1}). These particular speeds were selected as they have been shown to be typical speeds of a good standard of soccer player.

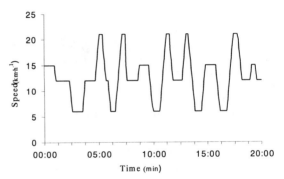

Figure 1. Changes in treadmill speed during the soccer specific intermittent protocol.

Skin temperature was measured at three sites on the foot (toe, instep, ankle) and four other sites on the body (chest, arm, thigh, shin) to calculate a weighted mean skin temperature according to the method of Ramanathan (1964). All measurements were taken on the left hand side of the body and thermistors were attached using white surgical adhesive tape.

The subjects were clothed in shorts, T-shirt, ankle socks and underwear. Prior to the test start, thermistors were allowed to equilibrate for 30 min while the subject rested. All subjects were familiar with running on a motorised treadmill. Environmental conditions were held relatively constant (22 ± 2°C and 45 ± 5% relative humidity).

3 Results

Results of the study were examined for the possible significant effects of intermittent exercise compared to continuous exercise on foot skin temperature. Physiological measurements recorded showed similar responses within conditions. Results are presented as mean values of the two groups. The details of subject characteristics are displayed in Table 1. Analysis of the

Table 1. Mean (± SD) subject characteristics.

Condition	Age (years)	Height (m)	Mass (kg)	Shoe size
Intermittent	25 ± 3	178.5 ± 5.7	81.0 ± 8.2	8.5 ± 1.2
Continuous	25 ± 5	172.3 ± 4.8	69.9 ± 10.7	7.0 ± 1.5

results was carried out with two methods. Skin temperatures and heart rates were analysed using ANOVA and skin temperatures were also examined using analysis of serial measurements (Mathews et al., 1990) where the change from baseline to end-of-test was analysed using a t test.

3.1 Heart rate

Averaged 60-s data comparing heart rate values during intermittent and continuous conditions are displayed in Figure 2. At the onset of exercise the heart rate increased and during continuous exercise a steady state was attained within 2 min with a mean heart rate of 151 b·min^{-1}. During intermittent exercise heart rate fluctuated between a mean maximum of 176 and mean minimum of 127 with an overall mean heart rate (from 2 min to end of test) of 157 b·min^{-1}. Mean heart rate showed no significant difference between conditions (P = 0.07).

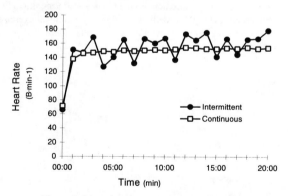

Figure 2. Mean heart rate during exercise tests.

3.2 Skin temperature responses

Typical temperature changes from a single subject are displayed in Figure 3. All subjects showed similar temperature responses.

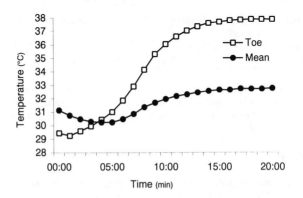

Figure 2. Typical changes in toe and mean skin
temperature.

The changes in foot and mean skin temperature from baseline to the end of exercise are displayed in Figure 4. Each site of measurement on the foot showed a significantly increased skin temperature above baseline ($P < 0.05$) with values during intermittent exercise being greater ($P < 0.05$). However, mean skin temperature did not alter significantly ($P > 0.05$) over the duration of the experiment.

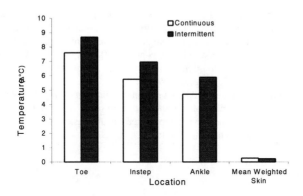

Figure 3. Change in temperature from baseline to end of test.

4 Discussion

The findings of the present study indicate that intermittent exercise compared to continuous exercise induces an increase in foot skin temperature of a greater magnitude. The high temperature of the foot compared to a lack of change of overall mean skin temperature would indicate that the foot maintains an altered thermoregulatory response not evident elsewhere on the human body.

As muscles in the legs and feet are working during running, the blood flowing to the foot increases in temperature. The toe is the furthest site of measurement from the opening of the training shoe followed by instep and lastly the ankle site nearest to the opening. As the site of measurement approaches the opening of the shoe at the ankle, the temperature increase is reduced. This response was evident in both intermittent and continuous exercise. It appears that the surrounding of the foot by footwear generates an enclosed microenvironment that attenuates the processes of radiative, convective, conductive and evaporative heat loss. Clothing and footwear also increase insulation by creating a trapped insulative layer of air between the skin and fabric or material layer. Evaporative heat loss is further impeded due to absorption of sweat into the sock and shoe. Areas of the foot nearer to the opening of the footwear are cooler due to more air movement and increased convective and evaporative heat loss. Adams et al. (1992) showed that skin temperature was significantly increased in conditions of low air movement compared to high air movement.

Temperature may also be elevated within the shoe due to friction. Friction can be caused by two mechanisms, movement of the foot within the shoe and also shoe-surface contact conducting heat into the shoe.

Soccer specific exercise is essentially intermittent with maximal sprints interspersed on a framework of lower intensity endurance exercise. Superimposed on this are the ball skills necessary to play the game of soccer. Laboratory simulations are unable to include ball skills. However, research has indicated that physiological responses are similar in magnitude and arrangement to those observed in match-play conditions (Drust, 1997). Indeed, heart rates recorded during intermittent exercise in this study (Figure 2) reflect heart rate profiles reported in previous work (Drust, 1997; Bangsbo, 1994).

This study demonstrates a greater thermal response of the foot during the soccer specific protocol. This may have ramifications for thermal comfort of the player and indicated that footwear manufacturers may need to incorporate the use of moisture and heat management materials into boot designs in order to facilitate heat loss.

5 Acknowledgements

Umbro International Footwear and Equipment is acknowledged for its contribution to this study.

6 References

Adams, W.C., Gray, W.L. and Ethan, R.N. (1992) Effects of varied air velocity on sweating and evaporative rates during exercise, *Journal of Applied Physiology*, 73, 2668-2674.

Bangsbo, J. (1994) The physiology of soccer – with special reference to intense intermittent exercise, *Acta Physiologica Scandinavica (Suppl)*, 619, 1-155.

Cable, N.T. and Bullock, S. (1996) Thermoregulatory response and in recovery from aerobic and anaerobic exercise, *Medicine and Science in Sports and Exercise*, 28, S202.

Day, R. (1967) Regional heat loss, in *Physiology of Heat Regulation and the Science of Clothing*, (ed L.H. Newburgh), Hafner Publishing Company, New York, pp. 240-261.

Drust, B. (1997) *Metabolic responses to soccer-specific intermittent exercise*, Unpublished Ph.D. Thesis, Liverpool John Moores University.

Mathews, J.N.S., Altman, D.G., Cambell, M.J. and Royston, P. (1990) Analysis of serial measurements in medical research, *British Medical Journal*, 300, 230-235.

Ramanathan, L.N. (1964) A new weighting system for mean surface temperature of the human body, *Journal of Applied Physiology*, 19, 531.

Reilly, T. (1996) Environmental stress, in *Science and Soccer* (ed T. Reilly), E. and F.N. Spon, London, pp. 201-224.

Sato, K., Kang, W.H., Saga, K. and Sato, K.T. (1989) Biology of sweat glands and their disorders. I. Normal sweat gland function, *Journal of the American Academy of Dermatology*, 20, 537-563.

Sawka, M.N. and Wenger, C.B. (1988) Physiological response to acute exercise heat stress, in *Human Performance Physiology and Environmental Medicine at Terrestrial Extremes*, (eds M.N. Sawka and C.B. Wenger), Benchmark Press, IL, USA, pp. 336-341.

PART NINE

Women and Football

60 MEASURES TAKEN BY THE JAPAN WOMEN'S SOCCER TEAM TO COPE WITH HIGH TEMPERATURE

T. KOHNO, N. O'HATA, H. AOKI and T. FUKUBAYASHI
Sports Medical Committee, Japan Football Association, Tokyo, Japan

1 Introduction

In sports activities, heat production occurs in muscles, but heat radiation from the skin can become a crucial matter due to the adjustment of the body temperature. In the high-temperature environment, since the efficiencies of radiation, conduction and convection tend to deteriorate, evaporation is the means of heat diffusion and it promotes a great deal of sweating during the sports activities. Unless adequate water-intake is achieved when it occurs, then dehydration symptoms will appear and athletic capacity will decline. Moreover, the heat diffusion efficiency deteriorates due to circulation failure caused by dehydration, heat transfer from the outer world into the body increases when the air temperature gets higher than the body temperature, and heat production is also raised by the increase of the radiated heat caused by the direct exposure to the sun. In such a high-temperature environment as this, unless the heat produced by exercise gets diffused as much as it is supposed to, it will lead to heat retention and it can be a cause of heat stroke. Therefore, for the protection against heat stroke, restrictions are placed on the sports activities or may even lead to cancellation under the high-temperature environment. Meanwhile, there are some cases where international matches are held despite such high-temperature environments. To deal with these sort of matches, some counter-measures for heat must be elaborated upon beforehand, such as acclimatization to the heat or water-intake to counter for fluid lost in sweat.

The 26th Olympic Games were held in and around Atlanta, USA in July 1996. According to the survey results on the climate of Atlanta, it turned out that it has a climate of high temperature and high humidity in July and often the temperature goes beyond 37°C. That accordingly required the Japan Women's Soccer Team to take counter-measures for heat and high temperature. In this report, the influences of the high-temperature environment on the women's soccer team players are discussed, and the results of the counter-measures for heat and high temperature taken for the Atlanta Olympic Games are presented.

2 Methods

Subjects in this study were 28 prospective soccer players preparing for the Atlanta Olympic Games from among the Japan Women's Soccer Team, with their ages ranging from 17–30 (mean = 22.5) years. The players stayed together in a camp for training in Kuala Lumpur, Malaysia, for seven days from 21 to 27 April, 1996. This study examined the weight at the time of wake-up, the weight before and after training or games, the amount of water-intake during training or games and blood tests for three days from the third to the fifth day after arrival. The blood tests were conducted before breakfast with the players having an empty stomach. The items measured for the blood tests were WBC, RBC, Hb, Ht, GOT, GPT, LDH, CK, BUN, Cr, Na, Cl and Fe. For the atmospheric conditions, the Wet Bulb Globe Temperature (WBGT) was measured.

3 Results

Figure 1 shows the changes of the WBGT during the camp training. 24 July was not included in the measurement since it was a day of rest. The WBGT showed a minimum of 27.7°C and a maximum of 32.7°C and its average for the six days measured was 30.7°C. Out of the seven days, there were four days that had over WBGT 31°C. According to the heat stroke guidelines by the Japan Physical and Health Association, any sports activities with the WBGT over 31°C are to be called off, yet the practice and games were enforced without being cancelled this time for the purpose of the counter-measures for heat and high-temperature.

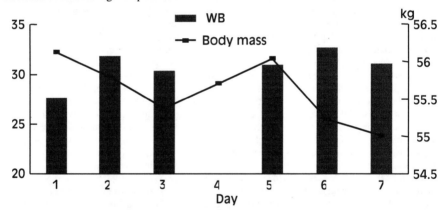

Figure 1. Change of WBGT and body mass in the morning.

The training schedule at this camp consisted of the morning practice from 10:00 hours to 12 noon and the afternoon practice mainly by games from15:00 to 17:00 hours, except the fourth day for a rest. The body mass at the time of wake-up showed 56.13 kg on the first day, i.e., the starting day of the camp training, and it gradually decreased compared to the start of the camp training to

55.80 kg on the second day, i.e., the following morning of the practice, 55.38 kg on the third day, and 55.71 kg on the fourth day. On the fifth day, i.e. the morning following the rest day, it recovered to 56.04 kg, the same level of the starting time, and it started to fall again with the restart of the practice to 55.23 kg on the sixth day, and to 55.01 kg on the seventh day. In other words, the correlation among the weight at the time of wake-up, the WBGT, and the practice schedule accounts for the facts that the practice under the high-temperature environment causes body mass to decrease and the one-day rest allowed it to recover to the normal on the next day.

Figure 2 shows the changes in body mass before and after the two games on the second day, 22 April, and the fifth day, 25 April, and the amount of water taken in. The environment for the games was almost equal on each day, exceeding 31.0°C in WBGT, 31.9°C on 22 April, and 31.0°C on the twenty-fifth. At the match on the twenty-second, the weight before the game on average decreased by 1.55 kg, which is 2.8%, after the game. The average amount of water-intake during the time from before the game to the after-game weigh-in was 0.95 l. In the 25th game, two water-intakes, i.e., 60 minutes and 30 minutes before the game, and water-intakes every 15 minutes during the game were added to the water-intake on22 April. As a result, as a matter of course, the amount of water-intake for this game increased to 1.85 l and the weight loss decreased to 0.85 kg on average, which is 1.5% of total body mass.

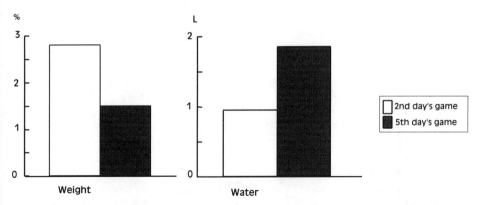

Figure 2. Changes in body weight before and after a game, water intake during the game.

Next are the results of the blood tests. Figure 3 shows the changes of BUN. The first day showed a slight increase to 22.8 ± 4.6 mg.dl^{-1} on average for all players, the second day it was 17.8 ± 4.1 mg.dl^{-1}, and the third day 16.3 ± 3.6 mg·dl^{-1}, with a steady decrease. For individual players, 24 out of 28 players, which is 83%, fell outside the normal range on the first day, but on the third day the number decreased to 17%. The average CK was 293 ± 213.1 IU·l^{-1}. It decreased to 246 ± 137.0 IU·l^{-1} on the second day, and came within the normal range of 137 ± 57.4 IU·l^{-1} on the third day (Figure 4).

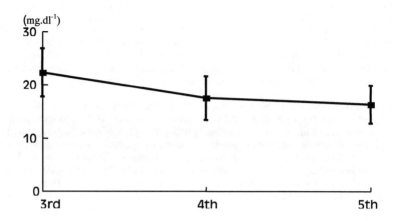

Figure 3. Change of blood urea nitrogen (BUN).

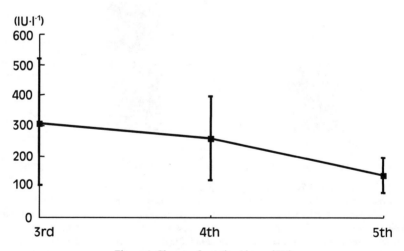

Figure 4. Change of creatine kinase (CK).

4 Discussion

The WBGT measured in this study was 30.7°C on average, and on four days it exceeded 31.0°C. The guidelines on the WBGT and exercise require exercise to be called off on the occasion of 31.0°C or over in principle, and strictly cautioned on the occasion of 28.0°C or over (intense exercise to be called off), and so this camp training was actually in the high-temperature environment where in principle training should be called off. The average body mass of the team at the time of wake-up on the next day of soccer training under such a high-temperature environment decreased on the next day when the training started, i.e., the second day, and on the third day. Even on the fourth day the mean body mass did not recover to the same weight as on the starting day. Since the fourth day was not spent on practice but allocated to rest, the weight had been recovered by the next morning. After that, with the restart of practice sessions, weight at the time of wake-up again decreased. Judging from the changes of the average weight at the time of wake-up observed so far in the overseas tours of the Japanese Women's Soccer team, since it has never experienced such a continuous decrease, it could be concluded that the weight changes over this time could be attributed to the training under conditions of a high-temperature environment.

The net weight loss caused by dehydration during the game was investigated by measuring weight changes before and after the game and the amount of water-intake. As for the influence of the weight loss during the game, if it exceeded 2.0% of the body weight it would be likely to influence athletic capacity, so it was the goal of this study to keep the weight loss during the game under 2.0%. The water-intake during a soccer game is usually allowed before the game and during half-time; other than that, water intake is allowed by the rules at the discretion of players from bottles containing water that are strategically placed around the ground. Concerning the water-intake in this study, the limited amount of water-intake before the game and during half-time, as in the first game, resulted in a weight loss of 1.55 kg, 2.8%, while as in the second game the additional water-intake, every 15 min during the game, resulted in the weight loss of 0.95 kg, 1.5%. In other words, it was possible to control the weight loss to under 2.0% even when in a high-temperature environment by adequately supplying water. Regarding the water-intake during exercise, since the amount that can be lost can exceed the amount taken in, it is necessary to take in water actively, while in other sport events where water-intake is only allowed at discretion by the event rules, it is crucial to discuss ways to take in water.

Next are the results of the blood tests that were conducted for three days in a row. The first and second days yielded the results of the blood tests on the morning following the practice, and the third day's results were on the following morning of the recess day. The changes of BUN were observed to indicate the influence of dehydration caused by the training under the high-temperature environment. Values for BUN increased on the first day to 22.8 ± 4.6 mg·dl^{-1}, confirming the influence of dehydration, but on the third day, it recovered to the normal range, 16.3 ± 3.6 mg·dl^{-1}. As for the changes of CK, measured as an index of muscular fatigue, CK values increased to 293 ± 213.1 IU·l^{-1} and recovered to 137 ± 57.4 IU·l^{-1}, within the normal range on the third day. With the changes in blood conditions of soccer players under a high-temperature environment, Kobayashi et al. (1997) reported that CK starts to increase right after a game, and it drops to the normal range on the third day. The results reported in this study confirmed this finding. These results indicate that a one-day rest is necessary after having participated in training under a high-temperature environment for two days.

5 Conclusions

The training programme was implemented for the Japanese Women's Soccer National Team under the high temperature environment with the WBGT over 31°C.
(1) The weight at the time of wake-up indicated a gradual decrease in consequence of training under   the high-temperature environmental conditions.
(2) The frequent water-intake before and during the game helped to get the weight loss before and after the game under control within 2.0%.
(3) The blood test results showed that a one-day rest is necessary after every two days of practice.

6 References

Kobayashi, M., Aoki, H., Ikela, S., Katsumata, T., Kohno, T., et al. (1997) High-school soccer summer tournaments in Japan – comparison of laboratory data of the players in summer and winter, in *Science and Football III* (eds T . Reilly, J. Bangsbo, and M. Hughes), E. and F.N. Spon, London, pp.190–195.

61 INFLUENCE OF PLAYING POSITION ON FITNESS AND PERFORMANCE MEASURES IN FEMALE SOCCER PLAYERS

C. WELLS and T. REILLY
Research Institute for Sport and Exercise Sciences, Liverpool John Moores University, Liverpool, UK

1 Introduction

Participation of females in soccer has increased enormously in recent years and in many countries is the fastest growing competitive sport. Not until the 1980s was the women's game actively promoted. It gained international credibility in 1991 when the first World Cup for women's teams was held in China. Whilst there has been a steep rise in women's soccer clubs throughout the 1990s, in many countries women's participation in sports, and particularly field games, is restricted by cultural, domestic and economic circumstances.

It is therefore unsurprising that research focused on women's soccer has lagged behind studies of the men's game. With the increased attention given to physical preparation for soccer, fitness assessment of female soccer players becomes relevant. This applies both to laboratory based tests of aerobic power, anaerobic power and muscle strength and their field-based counterparts where appropriate.

It has been established that the physiological demands of soccer, at least as played by males, varies with positional role with the highest work-rates being displayed by mid-field players. These positional differences are also reflected in fitness measures and mostly notably in aerobic indices (Reilly, 1997). Among male players a link has been established between dynamic muscle strength as recorded using isokinetic dynamometry and performance in the field as indicated by the distance the ball is propelled in a standardised place-kick (Cabri et al., 1988). This relationship was confirmed by Reilly and Drust (1997) in small samples of elite and recreational (both n = 8) female soccer players. It would seem that muscle strength as well as technique is relevant in one of the most fundamental soccer skills.

This study was based on female soccer players competing at University level. The two aims were:

(1) to explore the existence of positional differences in fitness and performance variables
(2) to investigate the relationship between kick distance and peak muscle torque at a range of angular velocities.

2 Methods

Forty-nine female soccer players of university standard participated in the study. Mean age was 19.0 ± 3.4 years, height 1.64 ± 0.09 m and body mass 60.7 ± 5.0 kg. They were self-classified into 11 centre-backs, 10 full-backs, 17 mid-field players and 11 forwards. These subjects participated in a series of laboratory and field based tests.

The maximal oxygen intake (VO_2 max) was estimated from a 20-m shuttle run test (Ramsbottom et al., 1988). A 'repeated sprint' test entailed ten 30-m sprints with a 10-s interval between sprints. The decrement in time over the repeated efforts was used as an index of 'fatigue'. A vertical jump test was used as a measure of anaerobic performance.

The percentage body weight as fat was determined using bioelectric impedance (Bodystat 500, Isle of Man). The lean body mass was obtained by subtraction from body mass values determined using Seca scales.

Kick performance was determined off a three-stride run up. Subjects were allowed three practice attempts, followed by six scored trials. The fullest distance from place kick to the first bounce of the ball was recorded as the performance. The ball used was size 5 (Mitre Astro pro, India), inflated to the manufacturer's recommendations. A marked area 15 m broad denoted the boundaries within which the ball had to land.

Peak torques for knee flexion and extension were measured by means of an isokinetic dynamometer (Lido Active, Davis, CA). Measurements were made for both concentric and eccentric actions at angular velocities of 1.05 and 3.14 $rad \cdot s^{-1}$. A sub-sample of 17 players attended the laboratory for these muscle performance assessments.

Differences between playing positions were examined using analysis of variance following checks for normal distribution of data for the 49 players. Tukey's post hoc test was used to localise significant differences. Data were examined using Kruskal-Wallis procedures when parametric analysis was contra-indicated. Relationships between kick performance, peak torque and lean body mass were investigated by means of Pearson product correlation analysis. In all cases a probability of 0.05 or less was taken to indicate statistical significance.

3 Results

There was a significant influence of playing position on the estimated value of VO_2 max ($F=9.84$; $P < 0.01$). Follow-up tests established that the difference lay between the mid-field players (mean values of 48 $ml \cdot kg^{-1} \cdot min^{-1}$) and the centre-backs (mean 43.7 $ml \cdot kg^{-1} \cdot min^{-1}$).

The forwards were the fastest in the 30-m sprint test but differences from the other playing positions were non-significant according to the Kruskal–Wallis test ($H = 1.63$; $P > 0.05$). Performances over the ten sprints displayed appreciable fatigue (averaging 83.1% of initial best time), which did not vary between positions ($F = 1.10$, $P > 0.05$). Similarly, although vertical jump performances were highest for centre-backs and forwards (see Table 1), and least for full-backs and mid-field players, the differences did not reach statistical significance ($P = 0.13$; $P > 0.05$).

Table 1. Field test performances (mean ± SD) according to positional role.

	Centre-back	Full-back	Mid-field	Forward
Vertical jump (cm)	35.4 ± 2.7	34.6 ± 3.6	35.0 ± 3.5	35.2 ± 3.6
VO$_2$ max (ml kg^{-1} min^{-1})	43.7 ± 3.0	45.7 ± 2.3	48.0 ± 1.8	46.3 ± 1.7
Sprint time (s)	4.81 ± 0.18	4.86 ± 0.19	4.84 ± 0.17	4.80 ± 0.25
Speed decrement (s)	3.79 ± 0.56	4.13 ± 0.38	4.05 ± 0.37	4.04 ± 0.54

Central defenders were significantly taller ($F = 7.08$; $P < 0.01$) than full-backs and mid-field players (Table 2). Forwards were intermediate and not significantly different in height from any of the other groups. The central defenders were also heavier than the midfield players ($F = 7.83$, $P < 0.01$) and had the greatest fat-free mass. There was no significant difference between the groups in body fat percent.

Table 2. Anthropometric measures and kick performance of players ($N = 49$) according to playing positions. Values are mean ± SD.

	Centre-back	Full-back	Mid-field	Forward
Height (m)	1.67 ± 0.01	1.62 ± 0.03	1.62 ± 0.04	1.65 ± 0.02
Body mass (kg)	62.6 ± 2.0	59.9 ± 2.5	59.0 ± 2.8	61.7 ± 2.7
Fat- free mass (kg)	47.1 ± 1.4	45.9 ± 1.6	45.5 ± 1.9	46.6 ± 0.9
Body fat (%)	22.8 ± 1.4	24.2 ± 1.4	23.1 ± 1.4	24.9 ± 1.1
Kick distance (m)	27.8 ± 2.0	27.1 ± 2.2	27.1 ± 2.2	26.7 ± 2.3

The values reported for peak torque during isokinetic movements (Table 3) are for the seventeen subjects who visited the laboratory for these tests. The centre-backs had consistently the highest values, being significantly different from full-backs for concentric knee extension ($F = 7.93$; $P < 0.05$) at 1.05 rad·s^{-1} and for eccentric knee flexion ($F = 4.33$; $P < 0.10$) at the faster angular velocity.

The centre-backs demonstrated consistently higher values than the other players for the isokinetic tests. Values were significantly different from those of the full-backs for extension (concentric) at 1.05 rad·s^{-1} ($F = 7.93$: $P < 0.05$) and flexion (eccentric) at 3.14 rad·s^{-1} ($F = 4.33$: $P < 0.10$).

Significant correlations for knee extensors were evident between kick performance and peak torques during concentric actions ($r = 0.62$ and 0.55 at 1.05 and 3.14 rad·s^{-1}, respectively) and for eccentric actions ($r = 0.75$ and 0.59 for the respective angular velocities of 1.05 and 3.14

rad·s^{-1}). The correlation coefficient between fat-free mass and kick distance was 0.69. The correlaton coefficients for knee flexors and kick performance were consistently lower than these values, being 0.52 and 0.49 for concentric and 0.47 and 0.43 for eccentric actions at 1.05 and 3.14 rad·s^{-1}, respectively.

Table 3. Peak torque (mean ± SD) for knee flexion and extension at two angular velocities.

	Centre-back	Full-back	Mid-field	Forward
Knee extension				
Con 1.05 rad·s^{-1}	109.0 ± 2.1	96.7 ± 2.1	105.6 ± 5.3	105.2 ± 2.5
Ecc 1.05 rad·s^{-1}	123.0 ± 9.8	116.6 ± 10.0	120.2 ± 5.3	118.0 ± 6.3
Con 3.14 rad·s^{-1}	76.6 ± 7.0	72.3 ± 5.1	75.8 ± 7.4	72.5 ± 2.0
Ecc 3.14 rad·s^{-1}	106.2 ± 8.5	100.6 ± 2.1	101.8 ± 4.0	101.7 ± 3.3
Knee Flexion				
Con 1.05 rad·s^{-1}	50.4 ± 7.1	42.0 ± 2.7	50.0 ± 5.5	47.7 ± 4.4
Ecc 1.05 rad·s^{-1}	55.4 ± 9.5	49.0 ± 1.0	54.6 ± 7.9	51.2 ± 3.0
Con 3.14 rad·s^{-1}	45.4 ± 5.3	35.0 ± 1.7	40.0 ± 6.6	43.0 ± 2.9
Ecc 3.14 rad·s^{-1}	50.8 ± 7.2	37.0 ± 1.0	46.6 ± 5.6	44.8 ± 3.2

4 Discussion

The estimated VO$_2$ max of the current players was intermediate between the English University players studied by Miles et al. (1993) and the England national squad (Davis and Brewer, 1992). The variability among out-field players was influenced by playing position, the highest values being found for the mid-field players and the lowest for the centre-halves. This separation is likely to reflect differences in physiological demands of playing in these positions, corresponding to observations in male players (Reilly, 1997). Even at the level of inter-university competition, it seems that specificity is evident in fitness measures. This specificity was not evident in the jump and sprint tests.

It appears also that anthropometric characteristics affected the orientation of individuals towards particular roles. The greater height of central defenders and forwards would be of benefit in contesting aerial possession of the ball. The greater body mass of players in these positions corroborates previous observations in male players (Reilly, 1979). Variability in body composition between individuals was greater than any effect of playing position.

The greater fat-free mass of the central defenders was reflected in their superior peak torque values. Peak torque in turn was related to kicking distance, demonstrating that gross muscle strength, particularly in the quadriceps is relevant when performing a kicking action. The values reported for knee flexors were below normal flexor–extensor ratios and suggest these players should train to improve knee flexor strength. The kick distances achieved by the players in the present study were practically identical to the University players and less than the scores of the elite players studied by Reilly and Drust (1997).

In the present investigation no data were collected to explore further anthropometric determinants of kicking performance. In future studies it would be helpful to estimate lean leg volume, as recorded by McCrudden and Reilly (1993) for male players. Such analysis would establish the potential for strength training to influence this particular aspect of soccer play.

The main observations in the present study were that positional differences among female soccer players were evident in body size and aerobic power variables. These differences were not evident in anaerobic and repeated sprint tests, or in body composition. Muscle strength measures were correlated with kicking performance, partly influenced by fat-free mass.

5 References

Cabri, J., De Proft, E., Dufour, W. and Clarys, J.P. (1988) The relation between muscular strength and kick performance, in *Science and Football* (eds T. Reilly, A. Lees, K. Davids and W.J. Murphy), E. and F.N. Spon, London, pp. 186–193.

Davis, J.A. and Brewer, J. (1992) Physiological characteristics of an international female soccer squad, *Journal of Sports Sciences*, 10, 142–143.

McCrudden, M. and Reilly, T. (1993) A comparison of the punt and the drop kick, in: *Science and Football II* (eds T. Reilly, J. Clarys and A. Stibbe), E. and F.N. Spon, London, pp. 362–366.

Miles, A., MacLaren, D., Reilly, T. and Yamanaka, K. (1993) An analysis of physiological strain in four-a-side women's soccer, in *Science and Football II* (eds T. Reilly, J. Clarys and A. Stibbe), E. and F.N. Spon, London, pp.140–145.

Reilly, T. (1979) *What research tells the coach about soccer*, AAHPERD, Washington.

Reilly, T. (1997) Energetics of high intensity exercise (soccer) with particular reference   to fatigue, *Journal of Sports Sciences*, 15, 257 –263.

Reilly, T. and Drust, B. (1997) The isokinetic muscle strength of women soccer players, *Coaching and Sport Science Journal*, 2, 12–17.

62 FITNESS CHARACTERISTICS OF ENGLISH FEMALE SOCCER PLAYERS: AN ANALYSIS BY POSITION AND PLAYING STANDARD

M. K. TODD*, D. SCOTT** and P. J. CHISNALL**
** Department of Exercise Science, University of Southern California, Los Angeles, USA.
** Faculty of Health and Exercise Sciences, University College Worcester, Worcester, UK

1 Introduction

The recent explosion in the popularity of female soccer in England is evident in the growth of the number of formally organised clubs in England from 263 to 700 in the past ten years with a concomitant doubling of the number of registered players from 7,000 to 14,000 (FA, 1998). The literature concerning the study of male players is abundant, but, much less is known about the female game, yet it would appear that the game places similar demands upon female players as it does upon male players. A Danish female international midfielder was reported to cover 9.5 km during an 80-min game, and time motion analysis revealed that percentage times spent standing, walking, jogging, running at various intensities, sprinting and backwards movement were similar to those for elite Danish male players in a league game (Bangsbo, 1993). A distance coverage of 8.5 km has been reported for Swedish female players during a league game with a distance of approximately 15 m per sprint (cited by Davis and Brewer, 1993). In a study of indoor four-a-side soccer, the intensity of play, as estimated by heart rate and blood lactate levels, was found to be the same for female players as for males in a similar previous study (Miles et al., 1993). Similarities between female and male team sports have been reported elsewhere, with trends in fitness scores related to position of female rugby players reflecting the trends seen in male players (Kirby and Reilly, 1993). There seems to be little data available on positional differences in female soccer players, though one study, which omitted statistical comparisons, showed trends observed for scores in fitness tests of the USA national team. This study indicates that the trend is for midfield players to have greater speed, aerobic endurance, and local muscular endurance compared to goalkeepers, defenders and forwards (Kirkendall and DiCicco, 1996). To our knowledge no study has examined physical fitness characteristics in different standards of female players.

2 Methods

A cohort of English players (N = 120) representing ten teams in either the National Premiership or two regional leagues were tested for measures of physical fitness. Teams were tested during a formal training session at the club practice facility within two weeks of the last game of the season. Players were measured for height (Seca stadiometer), body mass (Seca scales), percentage body fat (Harpenden calipers) estimated from skinfold thickness (Durnin and Womersley, 1974), aerobic capacity estimated from a 20 m shuttle run test (National Coaching Foundation), leg strength (TKK back and leg dynamometer, Japan), speed over 5 and 30 m (Newtest timing system, Sweden), and hamstring and lower-back flexibility from a sit and reach test (Bodycare, UK). All tests were performed in the same order, at the same time of the evening, on similar non-grass surfaces. All subjects gave informed consent and were screened prior to testing (BASES 1997).

2.1 Statistical analysis
Boxplots of the data revealed heavy-tailed distributions, and therefore a one-way ANOVA using 20% trimmed means (F_t) was applied to compare the groups (Wilcox, 1996). Confidence intervals for linear contrasts of the trimmed means were computed using a modification of the Welch–Šidák method to ensure that the probability of making a Type I error did not exceed $\alpha = 0.05$ (Wilcox, 1996, 1997). Relationship testing was analysed using a Percentage bend correlation (r_{pb}) which gives greater control over Type I errors than Pearson's correlation especially when data are skewed or heavy tailed (Wilcox, 1996).

3 Results

3.1 Whole group analysis
When assessing the physical fitness characteristics in the whole group, Percentage bend correlation analysis revealed significant relationships between 30 m sprint and the following variables: 5 m sprint, percentage body fat, vertical jump and endurance performance. Significant relationships were also found to exist between percentage body fat and the following variables: vertical jump, aerobic endurance, and 5 m sprint time (Table 1).

Table 1. Percentage bend correlation analysis for the percentage body fat, endurance, 5 m and 30 m sprints and vertical jump for all subjects (N = 120).

	Endurance	30 m sprint	5 m sprint	Vertical jump
Body fat		$r_{pb} = 0.45, P < 0.001$	$r_{pb} = 0.42, P < 0.01$	$r_{pb} = -0.26, P < 0.01$
	Endurance	$r_{pb} = -0.39, P < 0.001$	NS	NS
		30 m sprint	$r_{pb} = 0.79, P < 0.01$	$r_{pb} = -0.55, P < 0.001$
			5 m sprint	$r_{pb} = -0.45, P<0.01$

3.2 Playing standard

When players were divided into sub-groups (Table 2) of international representation within the current season (n = 25), Premier league players (n =4 4) and regional league players (n = 51), statistical analysis revealed differences (F_t = 5.0, P < 0.05) in the estimation of percentage body fat between international and regional players (22.9 ± 3.4% and 25.5 ± 3.5% respectively). The difference in playing standard was also reflected in significant differences in the total number of years spent playing competitive soccer between international players (with $\overline{X}_t$ =12.3 ± 4.9 years) and regional players ($\overline{X}_t$ = 9.0 ± 3.6 years, P < 0.05)

Table 2. Physical fitness and anthropometric data for all subjects (N = 120), presented by playing standard. Data from the present study are expressed as Trimmed means ($\overline{X}_t$ ± SD).

Standard	Age (years)	Height (cm)	Mass (kg)	Body fat (%)	VO$_2$ max (ml·kg^{-1}·min^{-1})
International (n = 25)	22.3 ± 4.3	162.8 ± 5.9	61.2 ± 5.2	22.9 ± 3.4*	46.8 ± 5.1
Premier League (n = 44)	23.4 ± 5.9	163.3 ± 5.5	62.1 ± 6.4	23.9 ± 4.2	45.0 ± 6.0
Regional League (n = 51)	21.3 ± 6.6	163.9 ± 6.3	61.6 ± 7.1	25.5 ± 3.5	43.9 ± 5.0

Standard	Leg strength (kg)	Speed 5 m (s)	Speed 30 m (s)	Flexibility (cm)	Vertical jump (cm)
International (n = 25)	122.0 ± 39.5	1.02 ± 0.05	4.62 ± 0.19	17.4 ± 5.9	47.8 ± 6.4
Premier League (n = 44)	113.7 ± 30.8	1.04 ± 0.06	4.64 ± 0.25	16.5 ± 8.9	49.0 ± 6.4
Regional League (n = 51)	116.5 ± 34.5	1.07 ± 0.06	4.70 ± 0.25	14.6 ± 8.0	46.5 ± 4.8

* Different from Regional players (P<0.05)

3.3 Positional differences

Positional differences between goalkeepers (GK; n = 9), defenders (DF; n = 45), mid-fielders (MF; n = 44) and forwards (FF; n = 22) were analysed to establish whether positional specificity was evident in the results obtained from the physical fitness measures (Table 3). Anthropometric differences were found for body mass (F_t = 6.5, P < 0.05) between GK and DF (26.3 ± 4.3 kg and 24.2 ± 3.9 kg respectively), between GK and MF (26.3 ± 4.3 kg and 24.0 ± 3.5 kg respectively) and between GK and FF (26.3 ± 4.3 kg and 24.3 ± 4.1 kg respectively); for height (F_t = 5.5, P < 0.05) between GK and MF (168.5 ± 4.3 cm and 161.6 ± 5.0 cm respectively), and between GK and FF (168.5 ± 4.3 cm and 162.5 ± 6.8 cm respectively), and between DF and MF (165.2 ± 5.6 cm and 161.6 ± 5.0 cm respectively), and for percentage body fat between GK and DF (26.3 ± 4.3% and 24.2 ± 3.9% respectively), and between GK and MF (26.3 ± 4.3% and 24.0 ± 3.5% respectively). A significant difference ($F_t = 2.4$, $P < 0.05$) was also found for 30 m sprint time between GK and FF (4.84 ± 0.36s and 4.58 ± 0.20s respectively). Trends were observed for aerobic capacity and flexibility but these were not significant at $P < 0.05$. The VO_2 max data for International, Premier League and regional league players were 46.8 ± 5.2 ml·kg^{-1}·min^{-1}, 45.0 ± 6.0 ml·kg^{-1}·min^{-1}and 43.9 ± 5.0 ml·kg^{-1}·min^{-1} respectively. Flexibility was greatest in the international players compared to the Premier League and regional league players (17.4 ± 5.9, 16.5 ± 8.9 and 14.6 ± 8.0 cm respectively)

Table 3. Physical fitness and anthropometric data for all subjects (N = 120), presented by playing position. Data from the present study are expressed as Trimmed means ($\overline{X}_t$ ± SD).

Position	No.	Height (cm)	Mass (kg)	Body fat (%)	VO_2 max (ml·kg^{-1}·min^{-1})
Goalkeepers	9	168.5 ± 4.3	68.9 ± 5.5	26.3 ± 4.3	40.5 ± 6.1
Defenders	45	165.2 ± 5.6**	62.7 ± 6.6*	24.2 ± 3.9*	45.3 ± 5.2
Mid-fielders	44	161.6 ± 5.0*	59.5 ± 5.0*	24.0 ± 3.5*	45.0 ± 5.5
Forwards	22	162.5 ± 6.8*	60.9 ± 7.3*	24.3 ± 4.1	46.2 ± 5.6

Position	Leg strength (kg)	Speed 5 m (s)	Speed 30 m (s)	Flexibility (cm)	Vertical jump (cm)
Goalkeepers	123.2 ± 41.0	1.09 ± 0.07	4.84 ± 0.36	17.2 ± 10.0	49.8 ± 7.0
Defenders	115.6 ± 29.3	1.03 ± 0.05	4.64 ± 0.23	15.6 ± 8.2	47.6 ± 4.7
Mid-fielders	114.6 ± 33.6	1.04 ± 0.07	4.69 ± 0.21	16.1 ± 7.6	46.6 ± 6.1
Forwards	119.2 ± 43.3	0.98 ± 0.06	4.58 ± 0.20*	15.6 ± 7.8	49.2 ± 6.9

* Different from GK (P<0.05), ** different from MF (P<0.05)

4 Discussion

4.1 General discussion

It may be assumed that experienced soccer players have adapted to the demands of the game (Bangsbo, 1993) and therefore the physical capacity of the female players in this study must reflect the demands of their specific positions and the standards at which they compete. Time and motion analysis of female players has revealed that in a FIFA World Championship game, the ball was in play for almost 51 min (Miyamura et al., 1995), while in an Italian first division game, 67.46% of the match was spent in the midfield and only 15.5% was spent in either attack or defence (Evangelista et al., 1992). Bangsbo (1993) reported the average heart rate for a Danish international player to be 171 b·min^{-1} in the first half and 168 b·min^{-1} in the second half of an 80 minute game, similar to the heart rates observed for male players. Any discussion on the physical capabilities of players must consider both the ergonomic requirements of the same position under different tactical situations and the differing technical capabilities of the player. Good anticipation, game awareness and skill coupled with positional freedom or restraint could lead to different adaptations of players broadly categorised as defenders, midfielders and forwards. When the players in this study are divided according to either international, Premier or regional playing standard, visual comparisons may be made with published findings of other female soccer players (Table 4).

Table 4. Physical fitness and anthropometric data for all subjects (N = 120), presented with those results from other published studies. Data from the present study are presented as trimmed means ($\overline{X}_t \pm$ SD).

Author	Standard	No.	Age (years)	Height (cm)	Mass (kg)
Present study	English International Premier and regional	120	22.6 ± 5.9	163.4 ± 5.9	61.8 ± 6.7
Colquhoun and Chad (1986)	Australian State and International	10	24.4 ± 4.5	158.1 ± 5.7	55.4 ± 6.5
Davis and Brewer (1992)	English International	14	24.5 ± 3.6	166.0 ± 6.1	60.8 ± 5.2
Jensen and Larsson (1993)	Danish International	10	24.7	169.0	62.2
Tumilty and Darby (1992)	Australian International	20	23.1 ± 3.4	164.5 ± 6.1	58.5 ± 5.7
Rhodes and Mosher (1992)	Canadian University	12	20.3	164.8	59.5

Author	Standard	VO$_2$ max (ml·kg^{-1}·min^{-1})	Flexibility (cm)	Vertical jump (cm)	Body fat (%)
Present study	English International Premier and regional	44.8 ± 5.8	16.0 ± 7.9	47.6 ± 5.8	24.4 ± 3.9
Colquhoun and Chad (1986)	Australian State and International	47.9 ± 8.0	9.4 ± 7.7		20.8 ± 4.7
Davis and Brewer (1992)	English International	48.4 ± 4.7	12.3 ± 6.9		21.5 ± 3.6
Jensen and Larsson (1993)	Danish International	57.6		37.8	20.1
Tumilty and Darby (1992)	Australian International	48.5 ± 4.8	12.8 ± 4.1	40.5 ± 4.5	19.7 ± 4.0
Rhodes and Mosher (1992)	Canadian University	47.1 ± 6.4			≈19.0[φ]

[φ]As estimated from reported sum of skinfolds

4.2 Anthropometry

There seems to be no ideal height requirement for soccer players (Bangsbo, 1993). However, taller players may be advantaged in some positions such as goalkeeping, or in challenging for an aerial ball, although explosive leg strength and the ability to time the jump may compensate the shorter player. There were no height differences between standards of player, though positional differences were observed. Goalkeepers were the tallest players in the current sample and were significantly taller than midfield players and forwards, while defenders were also taller than mid-field players (P < 0.05). Goalkeepers were also characterised by higher estimated percentage body fat levels than defenders and mid-field players. This is likely representative of a lower aerobic demand specific to the position. Similar conclusions may be drawn for the differences in percentage body fat between regional and international players, though time and motion studies are needed to confirm this hypothesis. The subjects in the current study, when considered as one sample (N = 120), were shorter than the Scandinavian national players (Jensen and Larsson, 1992), and reported similar body mass values to players in other studies, though Australian players had the least body mass of all reported studies.

4.3 Aerobic endurance

In the current cohort of players, there were no significant differences in estimation of VO_2 max between players when subdivided according to playing position or by playing standard, though trends were observed (Table 3), between goalkeepers and the outfield players, and also between international and domestic league players. The VO_2 max, however, may not be a sensitive enough measure to differentiate players, since Bangsbo (1993) also found no differences in VO_2 max between Danish first and second division male players. The USA national players displayed positional trends, with midfield players reporting greatest aerobic capacity; however, statistical comparisons were not reported (Kirkendall and DiCicco, 1996). The players in the current study had a lower aerobic capacity than the female elite touch football players reported by O'Connor (1995), and other comparisons may be gleaned from Table 4.

4.4 Strength, speed and vertical jump

Leg strength is important in kicking distance tests, but consideration must also be given to kicking technique (Bangsbo, 1993). Although the ball travels a greater distance during World Championship games than during other games (Miyamura et al., 1995), no significant differences in leg strength were found between players when analysed by position or standard in the current cohort of English players. Differences were found to exist between goalkeepers and forwards in the 30-m sprint test (P<0.05), though no significant differences were observed when players were grouped according to playing standard. The females in this study compared favourably with a study of professional German male players in the 5 m sprint test (Kollath and Quade, 1993). Male players recorded a time of 1.03 ± 0.08 s for a 5 m sprint and 4.19 ± 0.14 s for a 30-m sprint, compared to times of 1.01 and 4.62 s for the international players in the 5 m and 30 m sprint, respectively. When velocity was measured from 5 m to 30 m (a total of 25 m) in a small sample of the players in the current study, differences between GK (slowest) and DF, between GK and FF (fastest), and between MF and FF ($F_t = 5.79$, P < 0.05) were revealed. Since no significant differences were observed for the velocity over 5 m, it could be that increased rates of acceleration after 5 m are an important positional specific adaptation in the women's game. When the players were simply grouped according to the domestic league in which they participated (Premier or regional), significant differences were observed for vertical jump height between the two groups (P < 0.05). However, when the players were grouped according to position, or when the international representation was considered, the differences were no longer statistically significant. The current cohort of female players display greater explosive leg strength, as determined from the vertical jump test, than either Danish or Australian international players (Jensen and Larsson, 1992; Tumilty and Darby, 1992).

5 Conclusions

There were anthropometrical and speed differences existing in female players depending on the standard of play and positional responsibility within the team. Trends existed for aerobic endurance, flexibility, and explosive leg strength, whereas estimated percentage body fat correlates with aerobic endurance, speed and vertical jump indicating the interrelationship of fitness variables in multiple sprint sports. It is reported that the intensities of play and the percentage time spent in the various game related activities mirror those of male players though more research is required into time-motion analysis examining playing standard and positional differences.

6 References

Bangsbo, J. (1993) *The Physiology of Soccer – With Special Reference to Intense Intermittent Exercise*, HO+Storm, Copenhagen.

BASES (1997) *Physiological Testing Guidelines*, BASES, Leeds.

Colquhoun, D. and Chad, K.E. (1986) Physiological characteristics of Australian female soccer players after a competitive season, *Australian Journal of Science and Medicine in Sport* , 18, 9–12.

Davis, J.A. and Brewer, J. (1992) Physiological characteristics of an international female soccer squad, *Journal of Sports Sciences*, 10, 142.

Davis, J.A. and Brewer, J. (1993) Applied physiology of female soccer players, *Sports Medicine*, 16, 180–189.

Durnin, J.V.G.A. and Womersley, J. (1974) Body fat assessed from total density and its estimation from skinfold thickness: measurements on 481 men and women aged from 16 to 72 years, *British Journal of Nutrition*, 32, 77–97.

Evangelista, M., Pandolfi, O., Fanton, F. and Faina, M. (1992) A functional model of female soccer players: Analysis of functional characteristics, *Journal of Sports Sciences*, 10, 165.

F.A. (1998) *Talent Identification Plan*, English Football Association, London.

Jensen, K. and Larsson, B. (1993) Variations in physical capacity in a period including supplemental training of the national Danish soccer team for women, in *Science and Football II* (eds T. Reilly, J. Clarys and A. Stibbe), E. and F.N. Spon, London, pp. 114–117.

Kirby, W.J., and Reilly, T. (1993) Anthropometric profiles of elite female rugby union players, in *Science and Football* II (eds T. Reilly, J. Clarys and A. Stibbe), E. and F.N. Spon, London, pp. 27–30.

Kirkendall, D. and DiCicco, T. (1996) Measuring fitness, *NSCAA Soccer Journal*, May/June, 47–49.

Kollath, E. and Quade, K. (1993) Measurement of sprinting speed of professional and amateur soccer players, in *Science and Football II* (eds T. Reilly, J. Clarys and A. Stibbe), E. and F.N. Spon, London, pp. 31–36.

Miles, A., MacLaren, D., Reilly, T. and Yamanaka, K. (1993) An analysis of physiological strain in four-a-side women's soccer, in *Science and Football II* (eds T. Reilly, J. Clarys and A. Stibbe), E. and F.N. Spon, London, pp. 140–145.

Miyamura, O.S., Seto, S. and Kobayashi, H. (1995) A study of 'in-play' and 'out-of-play' time in the first FIFA World Championships for women's football, 1991, *Journal of Sports Sciences*, 13, 519.

O'Connor, D. (1995) Profile of elite female touch football players, *Journal of Sports Sciences*, 13, 505.

Rhodes, E.C. and Mosher, R.E. (1992) Aerobic and anaerobic characteristics of elite female university soccer players, *Journal of Sports Sciences*, 10, 143.

Tumilty, D.McA. and Darby S. (1992) Physiological characteristics of Australian female soccer players, *Journal of Sports Sciences*, 10, 145.

Wilcox, R.R. (1996) *Statistics for the Social Sciences.* Academic Press Inc, California, USA.

Wilcox, R.R. (1997) *Introduction to Robust Estimates and Hypothesis Testing*, Academic Press Inc, California.

63 GENDER DIFFERENCES IN STRENGTH AND ENDURANCE OF ELITE SOCCER PLAYERS

J. HELGERUD, J. HOFF and U. WISLØFF
Department of Sport Sciences, Norwegian University of Science and Technology, Trondheim, Norway

1 Introduction

During the last decade soccer for women has become a popular event. Differences between male and female athletes are obviously not only genetic, but are also influenced by level of selection, training and competition. Dimensional scaling must be considered when comparing groups with different body mass (Nevill et al., 1992; Wisløff et al., 1998). The major purpose of the present study was to examine gender differences in cardiovascular endurance capacity as well as muscular strength and power in the best male and female soccer teams in Norway. A further purpose was to investigate maximal oxygen uptake in proportion to body mass (m_b) for soccer players. A subsidiary aim was to establish normative data of elite female soccer players.

2 Methods

One male and one female team from the Norwegian elite soccer league participated in the study. Both Trondheims Ørn (women, N = 12) and Rosenborg (men N = 14) are the most successful teams in Norway over the last five years. The maximal oxygen uptake (VO_2 max) was measured in all the subjects and related to body mass. Performance measures included squats at 90° and bench press.

3 Results and Discussion

Results showed that maximal oxygen uptake did not increase proportionally to m_b in elite soccer players. This finding supports the argument that dimensional scaling should be used for soccer players. Mean results for the women's team were 54.0 ml.kg^{-1}.min^{-1} or 151.5 ml.kg$^{-0.75}$.min^{-1} for maximal oxygen uptake, 112.5 kg or 7.1 kg.$m_b^{-0.67}$ for 90° squats, 43.8 kg or 2.75 kg.$m_b^{-0.67}$ for bench press. Values of vertical jump height were 42.9 cm on average.

Considerable gender differences existed. Maximal oxygen uptake, squats and jump height were 20–25 % lower for women compared with men, and bench press values among women were 40 % lower.

4 References

Nevill, A. M., Ramsbottom, R., Williams, C. and Winter, E.M. (1992) Scaling individuals of different body size, *Journal of Sports Sciences*, 9, 427–428.

Wisløff, U., Helgerud, J. and Hoff, J. (1998) Strength and endurance of elite soccer players, *Medicine and Science in Sports and Exercise*, 30, 462–467.

PART TEN

Management and Organisation

64 THE MULTIDISCIPLINARY DELIVERY OF SPORT SCIENCE IN GAELIC FOOTBALL

C. MAHONEY
School of Sport Studies, Roehampton Institute London, West Hill, London, UK

1 Introduction

Irish sporting culture has been slow to acknowledge the usefulness of sport science, especially sport psychology. Commonly used references to 'bringing in the shrink' or 'sorting out the nutter' (Singer, 1978; Tutko et al., 1979; Patmore, 1986) are often heard in relation to the application of mental skills training by sport psychologists working with sporting teams or individuals. Such terms together with a reluctance to alter training behaviours have done little to enhance the reputation of sport science within academic or applied circles, but nonetheless the field is still the fastest growing discipline at undergraduate level in universities in the United Kingdom and Ireland. It is still ironic to note that its inclusion in football of all codes, within the UK and Ireland, appears invisible.

The role of the applied sport scientist is to educate athletes and coaches, by helping them improve and enhance their athletic performance through the use of strategies based on sound scientific evidence. This advice includes changes to nutritional regimes, altering exercise:rest ratios in anaerobic work, or empowering the athletes with psychological tools to achieve a zone of optimal functioning (Hanin, 1980). The combined effect of including sport science in training is to remove negative thoughts from performance and training contexts, prevent mental paralysis and its consequent fear of failure, and optimise personal potential in enabling peak performances to be achieved.

Few footballers would disagree that it is important to prepare themselves physically, technically, tactically and psychologically. All skills need to be developed and practiced regularly from an early age, and these include mental skills as well as physical and technical practice. The early identification of specific football requirements using a needs analysis, involves answering basic questions such as;

What are the important skills in football?

How can they be developed?

When they are developed, how can this be shown?

How can the enormous demands of football be encompassed through sport science?

Gaelic sports, and in particular Gaelic football, attract high levels of sports participation amongst children and people of sporting age throughout Ireland. Gaelic football is played and also watched by a large proportion of the population within Ireland, and has a most impressive worldwide following, with viewing figures for the annual All Ireland Championship final exceeding 30 million people world-wide. Within Ireland the viewing audience is comprised of males and females, both adult and children, which gives the sport a high family profile.

Being regarded as a family sport, Gaelic football is usually played on Sunday, regularly attracting crowds in excess of the safe limits of ground capacities in which matches are held, but attracting very little crowd trouble. Paradoxically, alongside this interest in playing, the sport has seemingly systematically failed to take notice of the advances in sport science by relinquishing outdated modes of training and performance delivery. Recent pressures from the Gaelic Athletic Association (GAA), the re-writing of Coaching Manuals for the sport and the recognised need to enable contemporary knowledge to infiltrate the sport, is beginning to promote change. The development of the National Coaching and Training Centre (NCTC) in Limerick has also enabled this process to occur more acceptably and with an appropriate sport science base. Nonetheless some counties have been proactive in embracing science knowledge used effectively by other sports and incorporated this into preparation of players for Gaelic football.

A systematic evaluation of sport science in Gaelic football has yet to be completed. However, a survey of County Managers showed an understanding of sport psychology with some examples in their coaching, but further assistance was needed (McMullen, 1986). The need for an integrated approach to the delivery of sport science was recommended. This advice has not gone unheeded as the NCTC, as part of its Masters Coaching Certificate in alliance with the GAA, has developed such a programme for all new coaches seeking accreditation.

2 Delivery of Sport Science in Gaelic Football

A representative of the management group who sought assistance in the team's preparation for the Championship, made the initial contact for the Doire team. The programme, which was described to the management, centred on the provision of physiological services integrated with a comprehensive mental skills training package.

The sport science package was designed to empower the athletes to become discerning consumers of sport science. Following an educational, developmental and practical application model of intervention, the sport scientist acted as a 'stretch coach' (Botterill, 1990) in which the players were assisted in identifying weaknesses. Interventions were then developed collectively and individually to enhance achievement and work towards the production of personal peak performances (Ravizza, 1984). Acknowledgement that mental fitness is as important as physical fitness and technical ability was recognised by the management, who were supportive of the need to integrate services and not have any aspect of sport science separate to other training aspects. Holistic programmes, integrating all sport sciences and practical coaching applications, were found by Rushall and King (1994) to develop any team more fully.

The work with the team was undertaken by following an integrated approach to the delivery of sport science and using a model adopted previously and outlined in Figure 1 (Mahoney, 1995). To enhance acceptance followed by accountability and evaluation, the players were first introduced to the principles of education followed by empowerment.

3 Type and Range of Services Offered

The first meeting with the players occurred following training where they were presented with a brief explanation of the sport science package that would be delivered. Prior to this meeting the players had been distributed with questionnaires to complete in an effort to provide some background information and make the initial stages of intervention more straightforward. Following this meeting a co-ordinated physical training (unrelated to skill development) and mental skills training package was put in place.

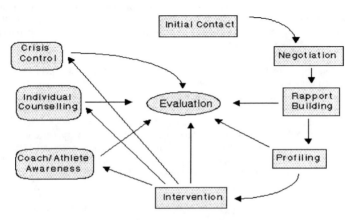

Figure 1. A sports science intervention model.

Player and management discussions on the specific needs of Gaelic Football and data from psychometric evaluation showed that general and specific fitness, together with communication, confidence, concentration, anxiety and choking were areas that reflected deficiencies in general team strategies. Open communication was promoted between players, management and all support staff involved in training and selection. Players were encouraged to take responsibility for communication strategies and to seek approval or knowledge related to any decisions of coaching or management team.

Preparation involved combinations of high intensity training, interval sessions and aerobic work through which players were able to establish a higher level of fitness than previously achieved. The resultant fitness improvements promoted individual and team confidence, reflected by greater training efforts, better practice performances, and improved competition results. By attending all training and competition fixtures, the fragility of confidence (Terry, 1989) often damaged by a bad refereeing decision, a missed point or a lost tackle, which necessitated crisis control strategies, was always available to the players when required.

Concentration strategies were not developed on a whole team basis, though several players sought individual assistance. This involved developing an awareness of focusing needs, identification of key concentration periods (Gordon, 1990) and focusing on the present, through which players were able to enhance their precise cue selection and minimise critical concentration periods. Players were assisted in the development of strategies similar to those established by Nideffer (1987).

Personal zones of optimal functioning were developed by assisting players to understand their arousal zones and establish how this affected their personal preparation, including developing pre-performance routines. The removal of the change room hype so often seen in football and the development of personal space for each player to develop his own arousal zone was a major aspect of the development process.

The team had been known for some years as fragile, 'choking' at the final hurdle. Having only once previously reached the All Ireland final, the team had a lack of success and a lack of expectation for success, though in the previous year its league record had been outstanding. To overcome choking at critical points, a range of confidence building strategies closely allied with fitness and appropriate 'what if' scenarios were developed with the team and individual players.

In Gaelic football, like many other combative team sports, goal-keepers require special attention. Playing as they do without a direct opponent, their concentration skills, confidence and coping strategies are of paramount importance. The team employed a specialist goal-keeping coach who worked exclusively with the two goal-keepers on the squad. While one goalkeeper was the 'regular starter', part of the sport science role involved maintenance of fitness and mental stability for the second goalkeeper if and when he might be required. Similar concerns existed for several players who were regular benchwarmers.

Involvement with the team was intense, three nights per week and every weekend with the team over a nine-month period. The sport science initiative was very much part of the team approach. As a consultant to the team it was possible to provide sport science with a 'shop window profile' throughout Ireland. It also resulted in consultant inclusion on the selection squad in an advisory capacity

4 Concluding Comments

The squad comprised 12 players with university degrees. Previous studies have shown that acceptance and participation in new initiatives are received more favourably by males with a higher level of education (McCarthy, 1994). It seems likely the 12 players with university degrees enhanced the acceptance of sport science. In addition several players had been part of a similar package during their university education and some players had been exposed to similar knowledge while overseas playing Australian Rules Football.

The use of a holistic approach to the delivery of sport science worked effectively with Gaelic football. Effectiveness seems to be closely linked with development of a close rapport with all players and support personnel and ultimately the education and cooperation of all concerned. The sport scientist should be a temporary support role in which empowerment, rather than reliance, is the outcome of intervention. Consultancy of this type is extremely time invasive, but the potential rewards of such programmes are measured by intrinsic satisfaction, player empowerment and performance rewards.

5 References

Botterill, C. (1990) Sport psychology and professional baseball, *The Sport Psychologist*, 4, 358-368.

Gordon, S. (1990) A mental skill training program for the Western Australian State cricket team, *The Sport Psychologist*, 4, 386-399.

Hanin, Y. (1980) A study of anxiety in sports, in *Sport Psychology: An Analysis of Athlete Behaviour* (ed W. Straub), Mouvement Publishing, New York.

Mahoney, C. (1995) Psychological interventions with an elite swimming squad: processes and products, *Unpublished PhD thesis*, The Queen's University of Belfast.

McCarthy, L. (1994) Patterns of Irish sport behaviour, *Unpublished PhD thesis*, The University of Oregon.

McMullen, M. (1986) Coaching in Gaelic football – a review of the contribution of the psychological aspects of sport to coaching, *Unpublished BSc thesis*, The University of Limerick.

Nideffer, R. (1987) Psychological preparation of the highly competitive athlete, *The Physician and Sportsmedicine*, 15, 85-92.

Patmore, A. (1986) *Sportsmen Under Stress*. Stanley Paul and Co Ltd, London.

Ravizza, K. (1984) Qualities of the peak experience in sport, in *Psychological Foundations of Sport* (eds J. Silva and R. Weinberg), Human Kinetics, Champaign, IL

Rushall, B. and King, H. (1994) The value of physiological testing with an elite group of swimmers, *Australian Journal of Science and Medicine in Sport*, 26, 14-21.

Singer, R. (1978) Sport psychology: an overview, in *Sport Psychology: An Analysis of Athlete Behaviour*, (ed W. Straub), Mouvement Publications, New York.

Terry, P. (1989) *The Winning Mind*, Thorsons Publishing Group, England.

Tutko, T., Pressman, M., Butt, D., Nideffer, R., Suinn, R. and Ogilvie, B. (1979) Who is doing what: viewpoints on psychological treatments for athletes, in *Coach, Athletes and the Sport Psychologist* (eds P. Klavora and J. Daniel), University of Toronto, Toronto.

65 MORE THAN A GAME: THE FAN–BUSINESS DICHOTOMY IN PROFESSIONAL SPORT

S. QUICK
School of Leisure, Sport and Tourism, University of Technology, Sydney, Australia

1 Introduction

When the article 'Paying to Win: The Business of the AFL' appeared in the Victoria University of Technology publications *NEXUS,* and the *Bulletin of Sport and Culture,* Australian Rules Football fans in Melbourne were incensed. Criticism emanated from quarters as disparate as journalists, interested academics and even the 'man-in-the-street' (Herald-Sun, 1997). The source of their ire was the suggestion that organisations such as the Australian Football League (AFL) who deemed themselves to be in the business of entertainment, should root decision-making processes in solid business principles and practices rather than populism. The article also suggested that 'rational debate had been derailed by rabid adherents to tribalism, community affiliation, history and tradition' (Quick, 1996a,b). The case was made that debates on rationalisation and commercialisation in the AFL, and by extension professional sport, should exclude die-hard fans. The rationale being that such fans may be incapable of arriving at an objective decision that is in the long-term interests of the game.

While the argument is sound that the management of the professional sport environment should include the creation of mission statements underpinned by strategic thinking, and framed by a vision for the sport that incorporates the fans, a dilemma quickly becomes apparent. That is, sport fans constitute a heterogeneous group with the die-hard being but one cohort of many. However it is this cohort which is the most identified, vocal and deified when it comes to popular discussion of the relationship between sport and its consumers. Such a position may be undeserved. In this report it is argued that in regard to professional sport the tribal, hard core fan is but a minor figure in the professional sportscape. Moreover this group, the hard core fan, may indeed have the least to offer.

2 Background

Despite Jones' (1997) assertion that there is little empirical research on sport fans, evidence would suggest otherwise. *Sport Marketing Quarterly, Journal for Sport and Social Issues, Journal of Applied Sport Psychology* and the *Journal of Sport Behaviour* are just four sport related publications that regularly feature research linked to sport fans. Furthermore journals in the mainstream disciplines of marketing, psychology, sociology and communication have regularly published research that has explored the sport – fan nexus. Irrespective of journal or the sport context, the fundamental underpinning of the majority of research has been the collection of data that determines who the fans are and their motivations for sport event attendance. While for the sport marketer such data assist in the establishment of strategies for continued and increased product consumption, such data do not differentiate fans based on type.

From a sport marketing perspective much of the published material has focused on socio-demography or motivations for attendance. Such cases are plentiful however the following few instances illuminate this trend.

Graham's (1992) study of spectators at professional tennis tournaments concluded that the demographic and economic characteristics of tennis fans were universal rather than a function of geographic location. Conversely Randl (1994) determined that the demographics of racetracks patrons alter depending on whether the track is a local race club or one of the large historic tracks. Ashley and Song (1995) undertook frequency distributions of attendance over three years at the Texas World Speedway which provided the demographic, economic and related information required by TWS management for the establishment of marketing strategies. This research was similar to that conducted by Randl (1994) in that it primarily addressed the question of who was in attendance rather than why they attended.

Hansen and Gauthier (1993, 1994) and Gauthier and Hansen (1993) systematically determined motivations for spectator attendance at PGA and LPGA golf events. They concluded that scenery was an important determinant in the non-golfing spectator's decision to attend events, as was the opportunity get fit and 'take in some fresh air'. For golfing spectators, the opportunity to get some tips, see a quality performance and 'big names' in action were the motivations for attendance.

Hofacre (1994) concluded that it was not 'sex appeal' that encouraged female consumption of professional indoor soccer but rather 'family appeal'. Interestingly Hofacre argued that while the indoor soccer marketing fraternity knew women comprised 50% of its audiences, they weren't sure of what to do in terms of promotional activity. While still focusing on demographics, Stotlar (1995) patronised the sports grill to discern the characteristics of patrons. The research concluded, 'sports viewing with friends was a major attraction to sports grill patrons'.

de Burca, Brannick and Meenaghan (1995) surveyed spectators at all-Ireland semi-finals in hurling and Gaelic football to identify motives for attendance and found that high attendance was directly related to product loyalty. Their major findings were that 'the game itself' and 'county team support' were the prime motivating attendance factors. They also suggested that spectators with a high degree of product loyalty were more likely to attend matches with friends than family.

Wakefield and Sloan (1995) concluded that team loyalty, the combination of history and performance, played the biggest role in determining spectators' decisions to be at the stadium, however they did not isolate such variables. While the issue of loyalty is critical to professional sports consumption, it is a dimension that is just starting to be seriously examined in its contemporary context.

The previous is but a small sample of the literature that has examined reasons for spectator attendance at, or consumption of professional sport. Without doubt the core area of concern has been a socio-demographic description of the fans, patrons and spectators. In such instances fan type and input into the decision-making process have not been variables for analysis. However loyalty is one variable that has been alluded to and is deserving of further discussion.

3   Fan Type

3.1 Fan categories

While traditionally sport fans have been differentiated on the basis of event attendees, and season tickets holders or members, in recent years the range of sport consumers has been widened to include fans that consume their sport through the media, merchandise purchase or sponsorship. Stewart and Smith (1997) devised a wide-ranging typology of the sport spectator that clearly illustrates the heterogeneous nature of sport fans. This typology is a useful starting point for an examination of fan type.

Stewart and Smith established five spectator categories, which provides nomenclature for the type of spectator, the motivation for sport attendance and the behaviour they exhibit. The categories are as follows.

1. The Aficionado which is the fan who seeks a quality performance. In the main, this fan is game loyal rather than team loyal.
2. The Theatregoer is the fan who seeks entertainment and wants a close contest. While they exhibit moderate team loyalty, the contest is key to their enjoyment.
3. The Passionate Partisan is that fan who wants the team to win and identifies strongly with team success and loss.
4. A Champ Follower exhibits loyalty that is related to team success. As long as the team is winning or competitive such fans will continue to consume the offered product.
5. Finally the Reclusive Partisan is the fan who strongly identifies with the team. However, this does not necessarily translate to game attendance as the fan can consume the product through other media such as television, radio, newspapers and magazines.

Given that it is now possible to differentiate clearly between sport fans based on type, a number of key questions need to be addressed. Furthermore how such categories link with the extant literature on sport fans is worth further exploration. Questions such as what are the factors that influence the fan's decision to continue to support an organisation, how far should/can an organisation go to accommodate different fan segments and, whether some fan segments are more important than others, need to be investigated.

It would appear from Stewart and Smith's typology that there are three major factors that contribute to the fan type. These are loyalty, performance (in this instance winning and losing) and identification. It is important to determine the importance of such variables before the preceding questions relating to fan influence, accommodation and importance can be resolved.

3.2 Loyalty

Sutton et al. (1997), Burca, Brannick and Meenaghan (1995), de Burca et al. (1995), Wakefield and Sloan (1995) and, as previously indicated Stewart and Smith, have all claimed that loyalty is one of the most important variables in understanding sport consumption. It could be argued that loyalty is very much an interpretative term. In most instances the lay person would interpret loyalty as an ongoing commitment; however, Macken (1998) suggested that loyalty tends to be a 'now' rather than a 'forever' concept. She argued that 'of all the human values to come under fire in our rationalist times, loyalty has suffered the most volleys' (p. 4). She further believed that 'the dismantling of bonds between employer and worker, between political party and true believer, between brand and consumer and between sport and fan has occurred without much discussion and seemingly without regret' (p. 4). The myriad of loyalty based programmes that target consumers almost daily gives credence to her belief that long term; ongoing loyalty is a concept under some pressure.

Perhaps even more significantly there is now little doubt that loyalty has become a divided concept. Consumers that demonstrate the trait have been classified as either rational or irrational customers. In the sport sense the rational customer is the one that wishes to be entertained while the irrational customer is the hard core fan, the passionate partisan. Pointedly Brendan Schwab, chief of the Australian Soccer Players' Association was quoted in the *Sydney Morning Herald* (1998) commenting that 'the rational customer as distinct from the irrational customer is the more desirable supporter' (p. 4). Fundamentally, Schwab was suggesting that the theatregoers might be a more valuable customer than Stewart and Smith's passionate partisan and definitely more valuable than the 'champ follower'. The football operations manager of a current National Rugby League (NRL) team in part supports Schwab's comments. He believes that 'the grassroots, die-hard supporter has no understanding of the commercial realities of professional sport' (Personal Interview, 1998).

3.3 Win/Loss

A cursory examination of the literature reveals that performance, or winning/losing is the variable that has been least examined and manipulated in the context of the professional sport spectator. While it could be argued that the analysis of such is problematic, performance is a sport consumption variable that cannot be ignored. While some research has included winning or 'the game' as an element for analysis, in the main it has not been tied into fan consumption. Nevertheless the recognition of 'the win' factor and the meaning associated with it does provide for a natural demarcation between the categories of research. A cursory exploration of such research is informative and provides components for a consumption framework.

It has long been argued that a winning team is the most important thing in professional sport. Vince Lombardi's original comment 'winning isn't everything, but trying is', has been mythologised into 'winning isn't everything, it's the only thing'. Similarly the legendary sport marketer Bill Veeck (1962) suggested that if he had a choice of great promotion and a losing team, or poor promotion and a winning team, he would always take the latter. In a recent edition of *Inside Sport* (October 1998) ex Sydney King, Tim Morrisey commented on the lack of on-court success of the King's team. He stated that, 'the average fan does not give a rat's arse if you've got a hundred sponsors or if your naming rights sponsor is Coca-Cola, the bottom line is if you are a fan you want to see your team win'. A more cautious comment is provided by Lapidus and Schibrowsky (1996) who argued that 'while the hot dogs may not taste better when the home team wins, the music sure improves'.

Mashiach (1980) conducted one of the earliest studies that briefly addressed the relationship between sport event attendance and winning. This research examined the factors that influenced American spectators to attend the 1976 Olympic Games in Montreal. While seeing the US teams compete against and beat international competition was of minor importance to the American male it was not a factor in the American female's decision to attend. The author concluded that behaviour is not determined by a single factor but by a myriad of factors.

Mawson and Coan (1994) did isolate the variables when they examined marketing techniques used by NBA franchises to promote home game attendance. They established that the marketing arms of the NBA franchises used 21 techniques to market their product, one of which was winning. Overall winning was a 14th priority ranking. Not surprisingly franchises with high attendances, strongly agreed with winning, as a marketing technique while those with low attendances didn't. The high attendance NBA franchises rated it 3 while the low attendance franchises rated it 21. Zhang et al. (1995) investigated reasons for spectator attendance at NBA games and argued that 'making spectators feel that they had contributed to the team's success was as important as winning'. This is an interesting yet untested observation.

Pan, Gabert and McGaugh (1997) surveyed season ticket holders for a NCAA Division 1A men's basketball team in an attempt to identify the constructs of motives that constitute the decision process of purchasing a season ticket. They readily indicated that winning was just one of a number of variables which influenced fan attendance and cited work by Kennedy (1980) who found that less than 25% of sport fans attend professional sport as a direct result of a team's winning record. However, their own research indicated that prior team record was an important consideration in the decision to purchase season tickets. Similarly Branvold, Pan and Gaber (1997) concluded that while a winning percentage was a significant factor in attracting attendances to Minor League Baseball, winning alone was not a valid predictor of attendance.

Kochman (1995) explored the relationship between winning and attendance and argued that maximal attendance at Major League Baseball was more closely linked to outcome certainty than uncertainty. The logic in this instance being that it costs too much to attend professional baseball to take the chance that the fan may see their team lose. While this in part begins to address the nexus between attendance and winning, it does not differentiate between fan type.

Cramer, McMaster and Lutz (1986) went further than Kochman and suggested that fans who had just witnessed their team win are far more likely to be generous towards others than fans who have seen their team defeated. Obviously winning in such instances is a contributing factor to fan enjoyment; however, its impact on future product consumption has not been discussed.

These few examples are indicative of the perception that links the importance of a team's win-loss record to organisational success and continued fan support. However, in what at first glance may appear to be a simple issue, i.e. winning teams have greater fan support, there is a need for some circumspection for two reasons. First, it has already been demonstrated in the literature that sport fans are a heterogeneous group and as such individually have very different expectations from the sport experience. Second, there appears to be little evidence that ranks, attributes and quantifies the importance of winning to this expectation.

In conclusion it can be argued that while there is a body of literature that includes winning as a variable and, as such assists in understanding of the impact of performance on consumption, it does not address the issue of continued product consumption by diverse groups of sport fans.

3.4 Identification

Wakefield and Sloan (1995) also acknowledged the importance of winning but agreed with Melnick (1993) and Stotlar (1995) who believed that many spectators also sought social interaction and enjoyment via their sport experience. While social interaction is one feature of the literature that discusses sport fans' motivations, the issue of identification is another. Examples of this are as follows.

The research undertaken by Wann and Shrader (1997) combined winning and identification. The authors concluded that a higher mean rating of sport spectating enjoyment was found for highly identified spectators who had just watched their team win. Conversely Sutton, McDonald and Milne (1997) suggest that 'as the degree of fan identification and involvement increases, the less likely the fan's behaviour will be impacted on by team performance' (p. 15). While they cited the Boston Red Sox as an example, within the Australian context the AFL's Collingwood Magpie fans are legendary for their support through both good times and bad. Similarly Sutton, McDonald and Milne (1997) suggest that 'people associate themselves in terms of interest, speech, appearance and behavior with winning sport teams to enhance their prestige in the eyes of others and perhaps even to increase their own self-esteem' (p. 18). They used the 37% increase in the sales of Houston Rockets' merchandise after their 1994 NBA Championship season as a case in point.

Kahle, Kambara and Rose (1996) suggested that 'identification with a winning team can be a viable vehicle for relationship maintenance as it allows group members to share the meaning of a win' (p. 53). This activity is commonly referred to as BIRGing (Basking In Reflected Glory – 'we won'). However, the authors also suggested there was another group of fans who identified more with a team or a region than winning and hence were less likely to engage in BIRGing or CORFing (Cut Off Reflected Failure – 'they lost'). The previous example of the Boston Red Sox or the Collingwood Magpies is once again appropriate. Kahle et al. (1996) found that there was no direct relationship between winning and attendance and that fans who attend matches purely for the purpose of being associated with winners are very likely to be brand switchers. Subsidiary research questions therefore are, when it comes to the fan's continued consumption of professional sport, how important is winning in the sport fan's decision to continue to support the organisation? Secondly, does the variable 'winning', have a greater or lesser impact on the decision-making process depending on the type of fan or consumer?

Wann et al. (1994) surveyed 105 undergraduate students at two universities in the American Midwest to determine the relationship between spectator identification, perceptions of influence, emotions and competition outcomes. A major finding was that spectators reported high levels of identification for successful teams but that game outcome did not significantly influence their level of identification.

As illustrated, the literature on sports fans is plentiful and diverse. At a basic level there appears to be two types of fans, rational and irrational. Similarly, irrespective of the elements that combine to give rise to sport consumption, the variables loyalty, performance and identity would appear to be the three major components related to fan behaviour. When Stewart and Smith's (1997) typology is adopted and cross-tabulated by the previously mentioned variables of loyalty, performance and identity it is obvious that the management of football fans, and the incorporation of their voice in the decision-making process is far more complex then first imagined. The following examples are illustrative of this point.

4 Football Fans in Australia

In recent years there has been an ever-expanding body of knowledge related to the various Australian football codes. Such research has aided the respective sport organisations in the establishment of more focused marketing strategies. While a feature of much of this data collection has been the ranking of game elements to determine the salience for the attending fans, in a number of instances it has been possible to cluster responses to create a sport fan typology. This in turn has the potential to provide a framework for plotting the fan's perceptions of their sports experience.

4.1 The Sydney Swans

In 1995 and 1996 two studies focusing on fan behaviour were undertaken on behalf of the Sydney Swans (Psychometric Marketing, 1995; Quick and Stinear, 1996). Given that the Sydney Swans is the relocated Victorian club South Melbourne, the surveys, although unlinked were conducted in both Melbourne and Sydney in an attempt to solicit information from the respective fan bases. The 1995 study was undertaken in Sydney and conducted by Psychometric Marketing who surveyed 510 football fans to elicit a range of information concerning behaviour and attitudes. The 1996 survey was undertaken in Melbourne by Quick and Stinear and involved a mail out to the Melbourne members of the Sydney Football Club.

While much of the information collected was demographic in nature, there were a number of responses in both instances that highlighted differences in fan typology. The 1995 Psychometric study divided the Sydney audience into light (those who attend 1–2 games), medium (those who attend 3—7 games) and heavy users (those who attend 8—12 games). In an interesting manipulation of the data they further divided the medium category into fragile and non-fragile. The fragile medium users were the only group where Swans' performance was a significant reason for non-attendance. Moreover 40% cited coming to watch a specific player as reason for attendance. This group constituted 15% of Swans' spectators. Finally, while both the medium and heavy users were game rather than team loyal, the light users' reasons for attendance revolved around entertainment. In this instance two types of fans have been identified, the aficionado and the theatregoers.

The Melbourne member base of the Sydney Swans once again provided distinctive segments within the one cohort. While 26% of respondents indicated that they had been members for only one year the same number had been members for more than 14 years. Given the rapid rise in membership in 1996, it could be argued that new members were performance oriented in as much as the club played in the championship game for the first time in nearly 50 years. The long-term members (14 years plus) were obviously South Melbourne members who remained loyal to the club when it relocated to Sydney in 1983. Another cohort had been members for 10 years and this corresponds with a finals appearance by the club in 1986–7. Once again it would appear that these members are performance based. In this instance the 'passionate partisan' and the 'champ follower' have been identified. Hence within one club across two cities, four fan types are clearly evident.

4.2 The Western Bulldogs (formerly the Footscray Bulldogs)

Similarly and also in 1996, 400 attendees at football matches involving the then Footscray Football Club (now Western Bulldogs) were surveyed at home games at both the Western Oval and the Melbourne Cricket Ground (MCG) to ascertain membership-related characteristics (Quick, 1996c). Of those interviewed, 31% had previously been members but had let their membership lapse. Only 6% of this group indicated that their decision not to renew was related to on-field performance. Other reasons for non-renewal included cost and work commitments. One additional major factor in the decision not to renew was related to the club's decision to play a number of home games at the MCG, Australian pre-eminent football stadium as opposed to the Western or Whitten Oval where the club had been traditionally based. The fans' decision to forego their membership due to this decision indicates a loyalty to the traditions of the club rather than solely the team. While it could be argued that in this instance the fans were club loyal as opposed to game loyal, and hence could be described as the 'passionate partisan', in many respects this group does not readily fit into one of the five categories of fans established by Stewart and Smith (1997). As a result an additional category may need to be established to cater for the fan that does not differentiate between product and place.

Of those that had never been members, 25% suggested on-field performance would sway their decision-making. Hence the win-loss ratio was a minimal factor in the decision to forego membership but a more significant factor in the decision to take it up.

4.3 The NSWRU Waratahs

1n 1998, 1000 Ruby supporters were surveyed on five occasions at Super 12 matches in Sydney (Quick, 1998). The number one reason for attending the game was to see the Waratahs play. The second most important reason for attending the game was to be with friends. While watching the Waratahs play is not necessarily linked to winning, there is a need to link the importance of winning and losing to such a response. It could be argued that implicit in such a response is the desire to see the home team win. It could even be argued that such a response is self-evident. However, a good number of fans were actually at the matches to watch the opposition, as was the case when New Zealand teams were playing.

5 Conclusions

While the preceeding is just a small sample of the research related to sport fan consumption there are some identifiable trends. Motivations for sport consumption appear to be psychologically, sociologically or, in certain cases yet to be illustrated, economically based. Furthermore the information obtained from such consumers is invariably demographic or psychographic.

Who the fans are, their likes and dislikes, what they watch and read and how well they identify sport sponsors are all facets of the professional sports environment that are becoming increasingly well known. While this may appear to be comprehensive, it is merely a starting point for successful marketing campaigns. A number of issues need to be considered. Without doubt a major one is the type of fans, product expectations and its impact on future sport consumption.

It would appear that while the research into motivations for attendance at professional sport contests is well established, that demographic profiles of such attendees is readily apparent and that a typology of fan segments has been identified, linkages between the three have been only superficially examined. Having established clusters of the factors that address the issue of professional sport consumption, there is a need to take the next step and systematically explore the impact performance has on the attendance patterns of different types of sport fans. Given that rational sport fans are increasingly becoming more important to sport organisations, their sensitivity or lack thereof to factors such as the club's win-loss record is an issue that needs to be clearly understood. While an assessment of the impact of various variables on ongoing fan attendance may indeed prove to be problematic, at the very least it is an issue worth exploring. If a framework can be created which has the potential to plot the nexus between the win-loss factor, the irrational and rational fan and identification, then sport organisations may be better placed to formulate strategies which will assist in the ongoing maintenance of fan support.

6 References

Ashley, F. and Song, C. (1995) Marketing the auto race: What can we learn from spectators, *Sport Marketing Quarterly,* 4, 27-32.

Branvold, S.E., D.W. Pan and Gaber, T.E. (1997) Effects of winning percentage and market size on attendance in minor league baseball, *Sport Marketing Quarterly,* 6, 35-42.

Cramer, R.E., McMaster, M.R. and Lutz, D. (1986) Sport fan generosity: A test of mood, similarity and equity hypotheses, *Journal of Sport Behavior.* 9, 31-37.

Critchley, C. (1997) Footy fans in firing line, *The Herald Sun.* January 8, p. 7

de Burca, S., Brannick, T. and Meenaghan, T. (1995) A relationship approach to spectators as consumers, *Ibar,* 16, 86-100.

Gauthier, R. and Hansen, H. (1993) Female Spectators: Marketing Implications for Professional Golf Events, *Sport Marketing Quarterly,* 2, 21-28.

Graham, P. (1992) A Study of the Demographic and Economic Characteristics of Spectators attending the US Men's Clay Court Championships, *Sport Marketing Quarterly,* 1, 25-30.

Hansen, H. and Gauthier, R. (1993) Spectator's views of LPGA Golf events, *Sport Marketing Quarterly,* 2, 17-26.

Hansen, H. and Gauthier, R. (1994) The professional golf product: Spectators views, *Sport Marketing Quarterly,* 3, 9-16.

Hofacre, S. (1994) The women's audience in professional indoor hockey, *Sport Marketing Quarterly,* 3, 25-28.

Jones, I. (1997) Mixing qualitative and quantitative methods in sports fan research, *The Qualitative Report,* 3, 1-6.

Kahle, L.R., Kambara K.M. and Rose, G.M. (1996) A functional model of fan attendance motivations for college football, *Sport Marketing Quarterly,* 5, 51-60.

Kochman, L.M. (1995) Major League Baseball: What really puts fans into the stands, *Sport Marketing Quarterly,* 4, 9-12.

Lapidus, R.S. and Schibrowsky, J.A. (1996) Do the hot dogs taste better when the home team wins? *Journal of Consumer Satisfaction/Dissatisfaction and Complaining Behavior,* 9, 1-11.

Macken, D. (1998) You can't trust anybody these days, *The Sydney Morning Herald,* October 10, 4s.

Mashiach, A. (1980) A study to determine the factors which influence American spectators to go to see the Summer Olympics in Montreal 1976, *Journal of Sport Behavior,* 3, 17-28.

Mawson, M. and Coan, E. (1994) Marketing techniques used by NBA franchises to promote home game attendance, *Sport Marketing Quarterly,* 3, 87-96.

Pan, D.W., Gabert, T.E. and McGaugh, E. (1997) Factors and differential demographic effects on purchase of season tickets for intercollegiate basketball games, *Journal of Sport Behavior,* 20, 447-464.

Psychometric Marketing (1995) *Report: Usage and Attitude Survey,* PM(Asia) Sydney

Quick, S. (1996a) Paying to Win: The Business of AFL, *Nexus,* 6, 4

Quick, S. (1996b) Paying to Win: The Business of AFL, *Bulletin of Sport and Culture,* 9, 1-2.

Quick, S. (1996c) Footscray Football Club Fan Survey, *Technical Report.*

Quick, S. (1998) NSWRU Super 12s Fan Survey, *Technical Report.*

Quick, S. and Stinear, E. (1996) Sydney Swans Melbourne Member Survey, *Technical Report.*

Randl, J. (1994) Demographic Characteristics of Racetrack Fans, *Sport Marketing Quarterly,* 3, 47-52.

Stewart, R.K. and Smith, A.C.T. (1997) Sports watching in Australia: A Conceptual Framework, *Advancing Sport Management in Australia and New Zealand,* SMAANZ, Deakin University, Australia, pp. 1-30.

Stotlar, D. (1995) Sports Grill Demographics and Marketing Implications, *Sport Marketing Quarterly,* 4, 9-16.

Sutton, W., McDonald, M. and Milne, G. (1997) Creating and Fostering Fan Identification in Professional Sports, *Sport Marketing Quarterly,* 6, 15-22.

Veeck, B. with Ed Linn (1989) *Veeck...As in Wreck.* Fireside, New York.

Wakefield, K.L. and H.J. Sloan (1995) The Effects Of Team Loyalty And Selected Stadium Factors On Spectator Attendance, *Journal of Sport Management,* 9, 153-172.

Wann, D.L., Dolan, T.J., McGeorge, K.K. and Allison, J.A. (1994) Relationships between spectator identification and spectators' perception of influence, spectators' emotions and competition outcome, *Journal of Sport and Exercise Psychology,* 16, 347-364.

Wann, D.L. and Schrader, M.P. (1997) Team identification and the enjoyment of watching a sport event, *Perceptual and Motor Skills,* June 84, 954.

Zhang, J., Pease, D., Hui, S. and Michaud, T. (1995) Variables Affecting Spectator Decision to attend NBA games, *Sport Marketing Quarterly,* 4, 29-40.

S. DICKSON
University of New England, Armidale, Australia

1 Introduction

An uneasy relationship tends to exist between game participants (players and coaches) and sports officials (referees and umpires). Historically, game participants and officials have viewed each other as a source of constant aggravation. This view is held because most interactions between officials and game participants are of an aversive or unpleasant nature. While players and coaches may view officials as a necessary evil in sport, many officials also hold a negative stereotype of coaches, and are inclined to be cynical of players' actions.

The extent of the divide between officials and game participants is well documented. There are recorded cases of players and coaches physically assaulting officials, throwing equipment at officials, yelling abuse and screaming profanities. Reasons for such behaviour and disrespect are unclear, although it is hypothesised that players and coaches are more likely to argue with an official who: (1) appears to lack self-confidence; (2) makes inconsistent rulings; (3) makes mistakes in positioning; and (4) shows a lack of concentration (Kaissidis and Anshel, 1993). Under such circumstances, the officials may need to examine the reasons why hostility is being directed towards them before blaming players and coaches for placing them (the officials) under undue stress.

From the referee's perspective, research has provided reasons why coaches, in particular, aggravate officials. These include: (1) intentional baiting of officials by players and coaches; (2) officials having to deal with players and coaches who do not have a thorough understanding of the rules; (3) coaches' influence on the selection and retention of officials; (4) performance ratings by coaches of officials; and (5) coach and player criticism of officials in the media. From a more global perspective, the roles which game participants and officials play in the sporting contest may help to explain the aggravation between groups. Obviously, players and coaches are most concerned with the outcome of the game, and consequently will take actions in an attempt to meet this outcome. In fact, the actions of officials are viewed by some participants as 'contrary to their [the participants] best interests' (Anshel, 1989, p. 32). Many players now hold the attitude that the official is responsible to catch them [the players] committing fouls. The recent soccer World Cup in France was a case in point. Numerous times players 'dived' with the intent of eliciting undeserved free kicks from referees. Such strategy and play, often supported and encouraged by the coach, can be interpreted as playing against the referee, rather than playing against the opposition. Juxtaposed to such motivations, officials are responsible for the fair and proper conduct of the game, and have little or no concern for the eventual outcome of the game.

Despite these apparent contradictions, all soccer players must play by the rules as determined by FIFA (and consequently, coaches need to frame strategy and tactics within the same context).

In tandem with this obligation, it is incumbent on the referee to ensure the match is conducted within these rules, and in doing so, apply the rules in a fair and consistent manner. In this regard, the ability of the referee to carry out his/her role has a major impact on player behaviour and game quality. When there is a discrepancy, between the referee and game participants, about the proficiency with which rules are applied, problems arise. This report is on a study, which assessed the proficiency of soccer referees in executing competencies essential to elite soccer refereeing. As such, opinions were sought from referees, assistant referees, referee inspectors, players, and coaches. Specifically, it was hypothesised that a significant difference of opinion exists between these groups.

2 Methods

A list of 37 refereeing competencies, seen as essential for effective soccer refereeing, was developed. This was achieved via a hybrid Behaviourally Anchored Rating Scale (BARS) technique (Anshel and Webb, 1991; Anshel, 1995). This technique has proved successful in gathering data for the purposes of occupational analysis, principally because it 'specifies concrete behaviors that must be performed by an individual on the job' (Blood, 1974, p. 513). Fundamentally, most occupations are somewhat complex and therefore require the interaction of many skills for the role to be performed effectively. Smith and Kendall (1963) acknowledged this complexity, and the associated need to recognise all facets of an individual's performance.

Subsequently, a five-point Likert Scale was attached to each competency, and respondents (N = 173) were asked to indicate the degree of proficiency referees demonstrate in executing each competency. Respondents included players (n = 92), coaches (n = 18), referees (n = 13), assistant referees (n = 23), and referee inspectors (n = 27). All respondents were involved at the highest level of Australian soccer (Ericsson Cup). For analysis purposes, the five groups were collapsed into two larger groups. Players and coaches formed the 'competitive' group (n = 110), while referees, assistant referees, and referee inspectors formed the 'officiating' group (n = 63).

3 Analysis

Due to the ordinal nature of Likert scale data, it was imperative to convert the data to interval scores before subjecting data for empirical analysis. One technique which has proved useful in this endeavour is Rasch Latent Trait Scaling (Wright and Linacre, 1989; Burton and Miller, 1998). Specifically, Rasch modelling is based on both order and objectivity, the two fundamental principles of measurement (Snyder and Sheehan, 1992; Hands, Larkin, and Sheridan, 1997). Accordingly, the model is able to provide estimates of item difficulty (analogous to item 'proficiency' in the present study) and respondent ability (i.e., case estimates, analogous to subject perception) for polychotomously scored items (Wright and Masters, 1982). Importantly, the estimates of item difficulty and respondent ability are expressed on a logit scale and hence as an interval/ratio measure (Hands et al., 1997). As a result, estimates can be utilised for subsequent empirical analysis techniques in assessing differences across perspectives and groups.

4 Results

One-way ANOVA indicated a significant difference between groups ($F = 66.58$, $P < 0.0001$). In an attempt to identify the specific competencies, which were most implicated in causing this difference, Cook's Distance (a form of regression analysis) was employed to identify significant outlying items. However, no single item proved to be significant, thus indicating that most, if not all, items were implicated.

As specific competencies could not be identified through parametric analysis, the rank order of competencies from each group was compared. Although rankings are without the statistical rigour of parametric techniques, they are able to provide a practical guide for discussion purposes. In order to develop the ranking structure, items estimates (derived from Rasch analysis) for each competency, from both groups, were obtained. Subsequently, each competency was given a ranking according to group. A full list of these rankings is provided in Table 1.

Table 1. Competency ranking by group.

	Competency	Officiating group	Player/ coach
1.1	Understands and interprets the Laws of Soccer correctly	2	11
1.2	Observes incidents and decides on appropriate action (e.g., fouls, player acting, free kick, booking, play-on)	13	24
1.3	Understands how the game is played (e.g., tactics and strategy, analyse patterns of play)	36	37
1.4	Anticipates players actions and reactions (eg., retaliation to over physical play)	25	31
1.5	Keeps a complete record of the game (e.g., bookings, goals, substitutions)	1	1
1.6	Works as a team with assistant referees	16	27
1.7	Observes play from the best position	23	13
2.1	Applies the Laws of Soccer consistently (within each game and over the season)	26	36
2.2	Observes, analyses and correctly interprets incidents	20	29
2.3	Reacts quickly and effectively to incidents	17	14
2.4	Distinguishes between fair and foul play	7	19
2.5	Interprets and discriminates between the severity of fouls	20	30
2.6	Encourages attacking play	19	32
2.7	Distinguishes between advantage and disadvantage	18	26
2.8	Manages conflict (communication with players, use of presence and personality)	29	35
2.9	Moves to obtain optimum positions (i.e., place with best view and close enough to react effectively)	14	10

Continued over

Table 1. *Continued*

	Competency	Officiating group	Player/ Coach
3.1	Effectively uses the whistle (e.g., volume, tone, timing, length, player reaction)	14	5
3.2	Uommunicates decisions with clear hand signals	6	6
3.3	Undertakes report writing and record keeping (e.g., send-off reports, administrative reports)	10	3
3.4	Mommunicates (verbal and non-verbal) with players on and off the field	30	34
3.5	Mediates disputes between opposing players	33	28
3.6	Mommunicates confidently with assistant referees	4	8
3.7	Communicates appropriately with coaches	34	25
3.8	Communicates with referee's inspectors (e.g., post match discussions, self reflection, constructive criticism)	35	17
4.1	Applies the Laws and sanctions (e.g., free kick, yellow and red cards)	9	18
4.2	Displays positive behaviours and attitudes	11	16
4.3	Manages disputes between players/coaches and match officials	31	23
4.4	Monitors player behaviour	20	33
4.5	Manages all aspects of the game (e.g., match control/players)	11	20
5.1	Prepares well in advance of the match (time of arrival at ground – on time, presentation on arrival, kit prepared)	3	2
5.2	Undertakes mental preparation for the match (e.g. visualisation)	24	15
5.3	Engages in post match activities (e.g., talk to coaches, attend post game functions)	37	20
5.4	Manages personal anxiety	32	22
6.1	Maintains required levels of fitness (e.g., fitness benchmarks/standards, hydration, warm up/cool down)	7	4
6.2	Maintains appropriate levels of personal health (eg., correct diet, weight control)	27	9
6.3	Undertakes appropriate risk management procedures (e.g., know legal responsibilities, state of the pitch)	28	7
6.4	Maintains concentration during the game	4	11

With respect to the proficiency of referees in demonstrating these competencies, only six showed a ranking differential of three or less, while 13 recorded a differential of 10 or more places. Just two competencies recorded complete agreement between the groups, these being competencies 1.5 *keeps a complete record of the game (eg., bookings, goals, substitutions)*, and 3.2 *communicates decisions with clear hand signals*. In both cases, the competencies were ranked highly by both groups.

The three competencies which showed the greatest discrepancy between groups were:

3.8   *communicates with referee's inspectors (e.g., post match discussions, self-reflection, constructive criticism)*;
6.2   *maintains appropriate levels of personal health (e.g., correct diet, weight control)*; and
6.3   *undertakes appropriate risk management procedures (e.g., know legal responsibilities, state of the pitch)*.

All three competencies where ranked lowly by officials and relatively high by the competitive group (particularly the latter two skills). However, it could be argued that opinions of the competitive group are not fully based on all available facts. This is because the fulfilment of these competencies are, in general, independent of the competitive aspects of the game, and as such, away from the evaluative glare of the players and coaches. Moreover, the practical significance of these results is tenuous, as the competencies are not seen to be relatively important by either group (the assessment of competency importance was a complementary focus of this study. 'Importance' was assessed, and ranked, using the same Rasch analysis technique described previously).

Of particular interest are those competencies which are seen as important, are ranked highly in terms of proficiency by one group, and conversely, low by the other group. To meet this end, decision rules are required to guide the analysis. Consequently, only competencies which were ranked in top half of proficiency by one group (i.e., ranked 1–18), which had a ranking differential of 10 or greater, are examined further. Table 2 provides and overview of competencies which met this criteria.

Table 2. Marked group differential on 'important' refereeing competencies.

Competency	Rankings		
	Officials	Player/ Coach	Differential
1.2 Observes incidents and decides on appropriate action (e.g., fouls, player acting, free kick, booking, play-on)	13	24	11
1.6 Works as a team with assistant referees	16	27	11
1.7 Observes play from the best position	13	23	10
2.4 Distinguishes between fair and foul play	7	19	12

Two main points arise from this table which are worthy of further discussion. First, the officials have provided higher rankings across all four competencies. Obviously, the officiating group sees the proficient delivery of these competencies in a more favourable light than the players and coaches. A reversal of this trend is not evident in the data. Second, three of the four competencies (1.2, 1.7 and 2.4) are related to the immediate decision making ability of the referee. The practical linkages between the competencies are of marked importance. In this regard, the execution of the three competencies will have an immediate impact on the actions of players, and subsequent conduct and flow of the game. In combination, the competencies require that the referee not just observe an incident, but be in the best possible position. Furthermore, the referee must decide if the incident represented fair play, and the consequential appropriate action (if any) to take. When both points are viewed in combination, it is apparent that competencies, which are fundamental to the referee's ability to make quick and accurate decisions, are over-rated by officials. Clearly, the players and coaches perceive a need for improvement in these essential competencies.

Of additional interest, although somewhat removed from the expressed purpose of this paper, is the perspective that players and coaches place on the combined proficiency and importance of each refereeing competency. In this regard, the results are illuminating. As can be seen from Table 3, a marked variation occurs between the importance of a number of key competencies and the proficiency of referees in performing these competencies.

Table 3. Major differences between the importance and proficiency of refereeing competencies – player and coach perspective.

	Competency	Importance	Proficiency	Differential
1.2	Observes incidents and decides on appropriate action (e.g. fouls, player acting, free kick, booking, play-on)	3	24	21
2.1	Applies the Laws of Soccer consistently (within each game and over the season)	2	36	34
2.2	Observes, analyses and correctly interprets incidents	4	29	25
2.5	Interprets and discriminates between the severity of fouls	8	30	22
2.7	Distinguishes between advantage and disadvantage	8	26	18
2.8	Manages conflict (communication with players, use of presence and personality)	8	35	27

5 Conclusions

There are a number of important implications that arise from these results. First, it is apparent that players and coaches bring a different perspective to the assessment of referee proficiency. While not unexpected, the value of their perspective should not be ignored. It is this group, particularly the players, which is most impacted by the actions and decisions of the referee. In this regard, players and coaches are not unlike consumers, or clients, of what may be called the referee's 'product'. Accordingly, their opinions provide unique and specialised augmented feedback.

Second, the triangulated nature of the feedback validates the assessment of referee proficiency. Traditionally, referees receive their most specific feedback from referee inspectors. However, the homogeneous nature of this group ignores potential feedback from other relevant stakeholders. The information provided by players and coaches may furnish a more complete picture of referee performance.

Third, the results provide an unambiguous focus for referee training and development. If it is accepted that player and coach feedback is valuable, then data provided by this group should feed into subsequent referee education programmes. It would be foolish of referee administrators to blindly ignore the opinions of players and coaches, particularly given the intimate 'working' relationship of these groups with the referee.

Last, the refereeing competencies, which were seen as most deficient by players and coaches, carry a common theme. They are generally competencies which relate to the application of rules and the decision making process. However, the development and maintenance of these competencies requires focussed training and is time consuming. Additionally, they are competencies that need to be practised and reflected upon with regularity. To meet this end, any move to establish professional referees should be encouraged. Professional footballers and coaches/managers have been part of the elite soccer scene for many years. Given the stated importance of referees to the quality of the match, it is timely for refereeing to also be accepted as a legitimate and full-time professional occupation. Such a move can only improve the standard of refereeing.

6 References

Anshel, M. (1989) The ten commandments of effective communication for referees, judges and umpires, *Sports Coach,* 12, 32-36.

Anshel, M. (1995) Development of a rating scale for determining competence in basketball referees: implications for sport psychology, *The Sports Psychologist,* 9, 4-28.

Anshel, M. and Webb, P. (1991) Defining competence for effective refereeing, *Sports Coach,* 15, 32-37.

Blood, M. (1974) Spin-offs from behavioral expectation scale procedures, *Journal of Applied Psychology,* 59, 513-515.

Burton, A. and Miller, D. (1998) *Movement Skill Assessment,* Human Kinetics, Champaign, IL

Hands, B., Larkin, D. and Sheridan, B. (1997) Rasch measurement applied to young children, *The Australian Educational and Developmental Psychologist,* 14, 11-22.

Kaissidis, A. and Anshel, M. (1993) Sources and intensity of acute stress in adolescent and adult Australian basketball referees: A preliminary study, *The Australian Journal of Science and Medicine in Sport,* 25, 97-103.

Smith, P. and Kendall (1963) Retranslation of expectations: An approach to the construction of unambiguous anchors for rating scales, *Journal of Applied Psychology,* 47, 149-155.

Snyder, S. and Sheehan, R. (1992) The Rasch measurement model: an introduction, *Journal of Early Intervention,* 16, 87-95.

Wright, B. and Linacre, J. (1989) Observations are ordinal; measurements, however, must be interval, *Archives of Physical Medicine and Rehabilitation,* 70, 857-860.

Wright, B. and Masters, G. (1982) *Rating Scale Analysis,* Mesa Press, Chicago.

67 INVESTIGATING COMPETENCIES FOR OFFICIATING: SOCCER – A CASE STUDY

S. DICKSON
University of New England, Armidale, Australia

1 Introduction

The continual growth and interest in sport has created enormous demand for sports officials. This has resulted in coaches and administrators, at all levels of sport, constantly seeking people who are competent at officiating (Clegg and Thompson, 1993). However, it would seem that the determination of what constitutes a competent official, and their subsequent identification and development, is far more complex than simply finding someone who exhibits a thorough knowledge of game rules (Brown, 1993).

However, a clear and concise description of what constitutes effective officiating is elusive. As noted by Anshel (1995, p. 9), 'relatively few attempts have been made to systematically quantify the criteria by which researchers and practitioners interpret and assess effective or desirable performance in sport'. In this regard, officiating is no exception. Part of the problem lies in defining the role of the official. Various definitions have ranged from 'crisis containment' (Grunska, 1995), to facilitation (Burke, 1991; Clegg and Thompson, 1993), to maintaining social order (Smith, 1982).

Notwithstanding this lack of clarity, perhaps the most recognisable role of the official is to uphold the laws of the game. Yet, in upholding the laws, their strict application may not necessarily be seen as 'good' officiating. It is maintained that officials should exercise some degree of discretion in their rulings (Reilly, 1996). However, such discretion can lead to numerous complications for the official, as officials are then expected to make arbitrary judgements which may be inconsistent with previous decisions. Moreover, the complexity of their role multiplies when controlling sporting contests which are contextually based, yet framed by specific and objective rules. Such demands are particularly relevant to football, where game dynamics ensure that very few, if any, decisions will be based on exactly the same set of circumstances.

Obviously, the complicated and ambiguous role of the official impacts on developing a clear description of refereeing skills. Perhaps the first step in developing and list is the recognition that officiating skills are sport-specific. Although it is recognised that some skills will underpin officiating across a range of sports (National Officiating Program, 1997), the demands placed on officials, within individual sports, requires a unilateral approach to skill identification.

Accordingly, this paper reports on a study that was undertaken to identify the essential competencies required by elite soccer referees. In addition, it aimed to determine which competencies were the most important. Subsequent results are discussed in terms of referee training, development and assessment.

2 Methods

2.1 Subjects

A hybrid form of the Behavioural Anchored Rating Scale (BARS) was employed for this investigation. Two main reasons are offered for the selection of this technique. First, it has been clearly established that BARS methodology encompasses extensive occupational analysis (Borman and Dunnette, 1975; Schwab, Heneman and DeCotiis, 1975; Grussing, Silzer and Cyrs, 1979; Jacobs, Kafry and Zedeck, 1980; Anshel, 1995), and second, the process requires the generation of qualitative data in the form of expressed competencies and skills related to occupational performance.

Two identical groups of representative 'experts' were selected to generate relevant skills. Each group contained the following: one Ericsson Cup (Australian National Soccer League) referee; one Ericsson Cup assistant referee; one Ericsson Cup referee's inspector; one Ericsson Cup player; one National Director of Coaching (current or former); one representative of the Australian Coaching Council; and, one academic from a cognate area.

Over a one-day period, the two groups identified essential skills required by elite soccer referees. A predetermined format was followed, and is outlined below:
1. generation of action verbs and phrases which described *essential* competencies for officiating soccer at the elite level, e.g., signal, organise, instruct, react;
2. generation of action phrases, e.g., react quickly;
3. writing *competencies*. This task was accomplished by transforming each of the relevant verbs and action phrases into a statement of specific behaviour/knowledge/ attitude/understandings, e.g., observes play from the best position; and
4. organisation of competencies into related or homogenous categories, referred to as *performance dimensions*, e.g., rule enforcement.
5. group responses amalgamated into one composite list.

The final list contained six performance dimensions, comprising 37 competencies. Based on this list a survey instrument was designed to assess the importance of the dimensions and competencies. As such, a five-point Likert scale was attached to each item (ranging from 'very important' to 'not important'), and sent for comment to a sample of national league referees ($n = 13$), assistant referees ($n = 23$), referee inspectors ($n = 27$), players ($n = 92$), and coaches ($n = 18$).

2.2 Analysis

The competency structure was assessed through principle components factor analysis with OBLIMIN rotation. For the purposes of this investigation, the 37 competencies represented variables, from which underlying constructs are sort (i.e., representative of competencies identified through the BARS procedure). Resulting factorial solutions are assessed via two methods, first the unambiguous eigenvalue criteria of greater than or equal to one, and second, the less objective scree criteria.

Rasch latent trait scaling was also used to assess performance dimension structure. Rasch modelling assumes homogeneity of data. Subsequently, if the data is shown to fit the Rasch model, it can be deduced that the data is unidimensional and thus devoid of latent traits. A second dimension to Rasch analysis is its ability to provide estimates of item difficulty (analogous to item importance in the present study), and subsequently expresses these estimates on a logit scale. Importantly, the analysis is able to detail the degree of separation of items on this scale as an item reliability index (Adams and Khoo, 1996). Measures greater than 0.7 are considered acceptable.

3 Results

Factor analysis on the 37 competencies yielded a six-factor solution, accounting for 63% of the explained variance. However, a number of concerns complicated the solution. First, a number of competencies ($n = 5$) loaded significantly onto more than one factor. Second, one competency did not load significantly onto any one factor. Third, the explained variance was far less than the recommended 75% (Stoskopf et al., 1992). Subsequent split-half tests, plus the application of the scree criteria, mirrored these problems.

Rasch analysis also demonstrated problems with the factor structure. Data were shown to fit the model (infit mean square = 0.99, infit t = -0.22), thus indicating that the data are essentially unidimensional and lacking in obvious latent constructs.

Consequently, all competencies were assessed for importance independent of the original dimension structure (item estimates and associated rankings are available from the author). Competencies were well separated (reliability of estimate = 0.88), indicating that their relative positions on the importance scale were stable.

The 10 highest ranked competencies are listed in Table 1. Of note is the stand-out threshold of competency, 1.1 *understands and interprets the Laws of Soccer correctly*. The threshold value of this competency is one standard deviation from the next highest ranked competency. No other highly ranked competency shows such a marked difference with respect to surrounding competencies.

The competencies listed in Table 1 show a certain degree of diversity. This includes competencies relating to rule interpretation and application, communication, fitness, and teamwork. However, the top 7 competencies are all related to the decision-making demands incumbent on referees. Yet the diversity of competencies, with respect to the types of decisions a referee is required to make, is illuminating. Technical aspects of rule interpretation, player behaviour and consistency of decisions are fundament skills, which, through their execution, have a profound effect on conduct of a match.

With respect to the least important competencies, the last ranked item is a more than one standard deviation below the 36th ranked competency (see Table 2). Obviously, this competency was not seen to be important by respondents, and as such calls into question the competency's relevance to elite officiating. Other competencies listed range from communication, to person management, to tactical understanding. Noticeably, none of the competencies listed in Table 2 relate to specific decision making processes during a match.

4 Conclusions

This study has shown that the competencies required by referees are highly diverse. In this regard, the role of an elite soccer referee is multi-skilled. Player management, rule enforcement, rule application, communication, health and fitness, legal responsibilities and record keeping are just some of the competencies which require attention and development. The notion of refereeing being multi-dimensional is a separate issue that cannot be fully explored here. Suffice to say that, although the empirical analysis showed otherwise, the referee's role can be considered multi-dimensional from a practical perspective. Further research and development is needed on the methodology employed to identify and categorise competencies into performance dimensions.

Table 1. Most important competencies by rank.

	Competency	Threshold	Rank
1.1	understands and interprets the Laws of Soccer correctly	1.32	1
1.2	observes incidents and decides on appropriate action (e.g., fouls, player acting, free kick, booking, play-on)	0.80	2
2.1	applies the Laws of Soccer consistently (within each game and over the season)	0.76	3
2.2	observes, analyses and correctly interprets incidents	0.6	4
6.4	maintains concentration during the game	0.56	5
4.1	applies the Laws and sanctions (e.g., free kick, yellow and red cards)	0.51	6
2.4	distinguishes between fair and foul play	0.49	7
2.8	manages conflict (communication with players, use of presence and personality)	0.47	8
6.1	maintains required levels of fitness (e.g., fitness benchmarks/standards, hydration, warm up/cool down)	0.45	9
1.6	works as a team with assistant referees	0.43	10

However, the methodology gains its major strength from the diversity of 'experts' used to generate the relevant competencies. This ensures that a widest possible perspective is brought to the development process. A perusal of the 37 competencies bears testimony to this diversity. Two distinct benefits accrue from this process. First, content validity is enhanced, and second, the final result is met with greater acceptance across all soccer stakeholders.

From an applied perspective, the results give a clear direction for referee training and development. Specifically, the 37 competencies and their resulting rank order represents one of the first (if not the first) studies to quantify refereeing performance criteria systematically. Consequently, referee educators are now provided with a range of competencies, ranked according to importance, that reflect the emphasis specific competencies require for training and development. In doing so, the results provide a framework for constructing educational programs which are relevant to the specific needs and requirements of elite refereeing.

Table 2. Least important competencies by rank.

	Competency	Threshold	Rank
3.2	communicates decisions with clear hand signals	-0.46	29
3.5	mediates disputes between opposing players	-0.49	30
5.4	manages personal anxiety	-0.58	31
5.2	undertakes mental preparation for the match (e.g., visualisation)	-0.6	32
3.1	effectively uses the whistle (e.g., volume, tone, timing, length, player reaction)	-0.61	33
3.8	communicates with referee's inspectors (e.g., post match discussions, self reflection, constructive criticism)	-0.61	33
3.7	communicates appropriately with coaches	-0.7	35
1.3	understands how the game is played (e.g., tactics and strategy, analyse patterns of play)	-0.81	36
5.3	engages in post match activities (e.g., talk to coaches, attend post game functions)	-1.34	37

Last, the results derived from this study allow for more focused, specific, and accurate assessment feedback. Moreover, assessment can heed the weightings, or ranking, given to each competency. The approach represents a shift away from the subjective criteria normally applied to referee assessment. Obviously, the referee who performs the most important competencies well (and the poorer one not-so-well), should be judged in a better light than the referee who excels at the least important competencies only. Assessments can be more considered, and freed from the subjective and emotional biases, which can often pervade judgement. The benefits of such an approach can only lead to an improved overall standard of refereeing.

5 Acknowledgement

Soccer Australia facilitated this research.

6 References

Adams, R. and Khoo, S. (1996) *Quest – The Interactive Test Analysis System (Manual), Version 2.1,* Australian Council for Educational Research, Camberwell, Victoria.

Anshel, M. (1995) Development of a rating scale for determining competence in basketball referees: implications for sport psychology, *The Sports Psychologist,* 9, 4-28.

Borman, W. and Dunnette, M. (1975) Behavior-based versus trait-oriented performance ratings: an empirical study, *Journal of Applied Psychology,* 60, 561-565.

Brown, J. (1993) Umpires are human ... aren't they? *National Netball Magazine,* July/August, 16-19.

Burke, K. (1991) Dealing with sport officials, *Sport Psychology Training Bulletin,* 2, 1-6.

Clegg, R. and Thompson, W. (1993) *Modern Sports Officiating. A Practical Guide,* WCB Brown and Benchmark, Dubuque.

Grunska, J. (1995) Better living through officiating, *Referee,* 20, 36-41, 61.

Grussing, P., Silzer, R. and Cyrs, T. (1979) Development of behavioural-anchored rating scales for pharmacy practice, *American Journal of Pharmaceutical Education,* 43, 115-120.

Jacobs, R., Kafry, D. and Zedeck, S. (1980) Expectations of behaviorally anchored rating scales, *Personnel Psychology,* 33, 595-640.

National Officiating Program (1997) *Officiating General Principles Syllabus for Referees, Umpires, Judges and other Sports Officials.* Australian Sports Commission, Belconnen.

Reilly, T. (1996) Special populations, in *Science and Soccer* (ed T. Reilly), E. and F.N. Spon, London, pp. 109-123.

Schwab, D., Heneman, H. and DeCotiis, T. (1975) Behaviorally anchored rating scales: a review of literature, *Personnel Psychology,* 28, 594-562.

Smith, C. (1982) Performance and negotiations: A case study of a wrestling referee, *Qualitative Sociology,* 5, 33-46.

Stoskopf, C., Glik, D., Baker, S., Ciesla, J. and Cover, C. (1992) The reliability and construct validity of a behaviorally anchored rating scale used to measure nursing assistant performance, *Evaluation Review,* 16, 333-345.

68 PRELIMINARY INVESTIGATION INTO THE SEASONAL BIRTH DISTRIBUTION OF ENGLAND WORLD CUP CAMPAIGN PLAYERS (1982 – 1998)

D.J. RICHARDSON and G. STRATTON.
Liverpool John Moores University, Liverpool, England

1 Introduction

Theoretically, the date of birth of international soccer players should be evenly spread throughout the year. Early maturing children or those born early in the academic/competition year (i.e., September to December) may have initial advantages in power, speed and strength. Brewer et al. (1995) reported that over 50% of England U-16 Youth Internationals had their birthdays between September and December.

The complexity and speculative nature of talent detection and development has been previously recognised (Regnier et al., 1993). There are a number of prerequisites for football success (e.g., technique, determination, intelligence, speed). It would appear that at Youth International level the less mature players are not selected. One would assume that by full maturity there would be an even balance between the birth dates of 'elite' players across the year.

The aim of this study was to investigate the distribution of birth dates, in relation to the academic/competition year, for full England international football players.

2 Methods

The birth dates of senior England players (N = 139) involved in World Cup campaigns (ie., qualifying matches and tournaments) since 1982 were analysed. Players were placed into one of three categories according to their date of birth (September – December, January – April, May – August), and four categories according to their primary playing position (goal-keeper, defender, midfield, forward).

3 Results

A significant difference, (χ^2, (2) = 17.703; P < 0.01), was evident between the dates of birth of the players. Fifty percent of players had birthdays between September and December compared to 28% between January and April and 22% between May and August. Figure 1 shows that this trend is consistent throughout each of the campaigns since 1982.

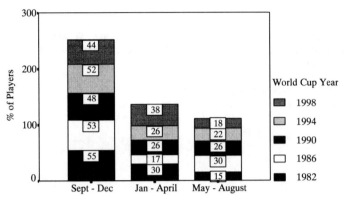

Figure 1. Birth dates for England World Cup campaign players (1982–1998).

Figure 2 highlights the differences across playing positions. Sixty nine percent of goalkeepers, 51% of forwards and 48% of defenders had their birthdays between September and December. The birth dates for midfield players were more evenly spread across the three calendar periods, 44%, 21% and 36% respectively.

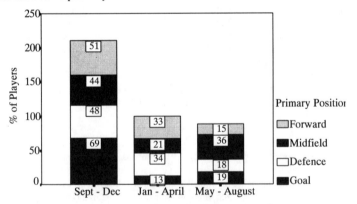

Figure 2. Birth dates of England World Cup campaign players (1982–1998) with respect to their primary playing position.

4 Discussion

Results indicate an over-representation of elite players with birthdays between September and December. These results concur with Dudink (1994) who reported that significantly more English and Dutch football league players were born in the first quarter of their respective competition year (P < 0.001). Moreover, Edwards (1994) identified a birth-date effect for 'early-year' (i.e., April - September) UK county cricket fast bowlers, but not for spin bowlers. The skewed effect was attributed to physical rather than psychological factors. Similarly, the more even nature of the birth-date distribution for midfield players could be attributed to the less dominant physical requirements of the position (i.e., compared to goalkeepers and defenders (e.g., height) and forwards (e.g., speed). However, these results contrast Edwards' (1994), who concluded that all footballers, regardless of height, show a birth-date effect.

On the basis of probability one would assume an equal distribution of players' birth dates at full maturity. The trend indicates a bias toward the 'early bird players' (ie., born early in the academic/competition year) during the identification and selection process. This may be from as early as eight years of age, thus, giving them a 'head start' over other players. These players may be exposed to a higher playing standard and better coaching, and are therefore more likely to graduate to the elite level than late maturing players.

If the birth-date theories are to be dismissed, then what happens to the other talented young footballers in the same academic/competition year (i.e., those born between January and August)? This study indicates that for a young player, with a birthday between January and August, to progress to the elite level they must either be exceptionally talented (e.g., Alan Shearer, born 13th August, 1970), or that managers, coaches and talent scouts alike, must begin to recognise an individual's rate of maturity.

A solution may lie in a more flexible approach to the age-banding of training and competition. Thus, facilitating a more progressive approach to an individual's upward mobility. The game cannot afford to prevent talented youngsters progressing just because their biological clock runs more slowly. Coaches and teachers must develop a greater understanding of the growth and development of young footballers and look more closely at mechanisms to reduce this relative-age advantage.

5 Conclusion

There was an over-representation of England World Cup campaign players (1982–1998) born early in the academic/competition year (ie., September through December). The discrimination effect was greater for goal-keepers, forwards and defenders. Talent identification and selection procedures must place more emphasis on talent and less reliance on physical attributes (i.e., size and strength). More flexible approaches to young player development must be adopted.

6 References

Brewer, J., Balsom, P. and Davis, J. (1995) Seasonal birth distribution amongst European soccer players, *Sports Exercise and Injury,* 1, 154-157.

Dudink, A. (1994) Birthdate and sporting success, *Nature,* 368, 592.

Edwards, S. (1994) Born too late to win, *Nature,* 370, 186.

Regnier, G., Salmela, J. and Russell, S. (1993) Talent Detection and Development in Sport, in *A Handbook of Research on Sport Psychology* (eds R.N. Singer, M. Murphey and L.K. Tennant), Macmillan Publishing, New York, pp. 290-313.

69   CHANGES IN PROFESSIONAL SOCCER IN GERMANY SINCE 1990

W KUHN
Institute of Sports Sciences, Free University of Berlin, Germany

1 Introduction

The purpose of this study is to report important changes in professional soccer in Germany since the opening of the border in 1990. Diverse documents such as the official communications of the German Soccer Federation, weekly journals, sports journals and law texts were used as sources to prepare this report. First, the migratory movements of male professional or licensed soccer players within Germany and Europe will be discussed. Second, the transfer and compensation system before and after the Bosman-verdict will be dealt with. Third, the problems pertaining to young developing soccer players in Germany will be analysed from two points of view: (i) What are the consequences of the extensive hiring of foreign players? (ii) What is the level of quality of German youth development? In the last section attempts to solve some of the problems will be considered.

2 Migratory Movements within Germany and Europe

Since the beginning of organized soccer competition in Germany, foreign players were attracted for several reasons. In most cases political constraints (such as discontent with a social system) and commercial advantages (generous transfer money and high salaries) were dominant, to a lesser degree professional incentives, or playing in a well-organized league or getting to know a new culture and a new language. Many outstanding national players from East-European countries (e.g., Cajkovski, Csernai, Horvath, Lorant, Zebec) played and coached in German soccer clubs after the 2nd World War. However, these were single cases and not mass movements.

After the wall was brought down, important East–West movements have taken place within Germany. In 1990–91, 13 top players from the former German Democratic Republic (including representative players such as Doll, Kirsten, Sammer, Thom) as well as many talented junior players could not resist the wooing of the rich Western clubs (Kicker-Sportmagazin Sonderheft 1990, S. 176 f.). These players were – in spite of initial adaptations in training – easy to integrate, as there was no language barrier. Unfortunately Eastern clubs lost valuable talent and role models whereas some Western clubs profited from a very cheap face-lifting.

In harmony with the unification of both German states, the first Federal Soccer League was enlarged to 20 teams in 1991–92 by giving a berth to the two first placed finishers of the North East Oberliga in 1990/1991 (Kicker-Sportmagazin Spezial 1998). The second Federal Soccer League was increased in 1992/1993 by six East-German teams to 24 teams. In the season 1998/99 only one Eastern club was a member of either league with 18 teams. Thus, from the original eight teams only two have survived the fierce competition.

An overview of migratory movements of foreigners in Germany is given in Tables 1–4 (Kicker-Sportmagazin Sonderheft 1979, 1989 – 1998). Table 1 demonstrates that the number of foreign players in the first Federal Soccer League constantly increased from 28 in 1979–80 to 165 in 1998–99. These figures underestimate real numbers as 1 July was taken as the deadline for player movements for the following season in the middle of August. In Germany transfers can be conducted until 11 January. This development which has probably not yet come to an end was influenced by two incidents: the fall of the Iron Curtain and the Bosman-verdict. In 1998-99 it happened for the first time that of all newly hired players more than half of the contracts were given to foreign players. In 1998–99 each first division club had an average of nine players under contract, which means that about 35% of the rosters were of foreign origin. The difference between teams is enormous. The number of foreign players fluctuates between 4 and 13 players (17.39% and 50%). Thus, it would be feasible for a few German clubs to start and finish a match without even using one German player. A survey of the exporting countries reveals that in 1979–80 the majority of players came from Eastern and Central Europe (12 players each). This trend was increased in 1990 (22 and 14 players respectively) and 1998 (79 and 43 players respectively). Also players from overseas found German soccer clubs more and more attractive (see Table 2). Only the figures for West Europeans were dwindling. Data from the second Federal Soccer League are provided in Table 3. The tendency is similar to the first Federal Soccer League: Foreign players increased from 17 in 1979–80 to 118 in 1998–99. On an average six 'legionairies' per team are under contract. The fluctuation between clubs is just as high as in the first Federal Soccer League: three and 14 foreign players correspond to a percentage of 11.11% and 51.85% respectively. Table 4 reveals that the majority of players came from Eastern Europe (1998/99: n = 78), followed by players from overseas (n = 24) and Central Europe (n = 12).

Table 1. Absolute and relative number of foreign players in teams of the first Federal Soccer League in 1979/80, 1990/91, 1997/98 and 1998/99.

	1979/80	1990/91	1997/98	1998/99
Sum of foreign players	28	52	148	165
Arithmetic means per team	1.56	2.89	8.21	9.17
Minimum	0	1	6	4
Maximum	3	6	11	13
Average percentages per team	7.43%	12.83%	32.58%	35.08%
Minimum	0.00%	4.17%	25.00%	17.39%
Maximum	15.00%	24.00%	45.82%	50.00%

Table 2. Geographical location of foreign players in teams of the first federal Soccer League in 1979/80, 1990/91, 1997/98 and 1998/99.

	1979/80		1990/91		1997/98		1998/99	
	Total	Arith. Means	Total	Arith. Means	Total	Arith. Means	Total	Arith. Means
East Europeans	12	0.67	22	1.21	73	4.06	79	4.33
Central Europeans	12	0.67	14	0.76	30	1.67	43	2.33
West Europeans	2	0.10	6	0.32	18	1.00	4	0.22
Oversea[1]	2	0.10	10	0.56	27	1.50	39	2.17

[1]all confederations of FIFA, except UEFA

Table 3. Absolute and relative number of foreign players in teams of the second Federal Soccer League in 1979/80, 1990/91, 1997/98 and 1998/99.

	1979/80	1990/91	1997/98	1998/99
Sum of foreign players	17	50	97	118
Arithmetic means per team	0.42	2.50	5.39	6.56
Minimum	0	0	1	3
Maximum	2	5	11	14
Average percentages per team	2.07%	11.56%	22.73%	26.46%
Minimum	0.00%	0.00%	5.25%	11.11%
Maximum	13.33%	23.81%	45.82%	51.85%

Table 4. Geographical location of foreign players in teams of the second Federal Soccer League in 1979/80, 1990/91, 1997/98 and 1998/99.

	1979/80		1990/91		1997/98		1998/99	
	Total	Arith. Means	Total	Arith. Means	Total	Arith. Means	Total	Arith. Means
East Europeans	8	0.20	34	1.70	63	3.50	78	4.33
Central Europeans	6	0.15	7	0.35	6	0.32	12	0.67
West Europeans	0	0.00	1	0.05	1	0.06	4	0.22
Oversea[1]	3	0.07	8	0.40	27	1.50	24	1.33

[1]all confederations of FIFA, except UEFA

In the lower leagues the number of foreign players has also risen sharply. One extreme example is the team line-up of Tennis Borussia Berlin. It had only one German player on its side when it played Hannover 96 in a promotion match for the second Federal Soccer League (Berliner Fußball Woche 22/1998, 3.).

Foreign players who do not fulfil expectations are frequently sold to other clubs in the same season they were hired. Importing countries are predominantly Belgium, France, The Netherlands,

Portugal and Spain. Some players from Eastern Europe also preferred to return home to their old clubs.

The German Soccer Federation and especially its coaching staff have often complained about the fact that too many mediocre players are recruited from abroad. The clubs have argued that the price–performance relationship is much better for foreign players than for native players. However, the German Federal League is also very attractive to top players (e.g., Balakov, Elber, Salou, Sforza). Table 5 shows that 205 of 704 internationals who were called up by their federations for participation in the 1998 World Cup, have earned their money in the top five European soccer leagues (Focus 27/1998, 221). Germany represents with 35 players about 5% of all participants (Focus 27/1998, 221). Table 5 (last column) also shows that national coaches frequently make use of the selected players. Altogether, 70.4% of these 35 legionaries played in the preliminary round of World Cup 1998. Only the Italian-based players displayed a higher percentage. However, when we take a look at the quarter-final, semi-final and final the number of top foreign players playing in Germany dropped sharply from 16 to 3 and 2 (see Table 6). The Italians comfortably take first place as they fielded 46% of the finalists, followed by England and Spain with almost 15% (Focus 29/1998, 209).

Table 5. Absolute and relative number of foreign players and their absolute and relative participation in the preliminary round of the World Cup 1998 in the five strongest European professional soccer leagues. (Adapted from FOCUS 27/1998,221).

Country	Foreign national players		Participation in prelim round of World Cup 1998	
	absolute	relative	absolute	relative
England	53	7.53%	85	53.46%
Italy	49	6.96%	106	72.11%
Spain	47	6.68%	91	64.54%
Germany	35	4.97%	74	70.48%
France	21	2.98%	27	42.86%
Arith. means	41.00	5.82%	76.60	62.28%

Table 6. Absolute and relative number of foreign national players and their absolute and relative participation in the quarter-final, semi-final and final of the World Cup 1998 in the five strongest European professional soccer leagues. (Adapted from FOCUS 29/1998, 209.)

	Quarter-final		Semi-final		Final	
	absolute	relative	absolute	relative	absolute	relative
England	8	7.14%	6	10.71%	4	14.29%
Italy	40	35.71%	14	25.00%	13	46.43%
Spain	11	9.82%	7	12.50%	4	14.29%
Germany	16	14.29%	3	5.35%	2	7.15%
France	1	0.89%	0	0.00%	0	0.00%
Total	76	67.85%	30	53.56%	23	82.16%

Germany is not only an importing, but also an exporting country. The number of players earning their money abroad has remained fairly constant over the last few years (Kicker-Sportmagazin Sonderheft 1998, 217). In 1998 11 of the 40 German legionaires played in Austria

(Kicker-Sportmagazin Sonderheft 1998, 217). The ratio of exporting to importing players is about one to four. Thus, the trade balance of Germany is largely import-oriented, reflecting the economic strength of the country and the high esteem of the first Federal Soccer League.

With regard to coaches, Germany is still an exporting country although the figures are slightly declining. Only four coaches work in one of the leading eight European soccer countries in 1998/99 (Kicker-Sportmagazin Sonderheft 1998, 217). On the other hand, there is a surplus of highly qualified coaches in Germany. In contrast to the past only one foreign coach was employed by a German first devision club in 1998/99.

The liberalization of the player market can also be observed internationally. Team line-ups of FC Barcelona, Real Madrid, Manchester United or Juventus (Turin) are known for their multicultural compositions. Chelsea FC started the European Cup Winner's Final in 1998 with three Italians, one Norwegian, one Dutchman, one Romanian, one Frenchman and one Scotsman. It is surprising that more and more Italians leave their country which has one of the strongest leagues in the world. In 1998–99 31 Italian 'legionairies' played abroad, mainly in England and Scotland (n = 20) and in Spain (n = 6) (Kicker-Sportmagazin 56/1998, 61). They took the place of the Scandinavians who represented in 1994/95, with 42.5%, the majority of foreign players in England (Maguire and Stead 1998, p. 66). The reasons for this exodus are manifold. On the one hand political changes in Italy resulted in loss of tax privileges for soccer clubs. On the other hand, clubs of the English Premier League gained substantial capital through merchandising and activities on the stock-market and the Spanish Primera Division increased its TV income. Hence, they could afford to pay higher transfer fees and salaries than their Italian counterparts (Kicker-Sportmagazin 56/1998, 61). This development is also reflected on the international scene. While for the European Championship in 1996 only one player was recruited for the Italian national team from abroad, selectors picked three for the 1998 World Cup (see Table 7). However, by far the largest loss in players was suffered by Jugoslavia. Around 450 players of the highest league left the war-ridden country in the last five years in order to earn their money abroad (Der Spiegel 25/1998, 98). From this point of view it was no surprise that Jugoslavia's World Cup team was made up of 19 'legionairies' and only three home players. Jugoslavia was surpassed only by Nigeria which had all its 22 players under foreign contracts.

Table 7. Absolute and relative number of national players in selected countries playing abroad in the World Cup 1990, 1994 and 1998.

Team	1990		1994		1998	
	absol. number	rel. number	absol. number	rel. number	absol. number	rel. number
Brasil	11	50.00%	11	50.00%	12	54.55%
Germany	5	22.71%	6	27.26%	4	18.17%
England	4	18.18%			0	0.00%
France					16	72.73%
Netherlands	5	22.73%	8	36.36%	13	59.09%
Italy	0	0.00%	0	0	3	13.64%
Spain	0	0.00%	0	0	0	0.00%

The migratory movements within Germany and Europe were presented in a quantitative manner. Exporting and importing players with respect to confederations can be clearly identified.

Prior to 1990 the main travel routes were easy to overlook. Italy was the main importing country for top players. After 1990 a lot of new avenues were opened up. Regarding the causes of migrations, we can only speculate since sound and comprehensive empirical data are lacking. There are only a few case studies (e.g., Maguire and Stead, 1998) in which foreign players were asked to talk about their motives and their experiences. According to these reports the motivational structure is complex. We must assume that there are individual types with regard to motives for transferring although financial rewards remain the principal incentive.

The influx of foreign players has heavily taxed the communication skills of coaches and players. Besides, according to FIFA rules national players have to be given leave for a certain amount of time for international matches. This can hurt a club badly in cases when the players concerned are needed most.

In view of tendencies of increasing globalisation in all areas of life, migratory movements in European soccer must be evaluated positively. They are witness of greater mobility, they also help to improve the culture of play and serve the endeavours of European integration. Such migratory movements are known not only in team sports, but also in individual sports
(e.g., boxing, fencing, wrestling, Olympic gymnastics, track and field, table tennis). It will become a problem, if there is an overload of foreign players. The consequences regarding the German national ice-hockey team should be a warning example for the German Soccer Federation.

3 The Bosman Verdict

The increase of migratory movements in soccer since 1990 cannot only be attributed to the opening of the border, but must also be allocated to the Bosman case (Flory, 1997). With the verdict by the European Court of Justice from the 15.12.95, ending a five-year long legal battle, the global player market was partially liberalized. The European Court of Justice decided that the compensation fee system, and the restrictions on the number of foreign nationals who could play for a team of a member state, contravened Article 48 of the Treaty of Rome (15.3.1957). This article specified no discrimination with regard to employment, remuneration and working conditions. In other words, a compensation fee is no longer required if three conditions are met:
– the contract of a player has expired
– the player is a member of the European Union (EU) and European Economic
Region (EER)
– the player intends to transfer within the EU or the EER.

It is important to note that these rules apply only for professional and licensed players of team sports within the EU and EER, i.e., if there is a transfer from one club of a member state to a club of another member state. The verdict has no influence on players who transfer after expiration of their contracts from one club to another within a member state.

The verdict was heavily criticized by clubs and federations both in Germany and abroad. Surprisingly only five months after the Bosman-verdict a row stirred. The Council of the German Soccer Federation went even further in its rulings (27.4.1996) than the verdict from Luxemburg in that no differences were made between European players from EU and non-EU countries, i.e., German professional clubs can hire and use in matches players from all member states of UEFA (Europe and Israel) in an unlimited fashion. In order to standardize competitive conditions UEFA also accepted this ruling. The Council of the German Soccer Federation further decided that an unlimited number of assimilated foreigners (resident players for five years, including three years as junior players) can be used in matches.

The previous rule that not more than three licensed players from all FIFA confederations (except UEFA) and three amateur players can play in a professional team was maintained. The Council of the German Soccer Federation also decided that a minimum of 12 players of German origin has to be under contract. The reasons for this move are obvious but it is inevitable that the ruling will be contested by the EU.

Table 8 displays an overview of the development of foreign player quotas in German professional soccer. Since the 1960s two foreigners were tolerated, but hardly any club made use of this rule. When the German Federal Soccer League was inaugurated in 1963-64 only three foreign players were under contract. Starting from 1978 an additional three players who played in German clubs for five years could be used in matches. From 1991 onwards the rule '3 + 2' was applied which meant that in addition to three foreigners, two players could be contracted who had been active in German clubs the last five years including three years as junior players (so-called assimilated foreigners). As a consequence of the Bosman-verdict, more than five foreigners can now play at the same time for a German club.

Table 8. Modalities with regard to foreign players in German professional soccer clubs.

Year	Modalities
From 1960	Two foreign players
From 1978	Two foreign players and two assimilated players
From 1992	Three foreign players and two assimilated players
From 1996	An unlimited number of foreign players from counties affiliated with UEFA;
	a maximum of three players from other confederations; a minimum of 12 licensed players of German nationality

The missing transition period was a major point of critique of the Bosman-verdict. The German Soccer Federation followed the lawsuit in Luxemburg very closely and succeeded in reducing some of the consequences of the verdict by adding an important part to paragraph 11 (transfer regulations) of the work contract for professional players prior to the season of 1995/96. In anticipation of the Bosman-verdict, a player had to agree to extend his expiring contract by one year under the existing conditions if the club wished to do so. Likewise, a club had to agree to extend the player's contract by one year under the existing conditions if the player wished to do so. Club or player could make use of this option until 30 April. The German Soccer Federation sold this part of §11 as a safeguard for clubs and players. In reality, players willing to change clubs were forced to stay on for another year. Although courts have ruled §11 to be illegal and not compatible with working rights, the German Soccer Federation has so far refused to abide by the rule.

Originally it was thought that the Bosman-verdict would eventually eliminate the whole transfer system. However, top players signed well paid medium- and long-term contracts and agreed to astronomical transfer fees in case of premature cancellation of their contracts. The goal was to use the high transfer fees as a deterrent to potential buyers. Players who changed clubs without any transfer fees also cashed in. Therefore, the players were the big winners of the Bosman-verdict as their salaries soared. Hoeness, manager of Bayern Munich, commented on the situation as follows:

> 'The players make money whether they win or not. Bonuses make up only 5 to 10 per cent of the salary, sometimes even only 3 per cent. With bonuses they cannot be grasped.' (1998, 23).

A closer look at the transfer expenditures of the 18 German elite clubs reveals that the Bosman-verdict had two effects. First, in 1996–97, the year immediately following the end of the Bosman-case, clubs spent only 83 million DM, compared with 126 million DM in 1995-96. As many contracts had expired, players could transfer within Europe without compensation fees. Second, in 1997–98 and 1998–99 the clubs spent 121 and 113 million DM, almost as much as in 1995–96. The reason was that top players, having signed medium- and long-term contracts, had to be bought out of their existing contracts.

The Bosman-verdict is often held responsible for the influx of foreign players in Germany. This is not correct. As early as 1994–95 and 1995–96, eighty 'legionairies' (i.e., 18% of the players) from 35 countries were active in the first Federal Soccer League. Hence, the Bosman-verdict did not have a triggering, but only a reinforcing function.

Although the decision of the European Court of Justice had positive implications for European integration, new questions opened up and need to be answered. Will there be an exodus of relatively cheap East European players, flooding West and Central European soccer markets? Will this loss of talent hurt the performance level of East European teams? How can young native players be protected? Will German clubs spend enough time and money to develop and train native talents in the future?

4 The Misery of Young German Soccer Players

Vogts, the former Federal Head Coach, has always pointed out that Germany has a serious problem in the age-bracket 18 to 24 years. This is easy to prove. The Under-21 and Under-23 teams are currently not doing well internationally. In the season 1998–99 85% of all German professional players were past the age of 26. At the 1998 World Cup Germany fielded the oldest team, averaging almost 30 years. No German was amongst the 15 youngest players at the 1998 World Cup. The two youngest German players were already 24 years old and had only four and seven caps respectively. In former times Germany had players that started their international career at a very young age and also participated in World Cup competition very early.

Two reasons are frequently put forward to explain the misery of young German soccer players:
(a) Coaches from the German Soccer Federation have pointed out that hiring too many mediocre foreign players has had a snow-balling effect. Talented youngsters cannot get professional contracts and if they are lucky to obtain one, they do not receive any playing time or not enough for developing their skills at the highest level. The coaching staff of the German Soccer Federation is not totally against the boom of foreign players. High-class players are welcome as they create role models and improve the level of German game culture. This is important in a country in which football is quite often not seen as play but work.
(b) Licensed soccer coaches who work in the field of junior soccer have pointed out that compared with other countries the training of youngsters is not good enough. In all age classes, young players lack the criteria of modern soccer (such as ball control, surprising collective play, game intelligence, offensive thinking, individual creativity). The main problem is not the

age class from 14 to 18 years as all major clubs employ licensed soccer coaches. Most important is the training of the 6 to 12 year olds. Unfortunately, often groundsmen or parents look after these kids without having the necessary background in technical-tactical and pedagogical training. Success is for these people the only guideline. For the sake of winning early maturing kids get preference in club and regional teams because of their height and weight. The drop-out rate is especially high in the transition period from 10 to 12 years. In order to improve the quality of coaching, more licensed coaches have to be persuaded to work with young children. The German Soccer Federation has to promote programmes in co-operation with the schools and give thought to implementing a coaching licence entirely geared to juniors. The Federation of German Soccer Coaches also has to devote more attention to the problems surrounding training and development of youngsters. Specialists are paramount in furthering soccer talents as there is stiff competition with other sports.

5 Possible Solutions

1. Talented players must be given more playing time in professional clubs. An analysis of match day 1 in the first Federal Soccer League in 1998-99 revealed that 75 foreign players were amongst the 198 starting players (which corresponds to 37.9%). Assimilated Germans are not taken into account. The same picture repeated itself with the substituted players - 22 foreign players out of 52 (which corresponds to 42.3%). Foreign players were used in key positions (play-maker, striker) and scored 10 out of 23 goals. As coaches are under enormous pressure to win, they will not give inexperienced young professionals a chance to develop their skills at the highest level. Talents are on loan to teams of the second Federal Soccer League or they have to prove their skills in amateur teams. One way to solve the problem would to be ask the professional clubs to hire only top-class foreign players. Eventually all national teams would profit from this measure.
2. Identification and training of talented players have to be improved. Not only has money to be provided but also concepts have to be developed for the age-bracket 6-18 years. Specialists must be used throughout all junior ranks. A permanent scientific council should be established to advise the German Soccer Federation with regard to promoting young talent. The youth programmes passed on the 18.7.1998 by the German Soccer Federation serve more as an alibi for the failure of the national team at the 1998 World Cup.
3. The boom in salaries has to be stopped. Good and average players whose contracts expire ask for higher salaries as transfer fees are abolished. Clubs pay the money to maintain their standard of play although the league has debts of about 500 million DM. One way to solve the problem is to establish salary caps which exist world-wide in team sports. The German Soccer Federation or the league itself has to make salary caps mandatory for all clubs. They also help to make the league more balanced.
4. The first Federal Soccer League has to become more balanced in its competition. At the moment we have a three-class society: five to six teams are championship contenders or in danger of being relegated and the rest of the 18 teams are on the safe side. High interest of spectators and media is only guaranteed up to the last match day when a team can beat every other team. In order to make the league more balanced, thought should be given to establishing a draft system similar to one that exists in North American and Australian professional leagues. The weaker or newly promoted teams should be given first priority to draft players who are willing to change clubs.

5. Professional soccer clubs in Germany have to go to the stock market in order to optimize their income. An analysis in 1998 revealed that out of 15 teams only three clubs were ready for the capital market, six were almost ready and six not yet ready (Jacobs et al., 1998). Professional management is an important prerequisite to collect capital from investors. In this regard it might be helpful to set up an independent administration for the 36 professional teams with office, staff, logo and marketing strategies (within or outside the German Soccer Federation). At the same time further important sources of income such as merchandising or quality of stadium have to be improved.

6 References

Flory, M. (1997) *Der Fall Bosman. Revolution im Fußball.* Agon, Kassel.

Jacobs, J., Klimasch, R., Napp, A. and Scheffler, S. (1998) Der deutsche Börsenmeister, *Impulse,* 8, 14-18.

Kicker-Sportmagazin *Fußball WM 1990, 1994, 1998 (Sonderhefte),* Nürnberg.

Kicker-Sportmagazin *Bundesliga 1979/80, 1989/90, 1990/91, 1992/93, 1993/94, 1994/95, 1995/96, 1996/97, 1997/98, 1998/99 (Sonderhefte),* Nürnberg.

Maguire, J. and Stead, D. (1998) Border Crossings, *International Review for the Sociology of Sport,* 33, 59-73.

INDEX